ZnO Nanostructures
Fabrication and Applications

Nanoscience & Nanotechnology Series

Editor-in-Chief:
Paul O'Brien CBE FREng FRS, *University of Manchester, UK*

Series Editors:
Xiaogang Liu, *National University of Singapore, Singapore*
Ralph Nuzzo, *University of Illinois at Urbana-Champaign, USA*
Joao Rocha, *University of Aveiro, Portugal*

Titles in the series:

How to obtain future titles on publication:
A standing order plan is available for this series. A standing order will bring
delivery of each new volume immediately on publication.

For further information please contact:
Book Sales Department, Royal Society of Chemistry, Thomas Graham House,
Science Park, Milton Road, Cambridge, CB4 0WF, UK
Telephone: +44 (0)1223 420066, Fax: +44 (0)1223 420247
Email: booksales@rsc.org
Visit our website at www.rsc.org/books

ZnO Nanostructures
Fabrication and Applications

Yue Zhang
University of Science and Technology Beijing, China
Email: yuezhang@ustb.edu.cn

THE QUEEN'S AWARDS
FOR ENTERPRISE:
INTERNATIONAL TRADE
2013

Nanoscience & Nanotechnology Series No. 43

Print ISBN: 978-1-78262-741-8
PDF eISBN: 978-1-78801-023-8
EPUB eISBN: 978-1-78801-173-0
ISSN: 1757-7136

A catalogue record for this book is available from the British Library

The Royal Society of Chemistry is a charity, registered in England and Wales, Number 207890, and a company incorporated in England by Royal Charter (Registered No. RC000524), registered office: Burlington House, Piccadilly, London W1J 0BA, UK, Telephone: +44 (0) 207 4378 6556.

For further information see our web site at www.rsc.org

Printed in the United Kingdom by CPI Group (UK) Ltd, Croydon, CR0 4YY, UK

Preface

Nanoscience and nanotechnology have aroused tremendous and worldwide concerns in recent decades. In the early 1990s, developments in physics made it possible to realise revolutionary progress on characterization approaches in nanoscale, which refreshed researchers' understanding towards nanomaterials. Along with in-depth exploration of nanomaterial properties, and extensive attempts on their synthesis methods, nanoscience and nanotechnology were gradually developed into a cutting-edge inter-disciplinary research field.

Among the various material candidates for the construction of nano-structures, the growth of research interest in 3rd generation semiconducting materials, such as SiC, GaN, AlN and ZnO has brought a tremendous impetus to the high-tech industry from both a scientific technology and industry strategies perspective. A series of physical superiorities to the 1st (Ge, Si) and 2nd (GaAs, InSb) generation semiconducting materials, such as electrical properties and optical properties have been highlighted, especially the wide band gap which is suitable for high-frequency, high-power elec-tronic devices and circuits. Recently, studies on ZnO, which is one of the most important 3rd generation semiconductors, has entered a new stage of comprehensive use due to its multi-functional characteristics.

Since 2000, my group has been devoted to ZnO nanomaterials. Recently, along with the highly developed multi-disciplinary fusion and integration, considerable progress has been achieved in both fundamental research and technique applications of ZnO nanostructures. To date, we have published nearly 400 peer-reviewed journal papers and authored more than 40 patents in this field, which has significantly broadened the research and application areas.

This book is mainly based on the published works in my group. It covers our research results for ZnO nanostructures ranging from fabrication to

Nanoscience & Nanotechnology Series No. 43
ZnO Nanostructures: Fabrication and Applications
By Yue Zhang
© Yue Zhang 2017
Published by the Royal Society of Chemistry, www.rsc.org

characterisation and from prototype applications to practical applications. Certainly, the worldwide milestone achievements in the field have also been introduced. Therefore, this book could serve as a reference book for students, researchers or other relevant personnel to promote the development of this field. This book is divided into ten chapters, and overall guided by Prof. Yue Zhang. The detailed chapter theme and co-authors are as follows: Chapter 1, Overview; Chapter 2, Designing and Controllable Fabrication (Qingliang Liao); Chapter 3, Property Characterisation and Optimisation (Xu Sun, Yanwei Shen, Pei Lin, Zhuo Kang); Chapter 4, Electromechanical Devices (Zheng Zhang); Chapter 5, Photoelectrical Devices (Zhiming Bai, Yanwei Shen, Pei Lin, Guangjie Zhang); Chapter 6, Photoelectrochemical Devices (Zhuo Kang, Zhiming Bai); Chapter 7, Biosensing Devices (Yu Song, Zhuo Kang); Chapter 8, Self-powered Devices (Zheng Zhang, Zhiming Bai); Chapter 9, Service Behaviour (Peifeng Li, Qi Zhang); Chapter 10, Field Emission and Electromagnetic Wave Absorption (Qingliang Liao).

I would like to thank my current and former group members, as well as collaborators who have dedicated themselves to the development of ZnO nanoscience as well as its applications. I would also like to acknowledge the strong financial support from the Ministry of Science and Technology of The People's Republic of China, the Ministry of Education of The People's Republic of China, the National Natural Science Foundation of China, the State Administration of Foreign Experts Affairs, the People's Government of Beijing Municipality.

Lastly and most importantly, I thank my family for their continuous understanding and support. The research could not be carried out without their strong support.

Yue Zhang
University of Science and Technology Beijing, Beijing, China

Contents

Nanoscience & Nanotechnology Series No. 43
ZnO Nanostructures: Fabrication and Applications
By Yue Zhang
© Yue Zhang 2017
Published by the Royal Society of Chemistry, www.rsc.org

CHAPTER 1

Overview

YUE ZHANG

University of Science and Technology Beijing, Beijing, China
Email: yuezhang@ustb.edu.cn

1.1 Introduction of Nanomaterials

Nanoscience and nanotechnology, emerging research fields with pioneering theory and multi-disciplinary approaches, have raised wide concern since their inception. At the beginning of the 1990s, developments in physics brought revolutionary progress for small scale characterisation, as well as a novel understanding of nanosized materials. In the following decades, with an intensive study on physical and chemical properties of nanomaterials and extensive attempts at their preparation methods, nanoscience and nanotechnology have been gradually developed into the most popular frontier interdisciplinary subject. Their research content involves physics, chemistry, materials science, mechanics, microelectronics, biology, medical science and many other related subjects. More recently, the rapid development of nanoscience and nanotechnology have created great influences on social economy development science and technology advancement as well as daily life (Figure 1.1).

Since the beginning of the new century, European countries and the USA have successively launched various development programs focused on nanoscience and nanotechnology,[1–3] especially the National Nanotechnology Initiative (NNI) proposed by the USA, which raised the nanotechnology development to national strategy level. In 2006, the National Long-term Scientific and Technological Development Plan (2006–2020) issued by China has firmly claimed 'nanoresearch' as a key program for basic research and

Nanoscience & Nanotechnology Series No. 43
ZnO Nanostructures: Fabrication and Applications
By Yue Zhang
© Yue Zhang 2017
Published by the Royal Society of Chemistry, www.rsc.org

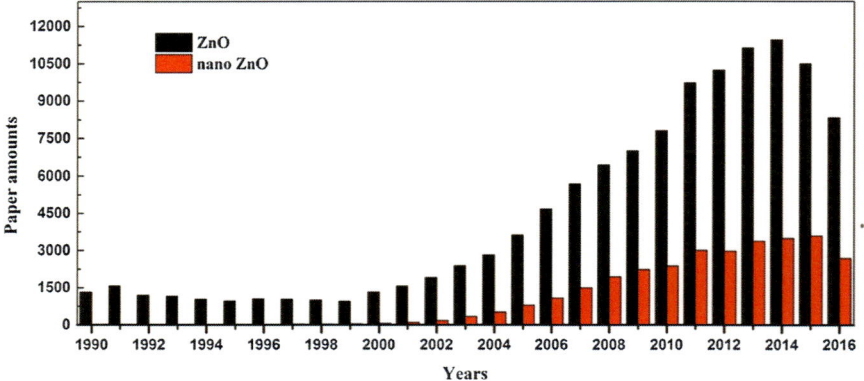

Figure 1.1 Statistics on ZnO-related published papers from 1990 to 2016.

identified nanoscience and nanotechnology advancement as a state major development policy on science and technology. China's Energy Development Strategy Action Plan (2014–2020) also listed micro-nano new energy as one of the nine key fields of innovation. The NASA 2015 technology roadmap also pointed out that nanotechnology was a necessary technical requirement to achieving NASA's goals in science, human exploration and science missions for the next 20 years.

Low dimensional functional nanomaterials remain a worldwide hot spot. They are ideal structural units for constructing functional nanodevices because of their special properties, and always lead to outstanding performance in multi-subject fields like energy, information, environment, micro-electronics, biology, medicine, national defence, *etc.* Their family includes advanced carbon nanomaterials, traditional single element semiconducting nanomaterials, metal nanomaterials, and organic nanomaterials, as well as emerging oxide semiconducting nanomaterials, *etc.*, which generally present excellent and unique mechanical, optical, electrical, thermal acoustic, magnetic properties that bulk materials do not have.[4–16] This is the essential foundation for the development of nanoscience and nanotechnology.

People have experienced a long-lasting period of investigating nanomaterials and nanotechnology. In the early of 20th century, Wilhelm Ostwald, the owner of the Nobel Prize in Chemistry and known as the father of physical chemistry, stated in his 'The World of Neglected Dimensions' that people had realised the importance of the mesoscopic field. The so-called 'mesoscopic' means the state between microscopic and macroscopic, namely the scale between nanometre and millimetre.[17] The conception of nanoscience and nanotechnology was first proposed by Richard Feynman in his speech at Caltech in 1959. After nearly 100 years of exploration until around 1990, along with the development of science and technology, the great breakthrough made particularly in terms of mesoscopic scale characterisation approaches, and great quantities of invented nanoscale characterisation facilities, such as transmission electron microscope (TEM), scanning electron microscope (SEM), scanning probe microscope (SPM), *etc.*, meant that people had truly entered the nanoworld.

The development process of nanomaterials can be divided into three stages: in the first stage (before 1990s), research was mainly focused on nanometre particles, such as nanometre crystalline, nanometre phase, nanometre amorphous as well as their characterisation methods and evaluation methods for their unique properties. In the second stage (1990–1994), research mainly concentrated on controllable synthesis for certain morphologies to acquire peculiar physical and chemical properties that originated from nanometre materials. Especially, exploration on synthesising the nanocomposites gradually became the main direction. In the third stage (1994–present), nanostructure assembling systems, or so-called nanoscale patterned materials, started to draw researchers' attentions. Based on the basic units like nanoparticles, nanowires and nanotubes, various complex nanostructures in one-, two- and three-dimensions, were successfully assembled. The third stage focused on designing, assembling and developing novel systems according to researchers' will, thereby realising the expected certain properties. For now, booming nanoscience and nanotechnology have stepped into a new era, where the modulation of structures and properties of nanomaterials have already tended to be developed and the nanomaterial based functional devices have already been realised for preliminary applications.

1.2 Introduction of ZnO Nanomaterials

The detailed physical property parameters of ZnO are as follows: the relative molecular mass is 81.37 g mol^{-1}, the density is 5.67 g cm^{-3}, the surface work function is 5.3 eV, the melting point of bulk ZnO material is 1975 °C, and the boiling point is 2360 °C. ZnO belongs to wurtzite, with a hexagonal crystal structure, and the space group is $P6_3mc$. The crystal lattice constants of ZnO are: $a = 3.2496$ Å and $c = 5.2064$ Å, and the actual measurement of these are: $a = 3.24$–3.26 Å and $c = 5.13$–5.43 Å.[8,18–20]

The structure of ZnO can simply be described as follows: countless closely packing O^{2-} and Zn^{2+} layers alternately stacked in the direction of the c axis, with the adjacent layers of O^{2-} and Zn^{2+} forming a tetrahedral structure. The structure of ZnO described in different ways is shown in Figure 1.2. The tetrahedral structure of ZnO is the intrinsic cause of its non-centrosymmetrical characteristic, which directly leads to its piezoelectric effect and thermoelectric effect.[21]

ZnO, a typical direct wide band gap semiconducting material (3.37 eV at room temperature) with piezoelectric and photoelectric characteristic, is similar to gallium nitride (GaN). In its energy band structure, the bottom of the conduction band (CB) is formed essentially from the 4s level of Zn^{2+} and antibonding sp^3 hybrid states, and the top of valence band (VB) from the occupied 2p orbits of O^{2-} or from the bonding sp^3 orbitals.[22,23] At room temperature, the defect level such as donor level and acceptor level are mainly distributed in the forbidden band of ZnO ranging from ∼0.05 to

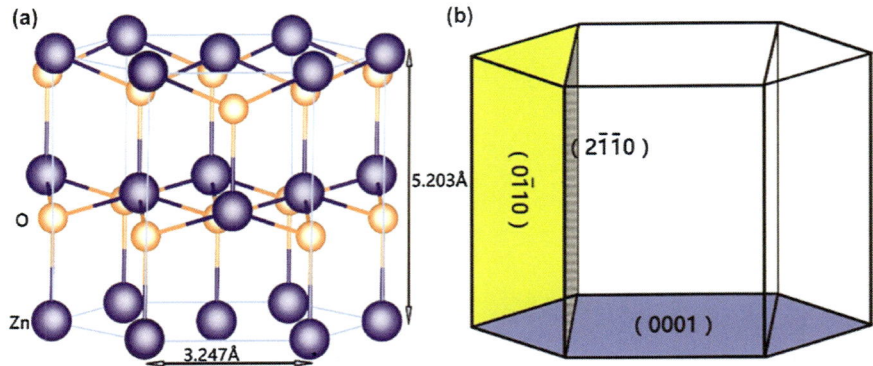

Figure 1.2 The ZnO crystal structure.

2.8 eV, and the band gap of ZnO is 3.4 eV. Besides, ZnO possesses a large exciton binding energy of 60 meV, which is much larger that of zinc selenide (ZnSe) (22 meV) and GaN (25 meV), and thus it is promising for various applications.

ZnO has unique electrical and thermal properties, as well as chemical stability. As a short-wave light-emitting material, its high stability demonstrates an enormous utility value. The traditional synthesis technology could hardly acquire favourable single crystal or thin film ZnO, which restricted its application as a light-emitting material. In recent years, with the development of synthesis approaches, the above-mentioned problem was gradually solved. Ever since people discovered the phenomenon of optical pumping simulated emission at near ultraviolet (UV) of thin film ZnO, it has played a very important role in improving optical recording density and the access speed of optical information due to the shorter wave length in the near UV photoluminescence of ZnO than the blue light emission of GaN. It is also expected to realise breakthrough results for application aspects such as surface acoustic wave, transparent electrode, optoelectronic devices and blue light emitting devices.

Various superior physical characteristics of ZnO make it widely adopted in the field of rubbers, ceramics, coatings and optoelectronics. At present, the research on photocatalytic performance for organic waste degradation and disinfection has been applied in practical waste water treatment; the research on UV absorption performance has been applied for anti-UV agents in textiles and cosmetics. In addition, the wide application prospects in the fields of photoelectric conversion, sensors, nanoelectro-mechanical systems, field emission devices and nanometre lasers have also been highlighted. Notably, the nanogenerator was invented by taking advantage of the piezo-electric effect of ZnO.[24-26]

In terms of morphology and scale size, ZnO nanomaterials can be divided into three types including zero-dimensional materials (ZnO nanoparticles), one-dimensional materials (ZnO nanorods, nanowires, nanobelts, nanocables), and two-dimensional materials (ZnO nanofilm). In terms of composition, there are pristine ZnO nanomaterials and doped ZnO ones, such as *n*-type semiconducting ZnO doped by elements such as In, Ga, Sn, Mn and Co, and *p*-type semiconducting ZnO materials doped by elements such as P, N, Li, as well as multi-element-doped ZnO nanomaterials.

Nowadays, in ZnO nanomaterials research, attention mainly focuses on their controllable and high-yield synthesis,[27-29] structure and property modulation,[22,30] functional nanodevice construction,[31,32] device performance evaluation and multi-field coupling effect,[30,33] as well as many effects induced by the nanometre scale, theory calculation and simulation and material and device damage as well as their service behaviours.[34]

This book is divided into three major parts: (1) synthesis and properties, (2) prototype device construction and (3) practical application exploration. The book has ten chapters covering topics such as fabrication approaches, property characterisation, prototype applications and practical applications of ZnO nanostructures.

References

1. T. Kalil, The national nanotechnology initiative, Supplement to the President's Budget for FY 2017.2016; [http://www.nano.gov/sites/default/files/pub_resource/nni_fy17_budget_supplement.pdf.].
2. R. W. Siegel, E. Hu, M. C. Roco, D. M. Cox and H. Goronkin, Nanostructure Science and Technology–A Worldwide Study, DTIC Document, 1999.
3. M. C. Roco, R. S. Williams and P. Alivisatos, *Nanotechnology Research Directions: IWGN Workshop Report*, Kluwer, 2000, pp. 279–287.
4. Y. Cui and C. M. Lieber, Functional nanoscale electronic devices assembled using silicon nanowire building blocks, *Science*, 2001, **291**(5505), 851–853.
5. H. Dai, J. H. Hafner, A. G. Rinzler, D. T. Colbert and R. E. Smalley, Nanotubes as nanoprobes in scanning probe microscopy, *Nature*, 1996, **384**(6605), 147–150.
6. S. S. Wong, E. Joselevich, A. T. Woolley, C. L. Cheung and C. M. Lieber, Covalently functionalized nanotubes as nanometre-sized probes in chemistry and biology, *Nature*, 1998, **394**(6688), 52–55.
7. Y. Cui, Q. Wei, H. K. Park and C. M. Lieber, Nanowire nanosensors for highly sensitive and selective detection of biological and chemical species, *Science*, 2001, **293**(5533), 1289–1292.
8. X. F. Duan, Y. Huang, R. Agarwal and C. M. Lieber, Single-nanowire electrically driven lasers, *Nature*, 2003, **421**(6920), 241–245.
9. Y. Zhang and Y. Dai, *Nanowires and Nanobelts–Materials, Properties and Devices*, Springer, New York, 2003.
10. Y. Zhang, Y. Dai, Y. Huang and C. Zhou, Shape controlled synthesis and growth mechanism of one-dimensional zinc oxide nanomaterials, *J. Univ. Sci. Technol. Beijing*, 2004, **11**(1), 23–29.

11. Y. Huang, Y. Zhang, X. Wang, X. Bai, Y. Gu, X. Yan, Q. Liao, J. Qi and J. Liu, Size Independence and Doping Dependence of Bending Modulus in ZnO Nanowires, *Cryst. Growth Des.*, 2009, **9**(4), 1640–1642.

12. Y. Yang, J. Qi, Q. Liao, H. Li, Y. Wang, L. Tang and Y. Zhang, High-performance piezoelectric gate diode of a single polar-surface dominated ZnO nanobelt, *Nanotechnology*, 2009, **20**(12), 125201.

13. X.-M. Zhang, M.-Y. Lu, Y. Zhang, L.-J. Chen and Z. L. Wang, Fabrication of a High-Brightness Blue-Light-Emitting Diode Using a ZnO-Nanowire Array Grown on p-GaN Thin Film, *Adv. Mater.*, 2009, **21**(27), 2767–2770.

14. W. A. De Heer, A. Chatelain and D. Ugarte, A carbon nanotube field-emission electron source, *Science*, 1995, **270**(5239), 1179–1180.

15. Z. W. Pan, Z. R. Dai and Z. L. Wang, Nanobelts of semiconducting oxides, *Science*, 2001, **291**(5510), 1947–1949.

16. M. S. Gudiksen, L. J. Lauhon, J. Wang, D. C. Smith and C. M. Lieber, Growth of nanowire superlattice structures for nanoscale photonics and electronics, *Nature*, 2002, **415**(6872), 617–620.

17. Y. Imry and M. Tinkham, Introduction to Mesoscopic Physics, *Phys. Today*, 1998, **51**(1), 60.

18. H. J. Yuan, S. S. Xie, D. F. Liu, X. Q. Yan, Z. P. Zhou, L. J. Ci, J. X. Wang, Y. Gao, L. Song, L. F. Liu, W. Y. Zhou and G. Wang, Characterization of zinc oxide crystal nanowires grown by thermal evaporation of ZnS powders, *Chem. Phys. Lett.*, 2003, **371**(3–4), 337–341.

19. Y. Dai, Y. Zhang and Z. L. Wang, The octa-twin tetraleg ZnO nano-structures, *Solid State Commun.*, 2003, **126**(11), 629–633.

20. C. J. Lee, T. J. Lee, S. C. Lyu, Y. Zhang, H. Ruh and H. J. Lee, Field emission from well-aligned zinc oxide nanowires grown at low temperature, *Appl. Phys. Lett.*, 2002, **81**(19), 3648–3650.

21. Z. L. Wang, X. Y. Kong, Y. Ding, P. X. Gao, W. L. Hughes, R. S. Yang and Y. Zhang, Semiconducting and piezoelectric oxide nanostructures induced by polar surfaces, *Adv. Funct. Mater.*, 2004, **14**(10), 943–956.

22. C. Klingshirn, ZnO: material, physics and applications, *ChemPhysChem*, 2007, **8**(6), 782–803.

23. L. Schmidt-Mende and J. L. MacManus-Driscoll, ZnO – nanostructures, defects, and devices, *Mater. Today*, 2007, **10**(5), 40–48.

24. Y. Qin, X. Wang and Z. L. Wang, Microfibre-nanowire hybrid structure for energy scavenging, *Nature*, 2008, **451**(7180), 809–813.

25. X. D. Wang, J. H. Song, J. Liu and Z. L. Wang, Direct-current nanogenerator driven by ultrasonic waves, *Science*, 2007, **316**(5821), 102–105.

26. Z. L. Wang and J. H. Song, Piezoelectric nanogenerators based on zinc oxide nanowire arrays, *Science*, 2006, **312**(5771), 242–246.

27. S. Baruah and J. Dutta, Hydrothermal growth of ZnO nanostructures, *Sci. Technol. Adv. Mater.*, 2009, **10**(1), 013001.

28. I. Gonzalez-Valls and M. Lira-Cantu, Vertically-aligned nanostructures of ZnO for excitonic solar cells: a review, *Energy Environ. Sci.*, 2009, **2**(1), 19–34.

29. Z. L. Wang, ZnO nanowire and nanobelt platform for nanotechnology, *Mater. Sci. Eng., R*, 2009, **64**(3–4), 33–71.

30. Y. Zhang, X. Yan, Y. Yang, Y. Huang, Q. Liao and J. Qi, Scanning probe study on the piezotronic effect in ZnO nanomaterials and nanodevices, *Adv. Mater.*, 2012, **24**(34), 4647–4655.
31. Y. Zhang, Z. Kang, X. Yan and Q. Liao, ZnO nanostructures in enzyme biosensors, *Sci. China Mater.*, 2015, **58**(1), 60–76.
32. L. J. Brillson and Y. Lu, ZnO Schottky barriers and Ohmic contacts, *J. Appl. Phys.*, 2011, **109**(12), 121301.
33. C. Liu, F. Yun and H. Morkoc, Ferromagnetism of ZnO and GaN: A review, *J. Mater. Sci.: Mater. Electron.*, 2005, **16**(9), 555–597.
34. Y. Zhang, Y. Yang, Y. Gu, X. Yan, Q. Liao, P. Li, Z. Zhang and Z. Wang, Performance and service behavior in 1-D nanostructured energy conversion devices, *Nano Energy*, 2015, **14**, 30–48.

CHAPTER 2

Designing and Controllable Fabrication

QINGLIANG LIAO AND YUE ZHANG*

University of Science and Technology Beijing, Beijing, China
*Email: yuezhang@ustb.edu.cn

2.1 Vapour Phase Deposition Methods

Among the fabrication methods for ZnO nanostructures, vapour phase transport and hydrothermal synthesis methods are the most commonly used to achieve the doping of ZnO nanostructures. The vapour phase transport method has been used to grow doped ZnO nanostructures *via* the vapor–liquid–solid (VLS) mechanism or the self-catalytic VLS mechanism. The vapour phase deposition methods of pure or doped ZnO nanostructures include physical vapour deposition (PVD) and chemical vapour deposition (CVD). Generally, PVD uses physical processes to produce a vapour of material, which is then deposited on the objects. There are no chemical changes during the phase or state transfer. The synthesis of ZnO nanostructures by PVD refers to two process: (1) evaporation and (2) deposition. Under high temperatures, the raw material (ZnO powders) firstly transfer to ZnO vapour. Next, ZnO vapour deposits and forms the solid nanostructures, such as nanowires, nanobelts, nanoarrays, *etc.* CVD, however, uses not only physical processes but also chemical processes. For instance, with zinc powder as a raw material, zinc vapour reacts with oxygen and generates ZnO. When the ZnO and carbon powder were used as raw materials, the ZnO generated Zn vapour by carbon thermal reduction reaction and then produced ZnO

Nanoscience & Nanotechnology Series No. 43
ZnO Nanostructures: Fabrication and Applications
By Yue Zhang
© Yue Zhang 2017
Published by the Royal Society of Chemistry, www.rsc.org

nanostructures by an oxidation process. During the synthesis of ZnO nanostructures by CVD, Zn powder, ZnO powder, and zinc compounds were used as the raw materials. ZnO nanostructures were fabricated by different physical and chemical process, such as the evaporation, the redox reaction, chemical decomposition and combination processes.

Controllable atmosphere tube furnaces are adapted to produce ZnO by vapour phase deposition. By precise control of the reaction temperature, atmosphere (gas type, atmosphere pressure, and flow rate), deposition temperature, catalyst type and state, and substrate type and position, *etc.* various morphologies and scales of ZnO, such as nanowires, nanobelts, nanorods, nanoarrays, nanocomb and self-assembly structures are produced. Usually, the carrier gases for the synthesis of ZnO nanostructures are Ar or N_2. Depending on the synthesis method and products, the source materials are Zn, ZnO, carbon or other powder materials.

2.1.1 Chemical Vapour Deposition by Thermal Evaporation

The process for chemical vapour deposited ZnO is through heating the raw materials (Zn, ZnO, carbon powders or other Zn compound powders) under a series of chemical changes, then generating ZnO vapour to deposit solid state ZnO nanostructures. Among the methods, using Zn powder as the raw material is one of the most important routes for ZnO deposition. Such a method includes two routes to achieve the ZnO deposition: (1) catalyst and (2) catalyst-free.

The reaction and phase transformation of Zn powder as the raw material are shown as followed:

$$Zn(g) + O_2(g) \rightarrow ZnO(g) \rightarrow ZnO(s) \tag{2.1}$$

$$Zn(g) + O_2(g) \rightarrow ZnO(g) \rightarrow ZnO(l) \rightarrow ZnO(s) \tag{2.2}$$

$$Zn(g) \rightarrow ZnO(l), \; Zn(l) + O_2(g) \rightarrow ZnO(s) \tag{2.3}$$

Eqn (2.1) indicates the catalyst-free V–S process, and Eqn (2.2) and (2.3) show the catalytic V–L–S process, in which (g) represents vapour, (l) refers to liquid, and (s) means solid state.

Zn vapour reacting with other vapours can also deposit ZnO nanostructures. For example, under certain conditions, the reaction $Zn(g) + O_2(g) \rightarrow ZnO(g) + H_2$ will happen, and ZnO(g) deposits as nanowires and nanobelts.

Besides Zn powder, the Zn contained compound can be oxygenised or decomposed to form and deposit ZnO, such as by ZnS oxygenation and ZnC_2O_4 decomposition.

2.1.1.1 Catalyst-free Thermal Evaporation Chemical Vapour Deposition

Chemical vapour transport is a simple and low-cost method of manufacture that produces no harmful waste. The characteristic of catalyst-free thermal evaporation deposition is the absence of liquid-phase transformation. During the catalyst-free process, there is neither external catalyst nor self-catalysis. By evaporation and oxygenation of the Zn powder, or oxygenation and deposition of Zn compounds (such as ZnC_2O_4), the vapour phase ZnO solidifies to nanowires and nanobelts.

In 2002, tetrapod-like ZnO (T-ZnO) nanostructures were first reported. Powders of 99.9% pure zinc were placed in a quartz boat and T-ZnO was deposited on Si substrate without catalyst.[1–3] The boat was inserted in a horizontal tube furnace, where the temperature and gas pressure were controlled. The temperature of the furnace was ramped to 850–925 °C at a rate of 50–100 °C min^{-1} and kept at that temperature for 1–30 min. White products were obtained on the Si substrate. Uniform T-ZnO nanorods were formed in high yield and the surface of nanorods were smooth. The length of the legs of the T-ZnO nanorods was 2–3 μm and the edge size of the centering nucleus was 70–150 nm. The diameter kept constant with the nanorod radial direction. The scanning electron microscopy (SEM) images of T-ZnO are shown in Figure 2.1. In addition, T-ZnO nanostructure with decreasing diameter from the center to the end of the leg was also reported. The length of the legs is 2 μm and the edge size of the central nucleus is ∼200 nm.[2,4–6]

T-ZnO has attracted significant interest due to its favourable optical and electrical properties. With the morphology modulation of T-ZnO, the T-ZnO nanostructures are suitable to fabricate different electronic devices. T-ZnO nanostructures were synthesised by a simple vapour phase oxidation method without any catalysts and used as a field emission cold cathode. T-ZnO nanostructures of high purity, uniform morphology and size and high aspect

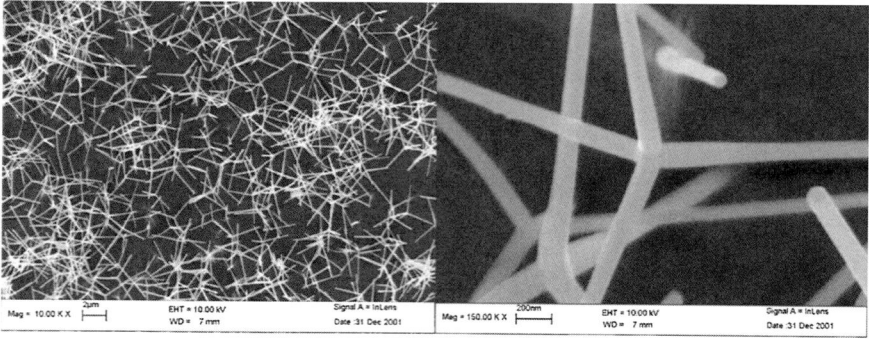

Figure 2.1   T-ZnO synthesised by CVD.
Reprinted from Chem. Phys. Lett., 358(1–2), Y. Dai, Y. Zhang, Q. K. Li and C. W. Nan, Synthesis and optical properties of tetrapod-like zinc oxide nanorods, 83–86. Copyright (2002), with permission from Elsevier.[1]

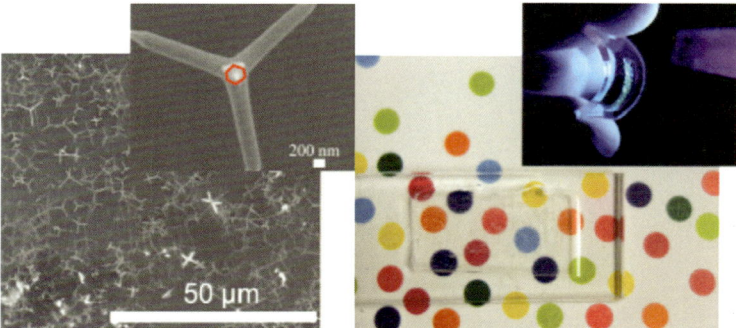

Figure 2.2 SEM image of ZnO nanotetrapods and the photograph of composite material exhibiting transparency and fluorescence under UV light. Reproduced with permission from ref. 8. Copyright (2016) American Chemical Society.

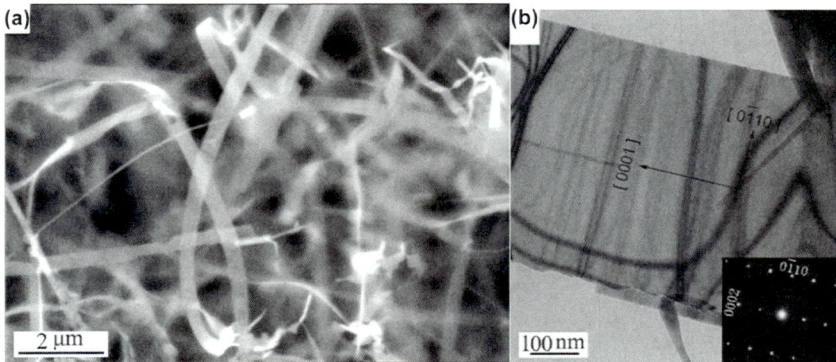

Figure 2.3 SEM and TEM morphologies of ZnO nanobelts. Reproduced from J. Mater. Sci., Morphology, structures and properties of ZnO nanobelts fabricated by Zn-powder evaporation without catalyst at lower temperature, 41(10), 2006, 3057–3062, Y. Huang, J. He, Y. Zhang, Y. Dai, Y. Gu, S. Wang and C. Zhou, (© Springer Science + Business Media, Inc. 2006) With permission from Springer.[9]

ratio have a low turn-on electric field, a large field enhancement factor and good field emission stability.[7] ZnO nanotetrapods were synthesised and used to construct the flexible light-emitting nanocomposite.[8] The SEM images and the fluorescence photograph of ZnO nanotetrpods are shown in Figure 2.2. T-ZnO is a good candidate for the application of flexible light-emitting materials.

The ZnO nanobelts were obtained at 650 °C with flow of 280 cm³ min⁻¹ (ratio of O_2 1–1.5%) by thermal evaporation of Zn powder. The SEM and transmission electron microscope (TEM) images of the obtained ZnO nanobelts are shown in Figure 2.3. The width is 400–900 nm and the thickness is 10–50 nm. The length of the nanobelts is more than dozens of micrometers.[9]

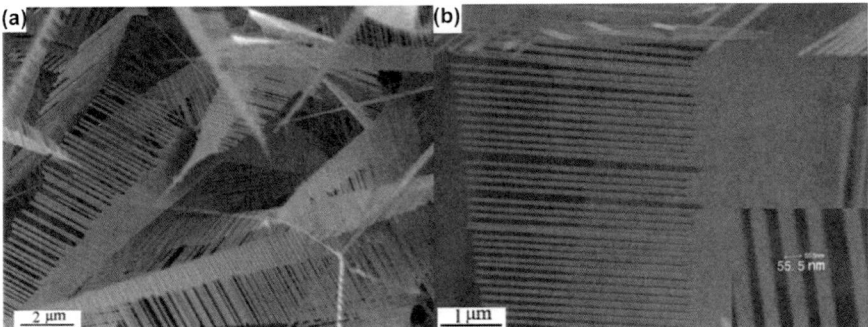

Figure 2.4 SEM morphologies of ZnO nanocombs.
Reprinted from Ceram. Int., 32(5), Y. Huang, Y. Zhang, J. He, Y. Dai, Y. Gu, Z. Ji and C. Zhou, Fabrication and characterization of ZnO comb-like nanostructures, 561–566. Copyright (2006), with permission from Elsevier.[8]

The high-quality comb-like ZnO nanostructures with various morphologies were produced by modulating the reaction flow rates, controlling the synthesis temperature, and thermally evaporating Zn powder.[10] Figure 2.4 reveals the inconsistent width of the backbones, which are non-uniform nanobelts. Along the opposite direction of the teeth, the backbones equivalent to the teeth are thick and other parts are rough and thin. The backbone of the ZnO nanocombs is about tens of micrometers in length and 40–50 nm in thickness. The teeth are 60–100 nm in diameter and 1–2 μm in length. Figure 2.4(b) shows ZnO nanocombs with consistent widths of the backbones equivalent to the thickness of teeth. The backbone of such ZnO nanocombs is about tens of micrometers in the length with 50–70 nm uniform thickness. Besides, the teeth are cone-like that possessed end diameter 30–50 nm, bottom diameter 50–70 nm, and length $\geq$2 μm. The first ZnO nanocombs were fabricated at 650 $^\circ$C with flow rates of 220–260 cm^3 min^{-1} (content of O_2 is 3–5%), and the second ZnO nanocombs were fabricated at 650 $^\circ$C with flow rates of 250 cm^3 min^{-1} (content of O_2 is 2–3%).

Single crystalline ZnO nanowires can be fabricated *via* thermal evaporation, and their cross-sectional geometry can be tuned by adjusting the gas flow rate.[11] The typical morphologies of two different ZnO nanowires are shown in Figure 2.5. By slightly adjusting the growth conditions, the cross-sectional shape of the nanowires can be tuned from hexagonal to circular.

ZnO nanowires with different tip morphologies can also be fabricated by a vapour-solid deposition process, as shown in Figure 2.6.[12] Nanocrystal facet evolution is critical for designing nanomaterial morphology and controlling their properties. Through a dynamically controlled deposition condition of ZnO growth, different crystal facets at the tips of ZnO were obtained. The tip facet evolution as a function of location and corresponding schematic drawing is presented. The relative area of these crystal facets continuously changed following the supersaturation.

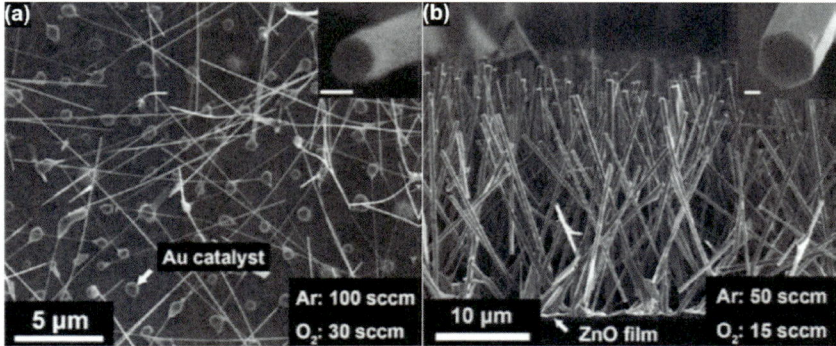

Figure 2.5 (a) SEM top view of the cylindrical NWs grown under higher carrier gas flow rate; (b) Side-view SEM image of hexagonal ZnO NWs synthesised under low carrier gas flow rate.
Reproduced with permission from ref. 11. Copyright (2015) American Chemical Society.

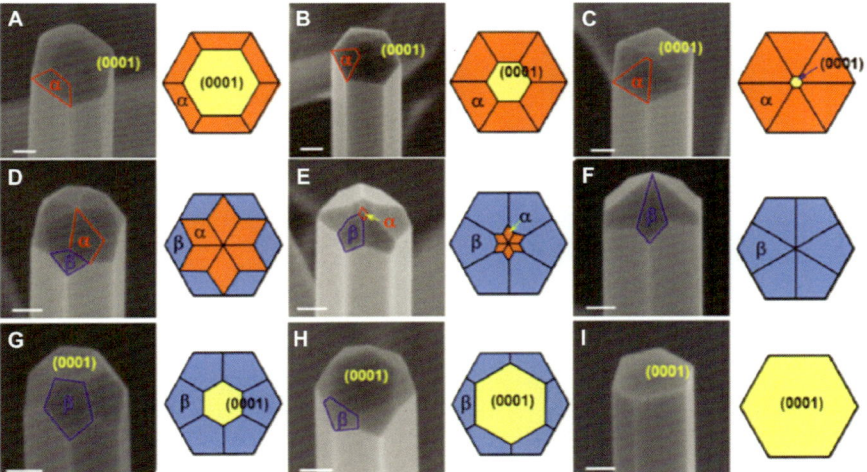

Figure 2.6 SEM images and corresponding schematic showing facets evolution at nanowire tips.
Reproduced with permission from ref. 12. Copyright (2016) American Chemical Society.

In summary, the process of catalyst-free synthesis of ZnO nanostructures by thermal evaporation uses a horizontal tube furnace, to evaporate ZnO powder with O_2 by controlling the temperature, flow rate, pressure, substrate position and location, to obtain the products. Beyond these, we can take some other representative examples to describe catalyst-free thermal evaporation synthesised ZnO.

Industry brass substrates (ϕ25 mm×1 mm) were used as the vapour source to produce ZnO nanobelts which were at 900 °C under 30 Torr.[13] The substrates were located at 450–500 °C. The fabricated ZnO nanobelts are 100–150 nm in width and 2–4 μm in length. Zn powder was used as the vapour source that reacted with H_2O to synthesise nanowires.[14] Preparation conditions are as follows; distilled water placed in a tube furnace entrance, Zn powder located at the central, and the substrate under the Zn vapour. At 730 °C, 50 cm^3 min^{-1} flow rate, the reaction time was 40 min. The obtained ZnO nanowires are 40–60 nm in diameter and more than tens of micrometers in length.

ZnC_2O_4 was pyrolysed at 900 °C to form ZnO vapour.[15] Subsequently, ZnO nanorods were obtained by a VS mechanism. The reaction equation is as follows: $ZnC_2O_4 \rightarrow ZnO + CO + CO_2$, and the diameter and length of the nanorods are 10–60 nm and 1–3 μm, respectively.

2.1.1.2   Thermal Evaporation Chemical Vapour Deposition with Catalytic Process

Usually, when Zn powder acts as source to fabricate ZnO nanostructures by CVD, Au, Co and Cu, or transition element compounds nanoparticles and films were deposited on the substrate as catalysts. The growth mechanism of ZnO nanostructures is a VLS mechanism. The VLS growth usually creates a coating of catalytic films or nanoparticles on a substrate. In the VLS growth of ZnO nanostructures, the Zn source and doping material were evaporated first, and then the gas-phase reactants were absorbed onto the catalyst, which created eutectic alloy droplets. The reactant concentrations would keep increasing in the droplet, and then exceed the saturation point. The supersaturated liquid drop led to the precipitation of the Zn, which was then mixed with oxygen to grow ZnO nanostructures. The catalysts can be achieved by magnetron sputtering, thermal evaporation coating or coating compound and pyrolysis. Besides, the Zn self-catalytic process can be also available to produce ZnO nanostructures.

Au nanoparticles and films are the common catalyst for ZnO nanostructures fabrication by CVD. Core–shell structured ZnO/SiO_x nanowires were synthesised on gold coated silicon (100) substrate by zinc powder evaporation at 550 °C.[16] A silicon substrate (100) was first cleaned ultrasonically in a mixture of HNO_3, HF and H_2O and rinsed subsequently in distilled water and alcohol. Then a thin gold layer about 20 nm in thickness was deposited on the cleaned silicon substrate by magnetron sputtering. The source materials were pure zinc and tin powders with a mass ratio of 1 : 1, which were mixed. The source materials were put in an alumina crucible and the silicon substrate was placed above it. Then the crucible was inserted into the center of the quartz tube in a horizontal tube furnace. Under a constant flow of Ar (98%)/O_2(2%) gas mixture (300 sccm), the temperature of crucible was raised to 550 °C at a rate of about 20 °C min^{-1} and kept at that temperature for 20 min, then cooled to room temperature. The synthesised core–shell structured ZnO/SiO_x nanowires are shown in Figure 2.7. The as-deposited

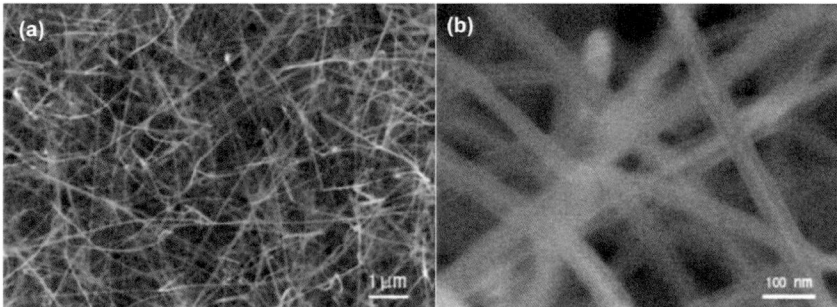

Figure 2.7 Core–shell structured ZnO/SiO$_x$ nanowires.
Reprinted from Mater. Lett., 60(2), J. He, Y. Huang, Y. Zhang, Y. Gu, Z. Ji and C. Zhou, Large-scale synthesis, microstructure and growth mechanism of self assembled core-shell ZnO/SiO$_x$ nanowires, 150–53. Copyright (2005), with permission from Elsevier.[12]

products were dense and wire-like nanostructures on a large-scale. With lengths up to several micrometers, the wires are ultra-fine and have a high aspect ratio of more than 50 and the thickness of the core–shell is 10–20 nm with 30–60 nm diameters. No Sn or other impurities were found in the samples. Sn powders served as an inhibitor to slow down the zinc vapour release rate.

ZnO nanoarrays were deposited on Au coated Si substrate by Zn powder at 500 °C.[17] The experiment method is shown as follows; first a Au layer (2 nm) was coated on the Si substrate (100). Then the substrate was placed above the Zn power with the Au layer facing the source (distance between substrate and Zn powder was 5 mm). Subsequently, the boat was inserted into the furnace and heated to 500 °C under a rate of 20 °C min^{-1}. Meanwhile, Ar and O$_2$ flowed into the furnace for 5 h. The synthesised nanorod was 70–100 nm in diameter and micrometers in the length.

By changing the synthesis conditions, such as reaction temperature, flow (flow type, pressure and rates), deposition temperature, substrate, and ways of placement, research has also produced ZnO nanostructures using Au catalytic Zn powder.[18–23]

ZnO nanostructures can also be synthesised by evaporating the raw materials directly. For example, ZnO nanowires were synthesised in bulk quantities by simple physical evaporation of Zn and Au powder at 900 °C. TEM observation reveals that the Zn/Au alloy catalysed the synthesis of ZnO at the ends. The Au nanoparticles were prepared by the reaction of gold salt (HAuCl$_4$) and sodium citrate. The immersed Zn powders were placed in an alumina boat placed at 900 °C with a flow of Ar and O$_2$ (volume ratio of 9 : 1).[24]

Zinc compounds can also be the raw materials for the catalytic synthesis of ZnO nanostructures. ZnO nanowires were successfully synthesised by thermal evaporation of ZnS powders onto silicon substrates in the presence of Au catalyst. The ZnO nanowires have diameters of about 20–60 nm and lengths up to several tens of micrometers.[25] Besides Au nanoparticles and films, Co is also a common catalyst for the synthesis of ZnO nanostructures. ZnO nanoarrays were fabricated on Co nanoparticle (6–8 nm) deposited Si substrate with a 50 nm diameter and a 13 μm length.[26] After being sulfuretted for 60 min in H_2S diluted with N_2 at 400 °C, the substrate was treated in Zn powder at 500 °C for the growth of ZnO nanoarrays.

The most successful result of NiO catalysis for the growth of ZnO nanostructures was that high quality ZnO nanoarrays deposited on Al_2O_3 substrate through Zn powder resource, NiO catalysis and low temperature (450–500 °C).[27,28] The ZnO nanowire array was fabricated by the following procedure. A nickel nitrate/ethanol solution was dropped onto an alumina substrate. After drying in ambient air, the substrate was loaded on a quartz boat filled with metal zinc powder. Then the quartz boat was inserted into a quartz tube under a constant flow of argon (flow rate: 500 sccm). The vertical distance between the zinc source and the catalysed alumina substrate is about 3–5 mm. The quartz tube was heated up to 450 °C under a constant flow of argon (flow rate: 500 sccm). After reaction at the temperature range of 450–950 °C for 60 min under the argon flow of 500 sccm, the substrate surface appeared white wax-like materials. The ZnO nanowires grown at 450 °C have uniform diameter about 55 nm and lengths up to 2.6 μm. At 500 °C, the diameter and length increase to 190 nm and 15 μm, respectively.

ZnO nanostructures can also be synthesised by self-catalysis. ZnO nanowires were synthesised by thermal evaporation of pure zinc powders under controlled conditions without the presence of catalyst.[29] The zinc powders were placed in an alumina tube that was inserted in a horizontal tube furnace, where the temperature, pressure and evaporation time were controlled. The temperature of the furnace was ramped to 850–950 °C at a rate of 50–100 °C min^{-1} and kept at that temperature for 10–30 min.

The deposition temperature has a great effect on control of the large-scale production and morphologies of ZnO nanostructure. In previous reports, the fabrication temperature of 450 °C for high quality ZnO nanostructures was very low.[27] ZnO nanoarray and ZnO nanowires by CVD further decreased the deposition temperature to 400 °C.[30,31] ZnO nanoarrays were deposited on stainless steel wires coated Au (10 nm) under 100 Pa at 400 °C. In such experiments, Zn powder was the source, the distance between Zn and stainless steel wires was 2–5 mm and Ar and O_2 were introduced into the furnace. The authors suggested that the Se powder located at the entrance of the quartz tube played an important role in the morphologies. The mechanism of the Se powder was still unclear. The ends of the nanorods were shaped and the radius of curvature was 5 nm. No Se element was found at the ZnO nanorods, while ZnSe nanoparticles were detected at parts of the stainless steel wires. In addition, synthesis of ZnO nanowires was reported using a mixture of Zn and Se (mass ratio 7 : 3) powder reacting at 1100 °C with a flow of 100 Torr.[32] This suggests that emanating ZnO nanowire grew along the ZnSe nanoparticles. Se powder catalyses the growth with a VLS mechanism and promotes liquid formation.

2.1.2 Thermal Evaporation Chemical Vapour Deposition with Carbothermal Reduction

During the fabrication process of ZnO nanostructures using carbothermal reduction, ZnO and carbon powder (including activated carbon powder, graphite powder or carbon nanotubes) are often used as reactants. At a controlled temperature and atmosphere, ZnO can be reduced to gaseous Zn vapour and then form ZnO nanostructures through the following possible processes: (1) gaseous Zn turned to liquid droplets (liquid Zn) with the catalyst and super-saturated liquid Zn; (2) gaseous Zn was oxidised to gaseous ZnO and then formed liquid droplets (liquid ZnO) with the catalyst, and solid ZnO was formed from super-saturated liquid ZnO; (3) gaseous Zn was oxidised to gaseous ZnO and then formed solid ZnO directly. Catalysis was needed in process (1) and (2) but not in process (3). Based on the above process, the carbothermal reduction method includes the following chemical reactions.

$$\text{Reduction reaction: } ZnO + C \rightarrow Zn\,(g) + CO/CO_2 \text{ or } ZnO + CO \rightarrow Zn\,(g) + CO_2 \tag{2.4}$$

$$\text{Oxidation reaction: } Zn(l) + O_2 \rightarrow ZnO(s) \text{ or } Zn(g) + O_2 \rightarrow ZnO(g) \tag{2.5}$$

To fabricate ZnO nanostructures by a carbothermal reduction method, catalyst is needed in some processes while in others it is not. Some commonly used catalysts include Au, Tin, Cu, CuO in the forms of particle or film, and they can be prefabricated on the substrate or added in the reactants. Substrates such as Si, Al_2O_3 and quartz are often used for reaction and a transition layer is needed sometimes in facilitating nucleation and growth.

Au is one of the most commonly used catalysts for producing ZnO nanostructures with carbothermal reduction.[33–40] Au particles or film can be evaporated or sputtered on the substrates. In 2001, ZnO nanowires were successfully fabricated through a carbothermal reduction method.[33–35] ZnO and graphite powder were mixed in the ratio 1:1 and reacted at 900–925 °C in an Ar flow of 20–25 $cm^3\,min^{-1}$ for 5–30 min.

ZnO nanorod arrays were deposited on Au coated Al_2O_3 substrates by evaporating ZnO and graphite powder at 950 °C in an Ar flow of 25 $cm^3\,min^{-1}$ and keeping the pressure at 300–400 mbar. The diameters of the ZnO nanorods are around 50–150 nm and the length is about 1.5 μm. Au particles on the tips of the nanorods were also observed, as shown in Figure 2.8.[36]

Sn can also be used as a catalyst for carbothermal reduction, which can be added in the source material instead of pre-coating on the substrates. By evaporating the mixture of ZnO, SnO_2 and graphite powders (atom ratio Zn:Sn:C = 2:1:1) at 1150 °C, ZnO nanorods, nanobelts and hexagonal structures were deposited on Al_2O_3 substrates, which were placed in the range of 550–600 °C with the Ar flow controlled at 20–25 $cm^3\,min^{-1}$ and pressure at 200 mbar.[11] Sn particles can be observed on the tips of the ZnO

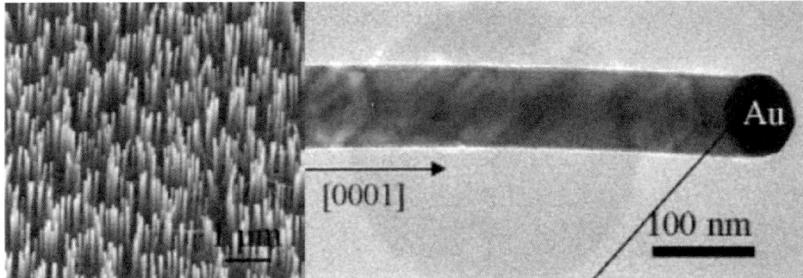

Figure 2.8 ZnO nanorod array fabricated by carbothermal reduction and the Au
particle on the top of nanorod.
Reproduced with permission from ref. 36. Copyright (2004) American
Chemical Society.

nanorods, indicating that SnO_2 is reduced to Sn and acts as a catalyst in the
process, which plays the key role in directional growth of ZnO nanorods.
When the pressure was changed to 2×10^{-3} Torr and the Ar flow was changed
to 25 $cm^3 min^{-1}$, ZnO featuring mesoporous structured polyhedral drums
and spherical cages and shells were formed.[42]

Cu is another commonly used catalyst for production of ZnO nanostructures with carbothermal reduction. For instance, ZnO nanopins were obtained
on Si substrates coated with Cu as the catalyst by evaporating ZnO and
graphite powders at 1150 °C.[43] ZnO nanorods were obtained on Cu coated
Si substrates by evaporating the source materials at 900 °C. CuO was also
reported to be used as a catalyst for the fabrication of ZnO nanostructures.[44]

Under some conditions, ZnO nanostructures can be synthesised through
carbothermal reduction without using a catalyst. In/ZnO nanoplates were
obtained by carbothermal reduction without any catalyst.[45–47] The fabrication process is as follows: Zn powder (purity > 99.9%), In_2O_3 powder
(purity > 99.9%) and carbon powder were ground and well mixed in a ratio
of $n(Zn) : n(In) : n(C) = 1 : 1 : 2$–$3 : 1 : 2$. The temperature was rapidly raised to
870–930 °C whilst controlling the Ar and O_2 flow at 250–350 $cm^3 min^{-1}$ and
10–15 $cm^3 min^{-1}$, respectively. The mixture was evaporated in the tube
furnace under 870–930 °C for 30 min and then cooled to room temperature
without changing the atmosphere. The obtained In/ZnO nanoplates have
two different morphologies.

Needle-like ZnO rods, nanobelts and nanowires were obtained with various morphologies and scales.[48] No metal catalyst, carrier gas or vacuum
conditions were involved in their fabrication process. The evaporation
temperature of ZnO and graphite powder was 1100 °C and the substrates
were put at a temperature of 800–750 °C, 750–650 °C and 650–500 °C.
Nanostructures with different morphologies and scales were obtained by
changing the position and temperature of the substrates. Needle-like ZnO
rods, nanobelts and nanowires were obtained at the above three different
temperature ranges, respectively, as shown in Figure 2.9.

ZnO, PbO_2 and graphite powders (mole ratio 3 : 10 : 15) were evaporated
at 10^{-3} Torr with a N_2 flow of 15 $cm^3 min^{-1}$ and finally deposited ZnO

Figure 2.9 ZnO nanostructures with various morphologies fabricated at different positions of Si substrate.
Reprinted from ref. 48 with the permission on AIP Publishing.

nanoflakes/belts structures at 900 °C on polycrystalline alumina substrates.[49] No evidence was found to support the existence of Pb, thus it was believed that Pb did not act as a catalyst in the process but had a great impact on the morphology of ZnO nanostructures, which is different from the catalysing effect of Sn in a similar fabrication process.[41,42]

Different catalysts have different effects on the nanocrystalline morphology and even nucleation and growth processes. The effects of pre-coated Au, Ag, Ni and Fe films on Si and sapphire substrates on the nucleation, growth and morphology of the products were compared.[50] The research suggested that ZnO nanowires were formed by a vapour–solid process on Fe films while a vapour–liquid–solid process was involved when using Au, Ag and Ni films as catalyst. The study also showed that these differences in catalyst are not only related to the differences in the size and aspect ratio of nanowires, but also the atomic composition ratio of Zn/O, as well as the relative intensity of the oxygen vacancy-related emission in PL spectra. Figure 2.10 shows ZnO nanowires grown on Au, Ag, Ni and Fe films, respectively.[51]

The effect of thickness on the thermally evaporated gold catalyst film on the density of aligned ZnO nanowires on $Al_{0.5}Ga_{0.5}N$ substrates was investigated.[52] The fabrication condition of the ZnO nanowire array was as follows. A mixture of ZnO and graphite powders (weight ratio 1 : 1) was quickly heated to 950 °C with a gas flow of 50 $cm^3\,min^{-1}$ (2% oxygen containing). The deposition process was conducted by placing the substrates coated with 1–8 nm Au film downstream in a temperature zone of ~850 °C under a pressure of 30 mbar for 30 min.

In addition, the influence of the total system pressure and partial pressure of O_2 on the morphologies of ZnO nanowire arrays under carbothermal reduction conditions was investigated.[53] It revealed that the total system pressure and partial pressure of O_2 have prominent influence on the morphology, size and distribution of ZnO nanowire arrays and there exists an optimal total system pressure and partial pressure of O_2 under a certain temperature.

Besides catalysts, different deposition substrates can be used for fabrication of ZnO nanostructures by carbothermal reduction. Beside Si and sapphire substrates, other commonly used substrates include graphite, AlN and W substrates.[54,55]

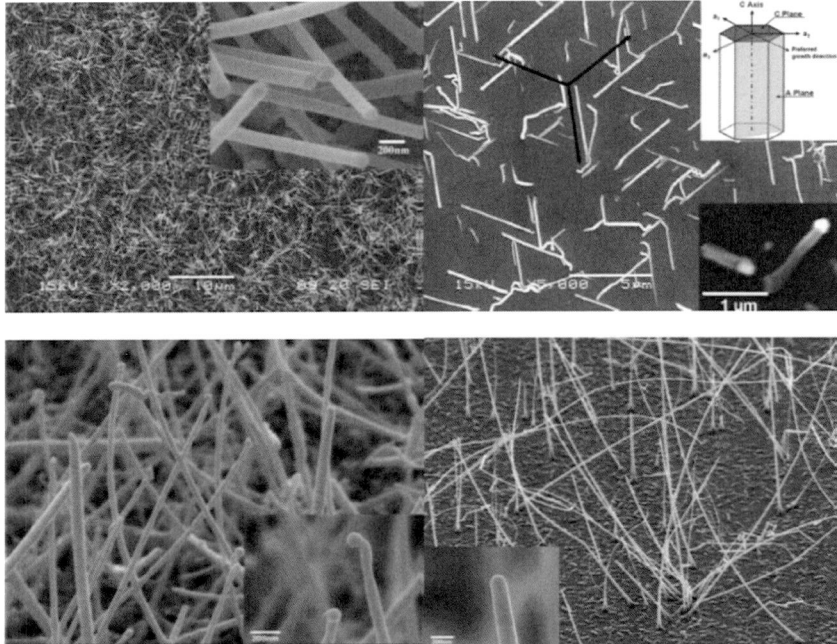

Figure 2.10 ZnO nanowires and nanorods fabricated on substrates coated with Au, Ag, Ni and Fe films.
Reproduced with permission from ref. 51. Copyright (2005) American Chemical Society.

2.1.3 Metal–organic Chemical Vapour Deposition

Metal–organic chemical vapour deposition (MOCVD) is a method that uses metal–organic compounds as source materials and includes the processes of evaporation, reaction and deposition. The low evaporation temperature of metal organics can lower the temperature of the fabrication process. The commonly used organic zinc compounds include zinc acetylacetonate ($Zn(C_5H_7O_2)_2$), diethylzinc ($Zn(C_2H_5)_2$, DEZn) *etc.* $Zn(C_5H_7O_2)_2$ is a commonly used source material for fabrication of ZnO nanostructures by the MOCVD method. ZnO nanorods were obtained on Si substrates at 500 °C by evaporating $Zn(C_5H_7O_2)_2 \cdot xH_2O$ at 135 °C and using O_2/N_2 as carrier gas. No catalyst was used in the fabrication process and the diameter of the nanorods could be controlled by adjusting the evaporation temperature of $Zn(C_5H_7O_2)_2$.[56]

The effect of substrates on the quality of ZnO nanorods during the MOCVD process was investigated.[57] $Zn(C_5H_7O_2)_2$ was heated up to 120 °C and carried by O_2 (500 $cm^3 min^{-1}$) and N_2 (200 $cm^3 min^{-1}$) to the substrates where the temperature was 500 °C. The use of sapphire (11–20) substrates has been demonstrated to result in the formation of ZnO films with better quality than that of ZnO films formed on sapphire (0001) substrates. $Zn(C_5H_7O_2)_2$ was evaporated to 105–125 °C and obtained ZnO whiskers on glass substrates at 500–600 °C.[58]

$Zn(C_2H_5)_2$ is another source material that can be used to fabricate ZnO nanostructures by the MOCVD method. $Zn(C_2H_5)_2$ has very low boiling point and is gaseous at room temperature and pressure, so it can flow into the reaction chamber with carrier gas and reactant gas. When $Zn(C_2H_5)_2$ is used as the Zn source, a catalyst may or may not be used. ZnO nanorods were fabricated on sapphire substrates coated with a ZnO transition layer at 400 °C by using $Zn(C_2H_5)_2$ (20–100 $cm^3 min^{-1}$) as source material and O_2 (0.5–5 $cm^3 min^{-1}$) as carrier gas.[59,60] The ZnO transition layer did not act as a catalyst, but facilitated nucleation and growth of ZnO. ZnO nanorods were deposited on Si substrates at 500 °C without any catalyst or transition layer.[61] ZnO nanorods were obtained on GaAs (002) substrates.[62] With the development of 2D nanostructures, ZnO nanostructures were grown on graphene layers by using catalyst-free low-pressure MOCVD. A selective growth of position-controlled ZnO nanostructures was obtained even without the aid of a growth mask.[63]

By using $Zn(C_2H_5)_2$ as a source material, a layer of Au particles with a diameter of 20 nm as catalyst was firstly coated on SiO_2/Si substrates and then fabricated ZnO nanorods at 400 °C, 550 °C and 900 °C, respectively. Au particles are typical catalyts for the formation of ZnO nanorods, which can be confirmed by Au particles observed on the tips nanorods by TEM. ZnO nanostructures were also fabricated with different morphologies through changing the density of Au particles and deposition temperature.[64] When the density of Au particles was $4\times10^9 cm^{-2}$ and temperatures were 400 °C, 500 °C and 550 °C, the products were nanorods, nanonails and nanowires, respectively. When the density of Au particles was $3\times10^9 cm^{-2}$, the corresponding products were nanorod arrays, nanowires and randomly distributed nanorods.[65]

The MOCVD method is similar to other vapour deposition methods where factors including reaction atmosphere, temperature, catalyst and substrates can influence the morphology and quality of the products.[66–71]

2.1.4 Thermal Evaporation Physical Vapour Deposition

The process of fabricating ZnO nanostructures by the PVD method is to evaporate ZnO powders into the gas state and then deposit into ZnO nanostructures. The fabrication method for ZnO nanostructures (or doped ZnO) using thermal evaporation physical vapour deposition can be catalogued into two ways: a process (1) without catalyst and (2) with catalyst.

2.1.4.1 *Thermal Evaporation Physical Vapour Deposition without Catalyst*

There are more reports on the fabrication of ZnO nanostructures by a PVD process without catalysts. Since the evaporation temperature of ZnO is very high (1975 °C), a high evaporation temperature is needed to fabricate ZnO nanostructures by this method. The commonly used temperature is 1350–1400 °C, while some studies also use 1300 °C in the process.[72–75]

ZnO nanobelts were fabricated on Al_2O_3 substrates by evaporating ZnO powders at 1400 °C without catalyst for 2 h and keeping the pressure at 300 Torr and Ar flow at 50 $cm^3 min^{-1}$.[72] The obtained nanobelts have a rectangle-like cross section with typical widths of 30–300 nm, width-to-thickness ratios of 5 to 10, and lengths of up to a few millimeters.

In addition, single-crystal complete nanorings of ZnO were also fabricated by similar methods.[73] The fabrication process is as follows. ZnO, In_2O_3 and Li_2CO_3 powders were mixed in a weight ratio of 20:1:1 and then heated to 1400 °C at 10^{-3} Torr for a few minutes. Then Ar carrier gas was introduced at a flux of 50 $cm^3 min^{-1}$ and the synthesis process was conducted at 1400 °C for 30 min. The products were deposited onto Si substrates placed in a temperature zone of 200–400 °C under Ar pressure of 500 Torr. The as-synthesised complete rings have typical diameters of 1–4 nm, thicknesses of 10–30 nm and widths of 0.2–1 μm.

One can also deposit a transition layer on Si substrates prior to the fabrication of ZnO nanostructures in facilitating nucleation and growth of nanowires. For instance, ZnO film was fabricated with a thickness of 200 nm on Si (100) substrates through pulsed laser deposition method and then ZnO powders were evaporated at 1350 °C and 10^{-3} Torr.[74] Without using any catalyst or carrier gas, they deposited ZnO nanorods with a diameter of 150–300 nm and width of 5.5 μm on the ZnO film.[75] The deposited ZnO nanowires on the Al_2O_3 film have diameters of 5–40 nm and lengths of 20–80 μm.

In a process of fabricating ZnO nanostructures by the PVD method, the typical source material is ZnO powders. For the fabrication of doped ZnO nanostructures, other oxide powders should be added into the source material in a ratio. The catalyst on the surface of substrates is typically gold or other elements in the form of nanoparticles or films, which can be achieved by sputtering, evaporation, spin-coating or thermal deposition of the coated compounds.

Au particles were fabricated as catalyst on sapphire and Si substrates and used to obtain epitaxial ZnO nanorods on catalyst particles by evaporating ZnO powders at high temperature.[76] High-quality pure ZnO nanowires and In-doped ZnO with different doping degrees were obtained.[77] The pre-coated ethanol solution of $HAuCl_4 \cdot 3H_2O$ on the Si substrates can be deposited into Au catalyst particles in a tube furnace. The chemical composition can be modified by adjusting the evaporation temperature. ZnO nanowires were obtained at 800 °C, $Zn_{0.85}In_{0.15}O$ nanowires at 900 °C and $Zn_{0.75}In_{0.25}O$ nanowires at 1000 °C.

The mixed powders containing highly pure ZnO and InO with mole ratio of 18:1 were evaporated, and In-doped ZnO nanobelts with atom ratio of In:Zn = 1:30 were deposited on Si substrates coated with Au film of 2 nm in thickness.[78] The width and thickness of the nanobelts are 100–500 nm and 10–30 nm, respectively, and the length of most nanobelts exceeds 10 μm. Other fabrication parameters include: evaporation time of 30 min, temperature zone for substrates of 800–1000 °C, Ar flow of 50 cm^3 min^{-1} and pressure of 50 Torr.

2.1.5 Pulsed Laser Deposition

Pulsed laser deposition (PLD) is a deposition method that uses a pulse laser beam of high energy source of rapid heating evaporation (target) is used to deposit nanometre materials. PLD can be used to the preparation of nano-structures. Through PLD preparation of one dimensional ZnO nanomaterial, ZnO is usually used as a heating material to prepare ZnO by PVD method. This method of nanorod preparation was. introduced early.[79,80] In pulsed laser deposition ZnO nanostructures, substrate types can be Si, GaN, Al$_2$O$_3$ and so on. The surface state includes a few classes such as catalyst layer, catalyst and prefabricated transition layer.

The nature of the pulsed laser size morphology of nanostructures will have an impact. When the target was heated by a nanosecond laser, the heated evaporated particle size could reach micrometer level, thus leading to an uneven size of the deposited products. A femtosecond (10^{-15} s) laser pulse has very high energy density, which makes the evaporation process of target a non-equilibrium process; the particle size is very small, achieving true nanometre level, and the particle size distribution range is small, which is advantageous to the sedimentary formation of tiny uniform products. A femtosecond laser (100 fs pulsed lase, wavelength of 800 nm) was used to prepare ZnO target material, and ZnO nanorods were deposited on the substrate of Au film.[81] Deposition pressure is 1 atm (1 atm $= 1.01 \times 10^5$ Pa), and the temperature is 900 °C. ZnO nanostructures were fabricated on GaN, Al$_2$O$_3$, and Si substrates using a high-pressure PLD method.[82] Vertically aligned hexagonal-pyramidal ZnO nanorods were obtained on the Al$_2$O$_3$ and Si substrates whereas interlinked ZnO nanowalls were obtained on the GaN substrates, as shown in Figure 2.11. A growth mechanism has been presented for the formation of ZnO nanowalls based on different growth rates of ZnO polar and non-polar planes.

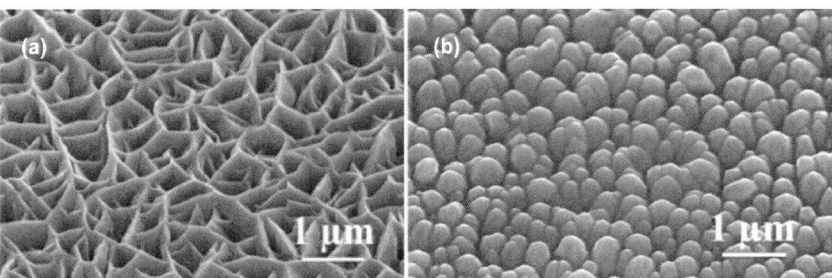

Figure 2.11 SEM images of ZnO nanowalls and nanorods.
Reproduced with permission from ref. 82. Copyright (2010) American Chemical Society.

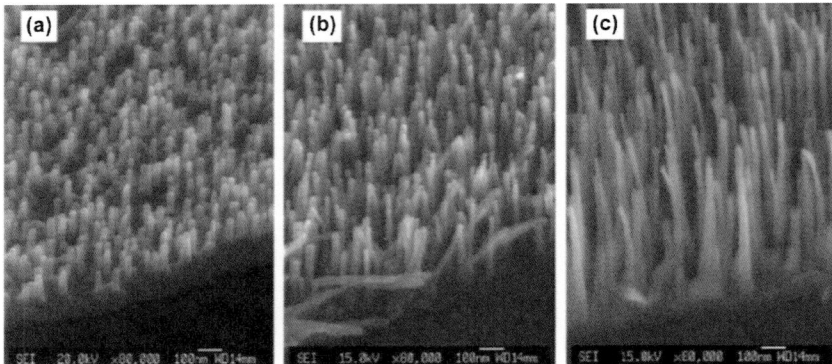

Figure 2.12 ZnO nanorod arrays fabricated by PLD method.
Reprinted from Chem. Phys. Lett., 396(1–3), Y. Sun, G. M. Fuge and M. N. R. Ashfold, Growth of aligned ZnO nanorod arrays by catalysts-free pulsed laser deposition methods, 21–26. Copyright (2004), with permission from Elsevier.[83]

The ArF laser heating ZnO target material was used to prepare ZnO nanorod arrays, as shown in Figure 2.12.[83] The deposition times of Figure 2.12(a), (b) and (c) are 15 min, 30 min and 45 min, respectively. The other preparation conditions are as follows: the energy density of the laser was about 100 J cm^{-2}, and the distance of Si substrate to target was 40–60 mm. Si substrate was heated by CO_2 laser, and the temperature was tuned from room temperature to 700 °C. The experimental deposition process, the substrate temperature was 600 °C, pure O_2 was added to the vacuum chamber with 10 cm^3 min^{-1}, the O_2 pressure was kept simultaneously at 10 mTorr.

Pulsed laser power density, reaction temperature and deposition temperature, and reaction of sedimentary chamber pressure, reaction sedimentary chamber gas ratio and the type of flow velocity and the substrate, surface state, substrate and the distance of the target material, *etc.*, will have a certain influence on the morphology and size.[84–87]

2.1.6 Molecular Beam Epitaxy

The molecular beam epitaxy (MBE) preparation process happens under ultrahigh vacuum (10^{-8} Pa above). The material molecular flow was sprayed onto the substrate directly at a certain temperature and a thin layer could be fabricated, which is also the method used in the fabrication of 1D ZnO nanostructures. MBE, including laser molecular beam epitaxy and microwave molecular beam epitaxy, can control components and the doping concentration and the atomic layer can be developed. However, because the current equipment and its maintenance cost is high, there is a need for ultrahigh vacuum and complex operation and product growth speed is slow, so the application range is not very extensive.

MBE of ZnO nanowires and related research has been reported.[88] ZnO nanowires were deposited on the Si/Al_2O_3 substrate that was covered with a discontinuous Au/Ag film layer. The preparation process is as follows: a preparation chamber pressure of about 5×10^{-8} mbar, O_3/O_2 as oxidation air, through the plasma ozone generator to keep the proportion of O_3/O_2 at 1–3%, with high purity metal Zn as the source of zinc, and O_3/O_2 partial pressure of gas mixture by ionisation gauge located in the front of the substrate, O_3/O_2 mixed gas pressure control at 5×10^{-6}–5×10^{-4} mbar, zinc pressure control at 5×10^{-7}–4×10^{-6} mbar, and Si substrate deposition of a 2–20 nm thick layer of Ag film. The substrate temperature was 300–500 °C, sedimentary 2 h, which can result in the growth on the substrate of ZnO nanowires.

In addition, the microwave plasma epitaxial growth method was used to prepare zinc nanowires, zinc/ZnO nanocables and ZnO nanowires.[89] ZnO nanorods were prepared by a microwave plasma epitaxial growth method.[90]

2.1.7 Magnetron Sputtering

Magnetron sputtering is a film preparation method that is commonly used; however, it is rarely used to prepare nanostructures. Sputtering is used to charge particles by bombarding the target material, causing the atoms or molecules to be sputtered on the target material and deposited on the substrate surface. Depending on the target material the process of deposition can be divided into common sputtering or reactive sputtering. In the preparation of thin films, for example, if the target is ZnO, a sedimentary process without chemical change is common sputtering. If the target material is zinc, zinc and the environment in the process of a deposition reaction of O_2 in atmosphere generated ZnO is reactive sputtering. The magnetron sputtering method requires a high vacuum, suitable sputtering power and substrate temperature, and protective gas with high purity (Ar commonly) for the O_2 reaction gas.

With regards to the working power supply, magnetron sputtering can be divided into different radio frequency (RF) magnetron sputtering and direct current (DC) magnetron sputtering. For radio frequency (RF) magnetron sputtering with a crystal as a general RF oscillator, the RF frequency is generally at 5–30 MHz, sputtering target materials are generally ZnO powder from sintering of ceramics, to ensure the stoichiometric ratio, on average mixed with a certain ratio of O_2 in a sputtering atmosphere. The DC magnetron sputtering with metal zinc acts as a target material, commonly with a mixture of Ar and O_2 gas for a sputtering atmosphere. In the process of sputtering, the launch of the electronic gun is accelerated by an electric field, bombarding the cathode target material, through the momentum exchange, and will target the material to atoms, ions and secondary electron detachment.

For nanostructures prepared by the magnetron sputtering method, generally by RF magnetron sputtering, the target material is ZnO without

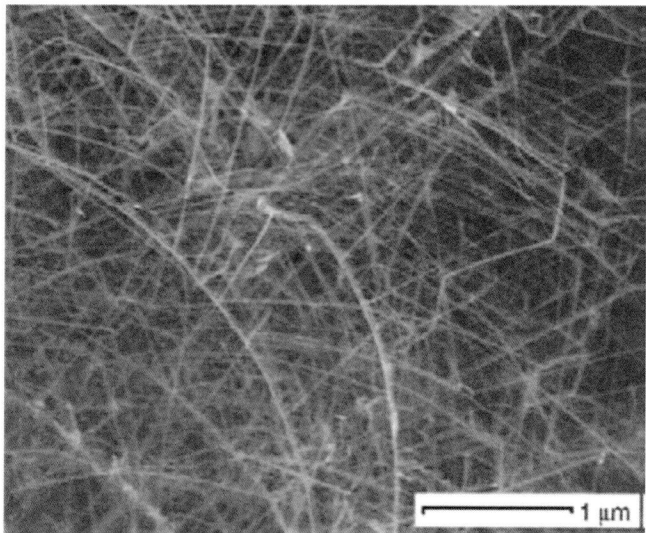

Figure 2.13   ZnO nanowire on Cu substrate fabricated by magnetron sputtering method.
Reprinted from S. Choopun, N. Hongsith, S. Tanunchai, T. Chairuangsri, C. Krua-in, S. Singkarat, T. Vilaithong, P. Mangkorntong and N. Mangkorntong, Single-crystalline ZnO nanobelts by RF sputtering, J. Cryst. Growth, 282(3–4), 365–369. Copyright (2005), with permission from Elsevier.[91]

chemical change. In the process of deposition, the sputtering parameters, such as deposition temperature, time, pressure, sputtering power and electrode distance will affect the appearance and quality of the product.

ZnO nanowires with diameters of 10–50 nm were fabricated on a Cu substrate by the magnetron sputtering method and the length of the ZnO nanowires reached a few microns.[91] The growth condition was as follows: pressure of 0.04 Torr, sputtering power of 300 W and deposition time of 60 min. The fabricated ZnO nanowires are shown in Figure 2.13. In addition, ZnO nanorods were deposited by the magnetron sputtering method on a Cu catalyst layer with different thicknesses.[92,93] An 80-nm thick layer of Ti was prepared on Si or glass substrate and then a layer of Cu was deposited as catalyst. The Ti layer is to prevent the spread of between the Cu and Si substrate.

2.2   Liquid-phase Reaction Methods

The synthesis of ZnO nanostructures has been reported *via* various solution-based methods, which can be classified as hydrothermal, solvothermal and electrodeposition methods as per the solvents used. The typical reagents for electrodeposition are like those used for hydrothermal growth, but the growth ratio is higher by several orders of magnitude than that of the hydrothermal method. Hydrothermal is the most common synthesis method for ZnO nanostructures out of the solution-based methods. Hydrothermal synthesis has a low cost, low growth temperature and can be used for mass production.

2.2.1 Liquid-phase Direct Reaction Method

2.2.1.1 Hydrothermal Reaction Method

The hydrothermal reaction is the method that uses chemical reactions to prepare nanostructures in aqueous solution under certain temperature and pressure conditions. Under the hydrothermal conditions, water can be used as a kind of chemical constituent, and participates in the reaction of chemical composition. It can also be solvent, mineraliser and transfer media of pressure at the same time. Under high pressure conditions, the majority of reactants can be partly dissolved in water, which is of great benefit to prompting reaction in the liquid- or gas-phase. The common types of hydrothermal synthesis include hydrothermal oxidation, hydrothermal reduction, hydrothermal synthesis, hydrothermal decomposition, hydrothermal precipitation, hydrothermal crystallisation reaction, hydrothermal chemical reaction, microwave hydrothermal method, ultrasonic hydrothermal method, *etc.*

The general procedures of synthesising crystal materials using hydrothermal reaction include the design and determination of reaction material ratio, batching and mixing, putting reaction materials into a kettle and sealing, adding pressure, confirming reaction temperature, time and condition, sampling, and sample characterisation. The size of the prepared materials will be influenced by the concentration of reaction materials, reaction temperature and time.

In 2001, the hydrothermal reaction method was adopted for the preparation of ZnO nanorod arrays.[94,95] The process was as follows: first the reaction solution (0.01–0.001 M) was obtained by mixing an equal molar ratio of hydrated zinc nitrate ($Zn(NO_3)_2 \cdot 6H_2O$) and hexamethylenetetramine (HMTA, $(CH_2)_6N_4$). Then, a substrate, such as F–SnO$_2$ glass, single crystal alumina, a Si/SiO$_2$ slice or one covered with ZnO film, was put into the reaction solution at 95 °C for several hours. Finally, ZnO nanorod arrays were obtained after cleaning and drying the reaction product in air. The diameter of the ZnO nanorod arrays can be controlled by the concentration of the reaction solution. When the concentration was 0.01 M, the diameter was 100–200 nm. A 10–20 nm diameter was obtained in the 0.001 M reaction solution.

Although development and optimisation was conducted, the hydrothermal reaction method was always adopted by many people as the simple prepared process.[96–98] In order to improve the quality of product and the controllability of the sedimentary process, the main development of the method was to prefabricate a seed layer on the substrate orto add surfactant in the reaction system of $Zn(NO_3)_2 \cdot 6H_2O$ and HMTA.[99–114]

In 2002, the hydrothermal synthesis process of ZnO nanorod arrays using a two-step hydrothermal reaction was further developed.[99] The method can be used to fabricate ZnO nanorod arrays onto every substrate, including 4 on Si substrate and 2 on plastic substrate. First, the homogeneous seed layer with 50–200 nm was obtained by sprinkling the ZnO nanostructure with 5–10 nm on 4 in Si crystal slice. Then, the sample was annealed at 150 °C to form strong adhesion between the nanostructure and the substrate. Then, ZnO nanostructures were synthesised in a reaction solution at 60 °C with stirring for 2 h. The reaction solution was obtained by 0.03 M NaOH dissolved in methanol solution dropped into another methanol solution of 0.01 M $Zn(NO_3)_2 \cdot 6H_2O$. The synthesised ZnO nanocrystal was a sphere and could be preserved in the solution for two weeks. The Si substrate coated with ZnO nanostructure was suspended in an aqueous solution of HTMA and $Zn(NO_3)_2 \cdot 6H_2O$ with a concentration of 0.025 M at 90 °C. During the reaction process (about 0.5 to 6 h), the substrate was suspended in the reaction solution. Finally, the product was cleaned and dried. It can be observed that when the growth time is 1.5 h, the diameter of ZnO nanorod is 40–80 nm, and the length is 1.5–2 μm. The diameter and the length can be increased up to 200–300 nm and 3 μm with a preparation time of 6 h.

The seed layer, with a thickness of 60 nm, can be prepared by the magnetron sputtering method in Si substrate of >2 in diameter.[99–102] The ZnO nanorod arrays were prepared by a hydrothermal reaction with a growth time of 6 h at 90 °C, which is shown in Figure 2.14. The concentration of the reaction solution was 0.05 M. A ZnO seed layer with a thickness of 50–100 nm was firstly prefabricated by a sol–gel method. Equimolar aqueous solutions (0.025 M) of HTMA and $Zn(NO_3)_2 \cdot 6H_2O$ were used for the growth of ZnO nanorod arrays at 90 °C for 3 h. The influences of the lower temperature heat treatment on the morphology of ZnO nanostructures were investigated.[101] In addition, bisectional ZnO nanowire arrays were fabricated. The growth process includes two steps: hydrothermal growth and CVD growth. The first ZnO nanowire array layer was fabricated by hydrothermal growth. Then, the second

Figure 2.14   ZnO nanorod arrays fabricated by hydrothermal method. Reprinted from ref. 99 with permission of AIP Publishing.

Figure 2.15 The aspect ratio of ZnO nanowire increased by adding of PEI. Reproduced with permission from ref. 111. Copyright (2006) American Chemical Society.

ZnO nanowire array was fabricated by a CVD method on the first ZnO nanowire array layer. The growth mechanisms of this special ZnO nanostructure were proposed.[102]

It should be noted that the ZnO nanorod arrays with a length-diameter ratio of >50 are hard to prepare by only using aqueous solutions of HTMA and $Zn(NO_3)_2 \cdot 6H_2O$. Figure 2.15 shows that the aspect ratio can increase to more than 125, when polyethylenimine (PEI) is added into the reaction solution.[111] The size of the product can be controlled by the concentrate of the reaction solution, temperature and growth time. Instead of PEI, sodium citrate with low concentration can also change the morphology of the ZnO nanorod into a prismatic shape.

In the procedure of the two-step method, the seed layer can be fabricated onto the substrate by sputtering, pulsed laser deposition, coating film or the sol–gel method.[104–109,115,116] The layer can also be deposited by hydrothermal reaction. Then, the ZnO nanorod arrays are prepared by another hydrothermal synthesis process.[110,112–114]

In addition, three-dimensional (3D) branched nanowire heterostructures could be fabricated using the hydrothermal growth method.[117] The 3D nanostructure are shown in Figure 2.16. The ZnO nanowire branches were grown on the Si nanowire cores using the hydrothermal growth method. The chemical bath deposition for the synthesis of ZnO nanostructures has the characteristics of low temperature and one step. By introducing ammonium citrate as a ligand and an ammonium halide as a buffering additive, dense, highly transparent, and conductive ZnO films can be fabricated.[118]

Usually, the hydrothermal method for fabricating ZnO nanostructures includes two steps: (1) the fabrication of a seed layer and (2) the growth of ZnO nanostructures using the hydrothermal method. There are many seed layer fabrication methods, such as sol–gel, magnetron sputtering, atomic layer deposition (ALD), *etc*. ZnO nanorods were synthesised *via* a two-step

Figure 2.16   SEM images of 3D ZnO/Si branched NW heterostructures.
Reproduced with permission from ref. 117. Copyright (2013) American
Chemical Society.

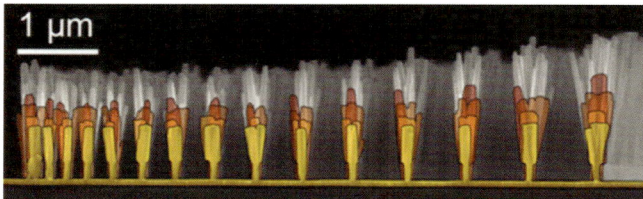

Figure 2.17   Cross-sectional SEM images of ZnO nanowire arrays with graded pitch
varying grown with increasing citrate concentration.
Reproduced with permission from ref. 120. Copyright (2016) American
Chemical Society.

hydrothermal method on Si nano-textured solar cells.[119] Hydrothermally
synthesised ZnO nanowire arrays are critical components in a range of
nanostructured semiconductor devices. The device performance is gov-
erned by relevant nanowire morphological parameters that cannot be fully
controlled during bulk hydrothermal synthesis due to its transient nature.
Cross-sectional SEM images of ZnO nanowire arrays with graded pitch are
shown in Figure 2.17. Independent tailoring of nanowire array dimensions
including areal density, length, and diameter was demonstrated by
employing continuous flow synthesis.[120] With the development of mixed-
dimensional nanostructures, hydrothermal growth is usually used to fabricate ZnO nanostructures on two-dimensional nanostructures.[121,122] The
hydrothermal growth is also an important method for fabricating ZnO
based composites.

The chemical reaction procedure of the synthesised ZnO nanostructures using HTMA and $Zn(NO_3)_2 \cdot 6H_2O$ is as below:

$$(CH_2)_6N_4 + 6H_2O \rightarrow 6HCHO + 4NH_3 \tag{2.6}$$

$$NH_3 + H_2O \leftrightarrow NH_4^+ + OH^- \tag{2.7}$$

$$2OH^- + Zn^{2+} \rightarrow Zn(OH)_2 \rightarrow ZnO + H_2O \tag{2.8}$$

In the above process, the ZnO was synthesised in both the solution and on the surface of the substrate. Homogeneous nucleation exists in the solution, and the heterogeneous nucleation appears on the surface of substrate.

The preparation of ZnO nanostructures using a hydrothermal reaction also has some limitations. First, the products are mainly nanorod and nanorod arrays, not as many as the ones synthesised by the vapour-phase method. The second is the limitation of the length/diameter ratio. It is hard for the length of the hydrothermally synthesised ZnO nanorods to reach dozens of micrometres.

The factors affecting the morphology and quality of ZnO nanostructures using a hydrothermal reaction are:

(1) Solution composition. The length/diameter ratio of the ZnO nanowire can be increased by adding polyethylenimin into the reaction solution of $Zn(NO_3)_2$ and HTMA. If sodium citrate replaces polyethylenimin, the cylindrical nanorods will become prismatic.[111]

(2) Concentration of reaction solution. The diameter of the ZnO nanorod decreases by reducing the concentration of the reagents.[123] The morphology of the ZnO nanorod will, in turn, become prismatic when the concentration of activator is at a low level in the reaction system of Zn^{2+}, OH^- and ethanediamine.[124]

(3) pH of the reaction solution. The morphology and size of the ZnO nanorod can be controlled by adjusting the pH of the aqueous solution of $Zn(CH_3CO_2)_2$ and ethanediamine/citric acid using NaOH.[125]

(4) Temperature and time. The diameter and length of the nanorods will increase as the reaction time and temperature increase.[99]

2.2.1.2 Solvothermal Reaction

The solvothermal reaction is the method that uses organic substances in the solution of chemical reaction to prepare nanostructures. The method expands the application range of the hydrothermal method, although the reaction principle is same.

Except for using $Zn(NO_3)_2$ in aqueous reaction solution, the ZnO nanostructures can also be synthesised by using a zinc salt in the organic reaction solution. ZnO nanowires with a diameter of 30 ± 5 nm and a length of 0.5–5 μm can be obtained by heating the solution of $Zn(CH_3CO_2)_2$ and trioctylamine at 300 °C.[111] A small amount of oleic acid will promote the process of the reaction. Co-doped ZnO nanowires have been synthesised by adding $Co(CH_3CO_2)_2$ into the reaction solution of $Zn(CH_3CO_2)_2$ and trioctylamine.

The solvothermal reaction was adopted to prepare superfine ZnO nanorods (ZnO quantum rod).[126] The process is: in a typical reaction, 4 mmol of dry zinc acetate $(Zn(CH_3CO_2)_2)$ is added to a mixture containing 15 mL of trioctylamine and 3 g of oleic acid (12 mmol) at room temperature. The resulting mixture is heated rapidly to 286 °C for 10–15 min, and the mixture changes to a yellowish solution. The temperature is maintained at 286 °C for 1 h under N_2 flow. The particles precipitate by adding ethanol after cooling the reaction mixture to room temperature and are then separated and cleaned by repeated precipitation of the hexane solution with ethanol. Finally, ZnO nanorods with diameters of 2 nm and lengths of 40–50 nm are prepared.

The reaction solution based on $Zn(CH_3CO_2)_2$, $ZnCl_2$, NaOH, methanol, and ethanol solution was reported.[127] The solution was stirred for 30 min. Then the autoclaves were airproofed and heated at a constant temperature for several hours in a regular laboratory oven. The products were washed by ethanol and water several times.

In addition to the above solution system for the preparation of ZnO nanomaterials using solvothermal reaction, common reaction groups include $Zn(CH_3CO_2)_2$-methanol-NaOH, $Zn(CH_3CO_2)_2$-methanol-ammonium hydroxide, $Zn(CH_3CO_2)_2$-paroline-stearic acid, $Zn(CH_3CO_2)_2$-$[C_2OHmim]^+BF_4$-NaOH, *etc.*[128–131]

2.2.1.3 Microemulsion Method

Generally, microemulsion is an isotropic, transparent or translucent, stable dispersion system, which is formed from two kinds of each phase solution under the action of surfactants. In common, two kinds of reactants that are soluble in the same two microemulsions are then mixed under certain conditions. Two reactants encounter each other by material exchange and then react. Through the overspeed centrifugal force, the reaction products and reacted microemulsion are separated. An organic solvent is used to clean and remove oil and surfactant, which are on the surface of the product. Finally, the required nanostructures are obtained by drying at a certain temperature.

Nanorods synthesised by the microemulsion method have been reported.[132,133] The microemulsions consisted of 2.77 g sodium dodecyl benzene sulfonate and 1 mmol of $ZnAc_2 \cdot 2H_2O$ both dispersed in 30 mL xylene by intensively stirring and sonicating until a homogenous slightly-turbid appearance of the mixture was obtained. Hydrazine monohydrate (four-fold diluted with ethanol) was added dropwise to the well-stirred mixture at room temperature by simultaneous vigorous agitation. To reach a thorough reaction, the stirring process must last for at least 1 h after completing the dropping of hydrazine monohydrate. The resulting precursor-containing mixture was subsequently heated to the boiling point of xylene for refluxing.

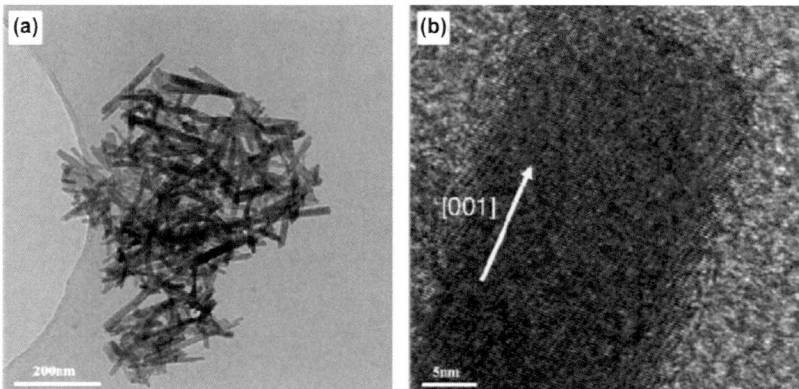

Figure 2.18 ZnO nanorods fabricated by the microemulsion method.
Reprinted from J. Lumin., 122, Y. Z. Lv, Y. H. Zhang, C. P. Li, L. R. Ren, L. Guo, H. B. Xu, L. Ding, C. L. Yang, W. K. Ge and S. H. Yang, Temperature dependent photoluminescence of ZnO nanorods prepared by a simple solution route, 816–818. Copyright (2007), with permission from Elsevier.[133]

After refluxing for 5 h, a milky-white suspension was obtained and centrifuged to separate the precipitate, which was rinsed with absolute ethanol and distilled water several times. Subsequently, the volatile solvent was evaporated under vacuum at 70 °C and finally a loose white powder was yielded. The diameter of the ZnO nanorods is 16 nm, and the length is 200 nm. Figure 2.18 shows the TEM and HRTEM images of the ZnO nanorods.[133]

Aside from this, ZnO nanorods synthesised by solvothermal methods have been reported.[134] The solvothermal method is a common method to synthesise nanopowders. When the method is used to prepare nanostructures, the product morphology is hard to control and the length-diameter ratio is generally low. Therefore, the solvothermal method is not a common method for preparing ZnO nanostructures.

2.2.2   Electrochemical Deposition

In electrochemical deposition (ECD) or electrodeposition fabrication of one-dimensional ZnO nanostructures, compared with general chemical solution methods (such as hydrothermal reaction), an electric field is introduced as an auxiliary. Normally in the three-electrode electrochemical cell (cathode, anode and reference electrode), nanorods or nanowires on the substrate are grown by negative potential deposition (electrolyte solution for zinc salt aqueous solution).

In 1996, ZnO film and ZnO nanorods were fabricated by an electrodeposition method.[135,136] The process of electrodepositing ZnO nanorods is as follows. First, a patterned Au film is fabricated on a Si substrate by

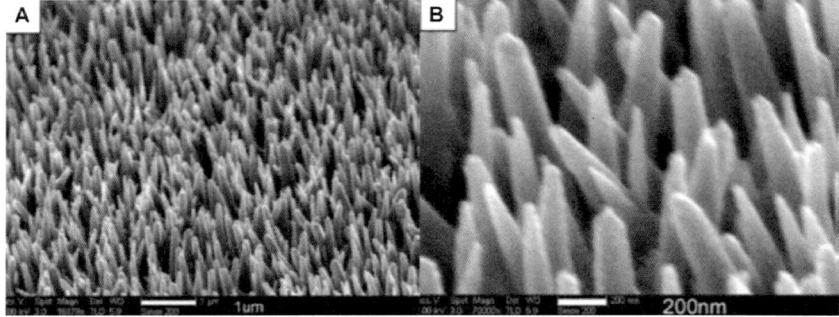

Figure 2.19 ZnO nanorod arrays fabricated by ECD method.
Reproduced with permission from ref. 137. Copyright (2006) American
Chemical Society.

photoetching. Second, ZnO nanorods are deposited on the patterned Au
film, at a temperature of 333 K; the concentration and pH of $Zn(NO_3)_2$
solution were 0.03 M and 5.3, respectively. The substrate for the cathode,
Ag/AgCl for reference electrode, Pt for anode, current density of deposition
process was kept 0.3 mA cm^{-2} for 2 h. The potential between the substrate
and the reference electrode is 0.45 V.

ZnO nanorods were fabricated by an electrochemical deposition
method.[137] 50 nm Au film was deposited on a Si substrate by thermal
evaporation and as an electrostatic cathode. The Zn sheet is the anode and
the Si sheet is the reference electrode. The fabricated product is shown in
Figure 2.19. The other preparation parameters are as follows: concentration
of $Zn(NO_3)_2$ solution of 0.05 M, pH of 6, deposition current of 0.9 mA,
temperature of 70 °C, and deposition time of 2 h.

When preparing one-dimensional ZnO nanostructures using a template
method, the electrochemical-assisted deposition is often adopted. For example, electrochemical-assisted deposition was adopted to help fabricate
ZnO nanorods using a porous polycarbonate template.[138,139] The electrochemical-assisted solution was KCl (0.1 M), $ZnCl_2$ (0.005 M), and H_2O_2
(0.005 M), the pH of solution was 6.85, deposition current was 0.1–0.3 mA for
different apertures templates, and the potential was 1.5 V. Changing the
solution system and increasing the ion composition in the solution can
prepare the doped ZnO nanostructures through the electrochemical assisted
deposition method. For example, Co-doped and Ni-doped ZnO nanorods
were prepared by electrochemical-assisted deposition method, through
adding $CO(NO_3)_2$ and $NiNO_3$ in the $Zn(NO_3)_2$-hexamethylenetetramine
solution.[140]

2.2.3 Template Method

The template method is defined as preparing one-dimensional materials or
specific morphology materials using nucleation and growth through the

channel in the material or the ability to limit other material. As early as 1870s, the template method was applied to prepare nanostructures.[141] The template method can be divided into (1) hard template synthesis methods and (2) soft template synthesis methods.

The hard template method mainly uses carbon nanotubes, porous alumina and the limited threshold deposition of the quantum well, and so on, as the template. The hard template method has a great advantage in controlling the size, morphology and distribution of nanostructures. The soft template synthesis method generally refers to the molecular system with no fixed organisation structure and can limit the threshold in a certain spatial range, such as micelle or reverse micelle template, single molecular layer template, biomolecular template, liquid crystal template, polymer flexible template, *etc.*

The synthesis of ZnO nanostructures by the template method usually uses anodic alumina membranes (AAMs), carbon nanotube template, and porous silica film template method and so on, as templates.

The anodic oxidation of porous anodic alumina template is widely used as a template, due to its high temperature resistance, good insulating property, uniform pore distribution, uniform pore size and controllable size. The ZnO nanowire array can be deposited in the AAM template by the electrochemical method.

ZnO nanorod arrays were prepared using porous alumina template, and similar research results were reported.[142,143] Porous alumina templates were also used for the preparation of Dy-doped ZnO nanorod arrays.[144]

In addition to the porous alumina template method for preparing one-dimensional ZnO nanomaterials, polymer templates, nickel phosphate nanopore templates and carbon nanotubes were successfully used in the hydrothermal method to prepare ZnO nanorods/nanowires.[145–148] Template synthesis is an important method for fabricating patterned ZnO nanostructures. Anodic aluminium oxide (AAO) template, with well-ordered hexagonal nanochannels, was fabricated by a two-step anodisation process at constant voltage in an oxalic acid solution. ZnO nanowires were fabricated in the AAO template by using a three-electrode system.[149]

It is an effective way to prepare all kinds of nanostructures by filling the template holes. However, it is difficult to obtain single crystal nanowires, as most are polycrystalline. The surface quality of the prepared nanowires hardly compares with that obtained by other methods.

2.2.4 Sol–gel Method

The sol–gel method refers to the easily dissolved precursor hydrolysis or alcoholysis in solvents (water or organic solvents), forming sol by hydrolysis or alcoholysis, and forming gel in the process of re-condensation and aging, and then drying, sintering and other processes are adopted to prepare the required materials. Precursors are generally inorganic salts, metal salts or alkyl compounds. The obtained inorganic materials could be particulate powder, film or fibre, *etc.* Generally, the sol–gel process can be divided into two types, (1) organic and (2) inorganic, as per the different materials used. In the organic process, a metal organic alcohol is usually used as a precursor, which is dissolved in an ester compound or metal alcohol salt in organic solvent to form a homogeneous solution. Then other components are added. The sol is prepared by hydrolysis and condensation reaction, and the gel is then obtained. Then the organic solution is removed by heat, and the nanometal oxide is obtained by drying. In the inorganic process, the sol is prepared by the hydrolysis of inorganic salts, and the formation of complexes with ligands, and adding a dispersant. In a certain way (such as heating dehydration), the sol becomes a gel, and the nanometal oxide is formed after drying and baking. The sol–gel method is widely used in the preparation of nanoparticles and thin films. Until now there have been some reports on the preparation of one-dimensional ZnO nanomaterials by the sol–gel method. The process of preparation of one-dimensional ZnO nanomaterials by sol–gel method is as follows: the precursor $(Zn(NO_3)_2 \cdot 6H_2O, Zn(CH_3COO)_2 \cdot 2H_2O)$ is dissolved in the solvent (such as hexamethylenetetramine, ethanol), forming a homogeneous solution. After hydrolysis (or alcoholysis) reaction of the solute and the solvent, the sol is formed with a certain spatial structure by evaporation and drying. Then ZnO nanomaterials are prepared by heat treatment.

When compared with one-dimensional ZnO nanomaterials prepared by the sol–gel method and hydrothermal method, the solution system and the hydrolysis reaction principle are all the same. The difference lies in the fact that for the hydrothermal reaction method the deposition product is a ZnO rod. But deposited by sol–gel method is a gel containing a zinc salt complex, and a ZnO wire or rod can be obtained after calcination. Based on the above difference, we must exclude some literature in the self-sol–gel method in this section. Here are some examples of the preparation of one-dimensional ZnO nanomaterials by the sol–gel method.

The sol–gel method was used to prepare ZnO nanowires, and the preparation process was as follows.[150] A mixture of 20 g 10% polyvinyl alcohol (PVA) and 1.5 g $Zn(CH_3COO)_2 \cdot 2H_2O + 2$ g H_2O aqueous solution was heated in water bath 60 °C for 6 h to form the sol. A 16 kV voltage was applied in the sol to heat the deposit on the aluminium foil as the electrode in a vacuum and at 70 °C for 8 h. The morphology of the product obtained after calcination 700 °C for 5 h is shown in Figure 2.20.

Y-doped ZnO nanorods were prepared by the sol–gel method using $Zn(CH_3CO_2)_2$ and $Y(NO_3)_2 \cdot 6H_2O$ as raw materials.[151] $Zn(CH_3CO_2)_2$ and $Y(NO_3)_2 \cdot 6H_2O$ were made into an ethanol–water solution of a suitable concentration. The pH of solution was adjusted using ammonia. The substrate was immersed in the solution, pulled up, and dried. The fabrication process could be repeated. Finally, a thin film was deposited on the substrate. Then the product was calcined at 450 °C for 1 h. The results showed that Y-doped ZnO nanorods with a diameter of 100–200 nm, long diameter ratio of 50–70 and Y mass fraction of 2% and 3% were prepared with the solution of pH = 4, as shown in Figure 2.21.

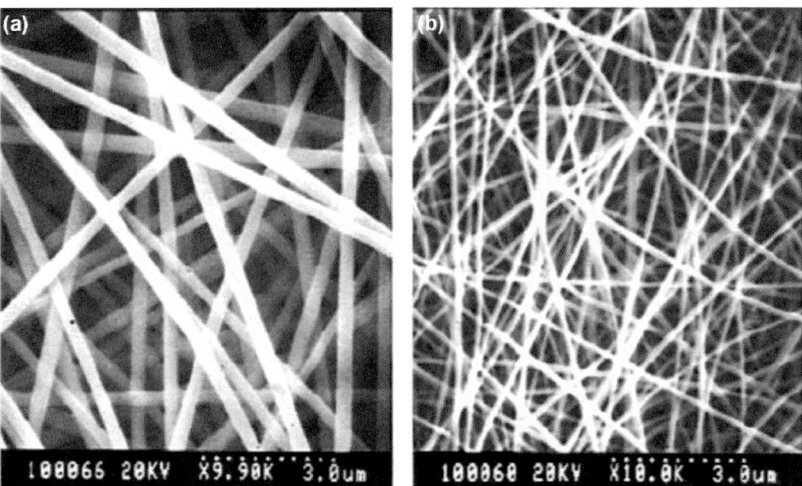

Figure 2.20 (a) PVA/Zn(CH₃COO)₂ fibre before calcination; (b) ZnO nanowires after calcination.

Reprinted from Inorg. Chem. Commun., 7(2), X. Yang, C. Shao, H. Guan, X. Li and J. Gong, Preparation and characterization of ZnO nanofibers by using electrospun PVA/zinc acetate composite fiber as precursor, 176–178. Copyright (2004), with permission from Elsevier.[150]

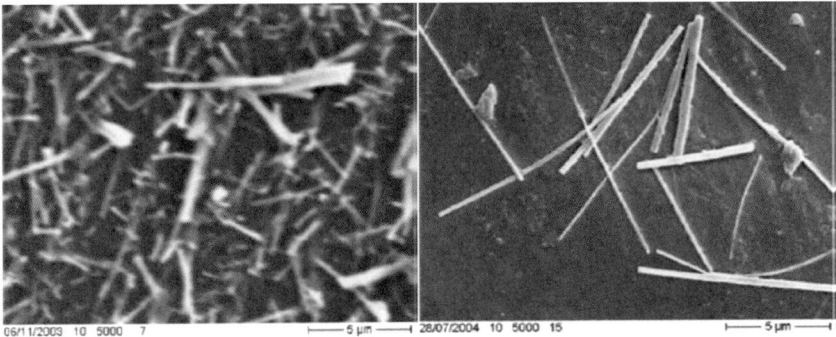

Figure 2.21 Y-doped ZnO nanorods fabricated by the sol-gel method.

Reproduced from J. Non-Cryst. Solids, 352(23–25), R. Kaur, A. V. Singh, K. Sehrawat, N. C. Mehra and R. M. Mehra, Sol–gel derived yttrium doped ZnO nanostructures, 2565–2568. Copyright (2006), with permission from Elsevier.[151]

As ZnO nanorods can be prepared using the sol–gel method, ZnO nanostructures are prepared by this method. Even compared with the hydrothermal process, which is similar, the morphology and quality of ZnO nanostructures made by this method are not satisfactory, and it needs further improvement.

2.3   Patterned Growth of ZnO Nanostructures

ZnO nanostructures, such as nanorod arrays (NRAs), have attracted a great deal of research interest over the past few years due to their unique properties and exciting potential applications. To meet their diverse needs of nanodevice structure and function, each of these applications requires a different arrangement, density and morphology of ZnO NRAs. For instance, vertically aligned ZnO NRAs are needed in piezoelectric nanogenerators to provide high output voltage and drive practical devices. Highly uniform ZnO NRAs are needed in near-UV/blue light emitting diodes for optical waveguide promotion and light extraction enhancement. Furthermore, large-scale highly ordered ZnO NRAs with long-range periodicity are needed for the integration of piezoelectric field effect transistors and self-powered nanosystems with multiple functions. Therefore, the ability to produce large-scale highly ordered ZnO NRAs with desired position, diameter, orientation, arrangement, density and morphology is essential for the integration and optimisation of related functional nanodevices and nanosystems.

In this regard, various patterning methods, such as optical lithography,[152,153] nanosphere lithography,[154–156] nanoimprint lithography,[157–159] electron beam lithography[160–163] and laser interference lithography[164–168] have been employed to provide templates and bring order to ZnO NRAs, which is called patterned growth. Among them, two-beam laser interference lithography (2BLIL) is proven to be a cost-effective and high-throughput technique for the fabrication of photoresist (PR) hole templates and patterned ZnO NRAs in a large area, usually combined with CVD or hydrothermal synthesis.[165–168] In this section, a fabrication method for large-scale patterned ZnO NRAs with tunable arrangement, period and morphology *via* 2BLIL and HTS was first introduced. The growth behaviours, crystal structures and optical properties of patterned ZnO NRAs grown on single-crystalline GaN film and a polycrystalline ZnO seed layer were both studied. Besides, the interesting relationship between orientation of PR template and morphology of resulting ZnO NRAs was investigated. In addition, the influence of solution pH on the top shape of the nanorod, and the height of the patterned ZnO NRAs and space between them as functions of growth time, were also studied. Furthermore, a simple approach was presented to fabricate large-scale highly ordered ZnO NRAs with uniform distribution. It combined three-beam interference lithography (3BIL),[169] top anti-reflective coating (TARC), and HTS. While maintaining most of the usual (2BIL + HTS) advantages,[168,170,171] it has many other merits. First, only a single exposure is needed to fabricate a PR hole template with hexagonal symmetry, without any sample rotation and multi-exposure. Second, developing and growth defects are easily reduced by introducing a water-soluble TARC, without any oxygen-plasma treatment. Finally, periods of the ZnO NRAs from several wavelengths of λ down to $2\lambda/3$ could be easily, continuously and precisely achieved by simply rotating the three-beam Lloyd's mirror interferometer.[166] These studies will be of benefit for a deeper understanding of the rules of patterned growth of ZnO NRAs, further enhancing the controllability of arrangement, period and morphology of the patterned ZnO NRAs on various substrates, and finally promoting the integration and optimisation of related functional nanodevices and nanosystems.

The fabrication sequence of large-scale patterned ZnO NRAs *via* 2BLIL and HTS consists of six steps, including substrate preparation, spin coating of PR and TARC, 2BLIL, developing, hydrothermal (HT) growth of ZnO NRAs and removing the PR template (Figure 2.22(a)).[171] The colour of the substrates will be changed in each step, especially when the substrates are covered with

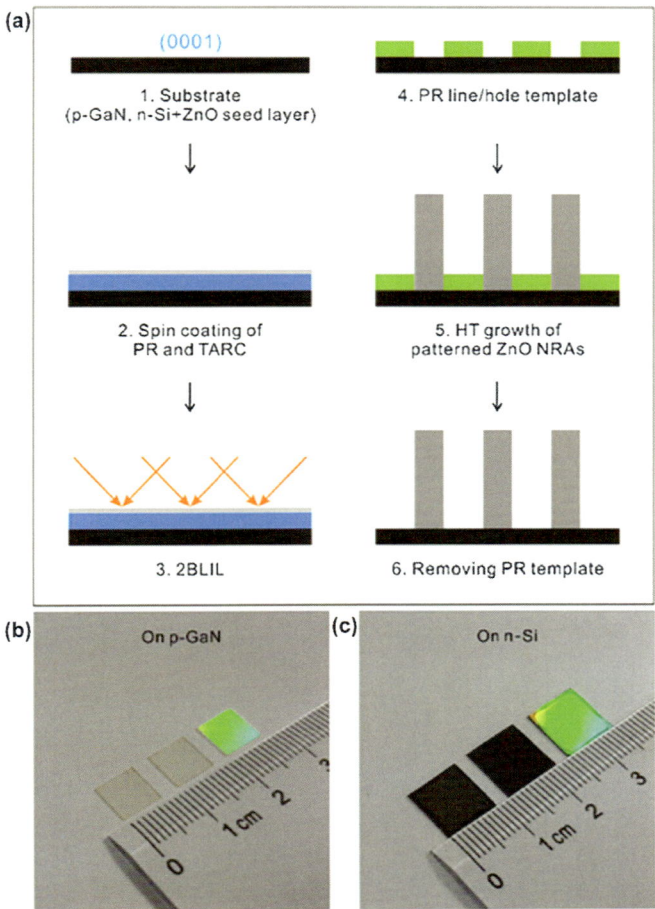

Figure 2.22 (a) Schematic illustration of the fabrication sequence of large-scale patterned ZnO NRAs based on 2BLIL and HTS; (b–c) Pictures of three *p*-GaN substrates (7 mm×7 mm) (b) and three silicon substrates (10 mm×10 mm).
Reproduced from ref. 171 with permission from The Royal Society of Chemistry.

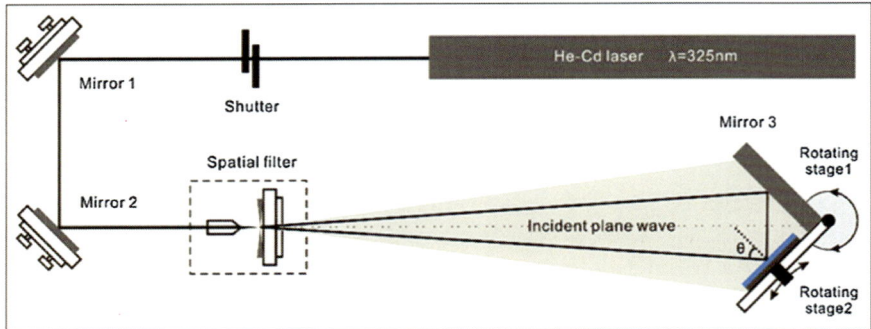

Figure 2.23   Schematic illustration of optical setup of two-beam Lloyd's mirror laser
interference lithography.

PR templates (Figure 2.22(b)–(c)). The bright colour caused by light diffraction indicates a uniform PR template over the whole surface, which is
essential for fabricating patterned ZnO NRAs in a large area with uniform
distribution.

For 2BLIL, a simplified diagram of the optical setup is shown in
Figure 2.23. A He–Cd laser (Kimmon IK5751I-G) with a wavelength of 325 nm
and an output intensity of 30 mW was used as the light source. It offers a
long coherence length (30 cm) and is at lower cost than other options, such
as argon-ion and excimer lasers. A spatial filter allowed the high frequency
noise to be removed from the laser beam to provide a clean near-Gaussian
profile. This diverging beam travelled 1.4 m over an optical table to a Lloyd's
mirror interferometer. The interferometer, which consisted of a dielectric
mirror fixed perpendicularly to the substrate, was used to provide two coherence beams: one travelled directly to the substrate and the other was
reflected onto the substrate by the mirror.

Besides the definition of arrangement, a far more economical and practical aspect of 2BLIL is that the period of PR template and resulting ZnO
NRAs can be precisely adjusted by simply rotating the two-beam Lloyd's
mirror interferometer to change the incident angle θ, for the period P as
given by:

$$P = \frac{\lambda}{2 \sin \theta} \tag{2.9}$$

with λ being the wavelength of the light source.

The heteroepitaxial growth behaviour of patterned ZnO NRAs on single-
crystalline p-GaN substrates was investigated. To fabricate large-scale individual patterned ZnO NRAs (a single nanorod in one position) with 2D
square and hexagonal symmetries on p-GaN, PR hole templates were adopted. Note that a PR hole template with circular holes, vertical hole sidewalls
and no residual PR particles at the bottom of the holes will benefit
the fabrication of perfectly hexagonal-faceted ZnO NRAs. Besides, the

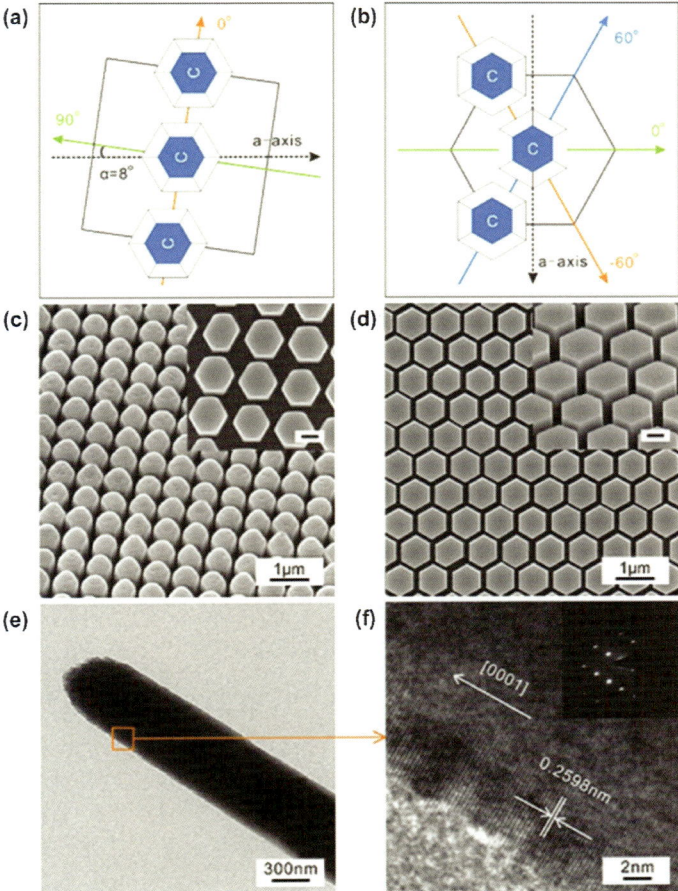

Figure 2.24 Individual patterned ZnO NRAs with 2D square and hexagonal symmetries fabricated on *p*-GaN by the use of PR hole templates; (a) schematic illustration of a square PR hole template prepared on *p*-GaN, where the angle of *a* axis to second exposure orientation (90°) was 8°; (b) schematic diagram of a hexagonal PR hole template fabricated on *p*-GaN, where the second exposure orientation was 0°; (c) 45° tilted view SEM image of the corresponding patterned ZnO NRAs; (d) top view SEM image of the corresponding hexagonal patterned ZnO NRAs; (e–f) TEM (e) and HRTEM (f) images of a *c* axis oriented ZnO nanorod. Reproduced from ref. 171 with permission from The Royal Society of Chemistry.

orientation of PR hole template will also affect the symmetry of resulting ZnO NRAs. For instance, a square PR hole template was prepared on *p*-GaN and the angle between a-axis and 90° exposure orientation was 8° in design (Figure 2.24(a)). Individual patterned single-crystalline ZnO NRAs in a period of 774 nm with 2D square symmetry were then fabricated. Thus, the arrangement of the ZnO NRAs follows the hole template, including the 8°

inclined angle, but the symmetry is broken (Figure 2.24(c)). To realise individual patterned ZnO NRAs with delicate symmetries, it is better to make the 0° exposure orientation perpendicular to the *a* axis, which is suitable for both square and hexagonal symmetries (Figure 2.24(b)). By doing so, perfectly individual patterned ZnO NRAs with hexagonal symmetry in the period of 774 nm were obtained (Figure 2.24(d)). What is more, an interesting phenomenon was observed that flat-top patterned ZnO NRAs (Figure 2.24(d)) could be turned into rough-top ones (Figure 2.24(c)) after the pH of the initial aqueous solution was changed from 8 to 10 by adding ammonia solution into it. TEM images of a single rough-top nanorod acquired from the patterned ZnO NRAs are presented in Figure 2.24(e)–(f). The high-resolution TEM (HRTEM) image of the nanorod shows that the distance between adjacent lattice planes is 0.2598 nm, which indicates that the nanorod is single-crystalline and the growth direction is *c* axis [0001] (Figure 2.24(f)).[165] The results show that pH = 8 is more suitable for fabricating individual patterned ZnO NRAs with a flat top, which are attractive for a number of applications that aim to integrate ZnO nanodevices and nanosystems, especially piezoelectric nanogenerators, electrically pumped lasers, and light emitting diodes.

Single-crystalline GaN film on sapphire has been proven to be suitable for patterned growth of vertically aligned ZnO NRAs. But for most applications, cost-effective substrates such like Si, glass, steel and polymer prefer to be adopted, and buffer layer (ZnO seed layer) are always required in HTS for reducing lattice mismatch and improving adherence between substrate and ZnO NRAs.[161] Based on the above considerations, the homoepitaxial growth behaviour of patterned ZnO NRAs on polycrystalline ZnO seed layer coated Si substrate was also investigated in this study. Large-scale patterned ZnO NRAs in a period of 1048 nm with 2D square symmetry were fabricated (Figure 2.25(a) and (b)). The nucleation density for each position is nearly the same due to the fixed hole diameter (~300 nm). It is obvious that multiple ZnO NRAs (more than seven nanorods) grow out of each hole, mainly because the hole was so wide that lots of small ZnO grains were exposed.[160] Moreover, the nanorods have relatively small diameters at the root due to the confinement of the PR hole template (Figure 2.25(c)). By increasing the growth time from 2 to 5 h, the height of the bunched ZnO NRAs was increased (from 0.8 to 2.1 μm), but the space between them was narrowed considerably (from 400 to 100 nm) (Figure 2.25(d)–(h)), which means that the ZnO NRAs are not perfectly vertically aligned and the growth is no longer strictly controlled by the PR hole template after the patterned ZnO NRAs grow out of the holes. This trend is like the one shown in ref. 160. Additionally, patterned ZnO NRAs may be broken down by the surface tension of residue water in drying process.[163] To realise large-scale patterned ZnO NRAs with uniform distribution, the substrate needs to be supercritically dried.[160] These bunched patterned ZnO NRAs, which have both high specific surface area and tunable density and size, could play significant roles in many areas, particularly in solar cells and bio-sensors.[172,173]

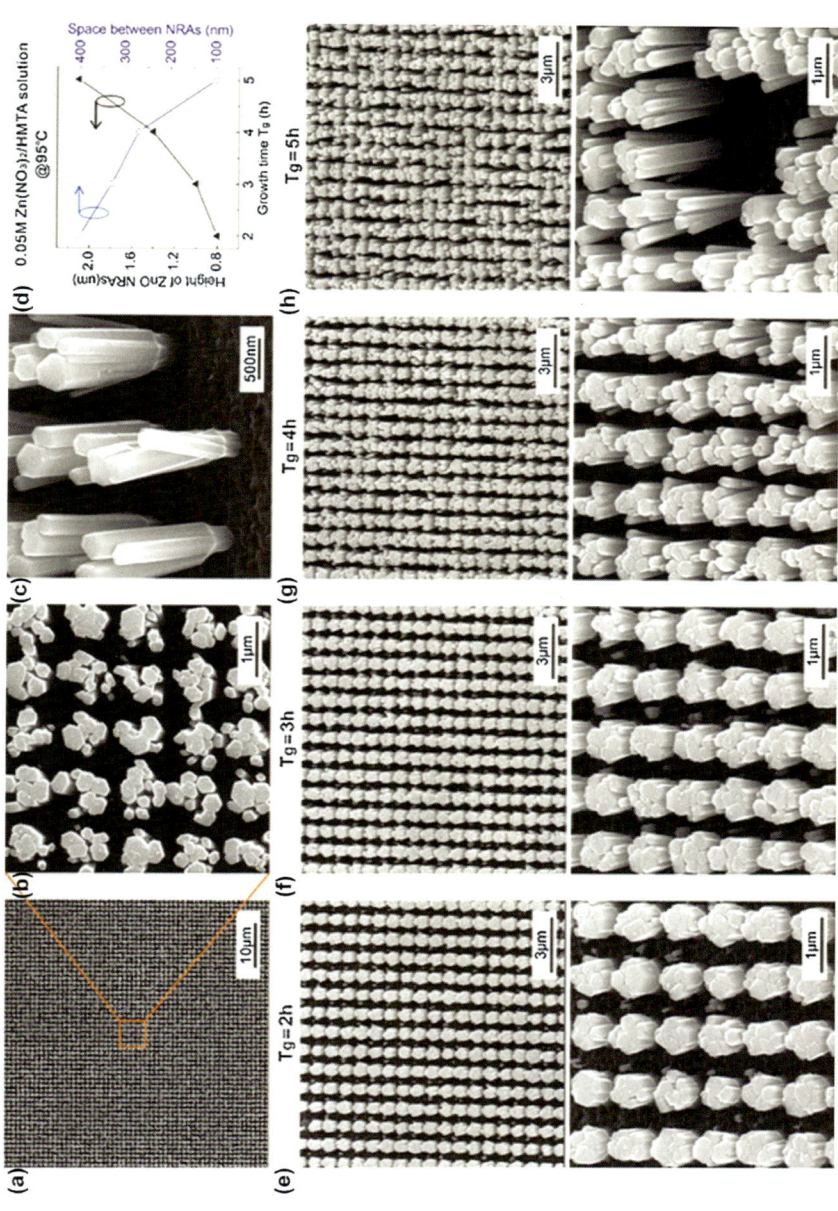

Figure 2.25 Bunched patterned ZnO NRAs with 2D square symmetry grown on ZnO seed layer coated Si. (a–c) Top view (a–b) and 45° tilted view (c) SEM images of the patterned ZnO NRAs in a period of 1048 nm at different magnifications; (d) height of ZnO NRAs and space between ZnO NRAs as functions of growth time; (e–h) 45° tilted view SEM images of the patterned ZnO NRAs under different growth times.
Reproduced from ref. 171 with permission from The Royal Society of Chemistry.

To reduce the fabrication time and make the patterned growth easier to handle, an efficient method includes 3BLIL and HTS. The schematic illustration of the 3BIL system was further developed as shown in Figure 2.26(a).[174] A He-Cd laser with a wavelength of 325 nm (Kimmon IK5751I-G, power: 30 mW, mode: TEM$_{00}$, coherence length: 30 cm) was chosen for the light source. The laser beam was continuously reflected by two dielectric mirrors (Daheng JGS1, diameter: 25.4 mm, reflectivity: >99%) and went through a spatial filter, which consisted of a focusing lens (Thorlabs LMU-10X-UVB, WD: 15 mm) and a 5 μm diameter pinhole. Then this diverging laser beam travelled 1.4 m over a pneumatic optical table (with an expanded diameter of 13 cm and a central power of 0.5 mW) to a three-beam Lloyd's mirror interferometer. The interferometer consisted of a sample holder, two dielectric mirrors and a rotating stage. The mirrors were both perpendicular to the sample holder and had an angle of 120° between each other (Figure 2.26(b)). In this way, the incident plane wave is divided into three: one travels directly to the sample holder and the others are reflected onto the sample holder by the mirrors. These three beams form a large diamond-shaped interference (exposure) area on the sample holder,

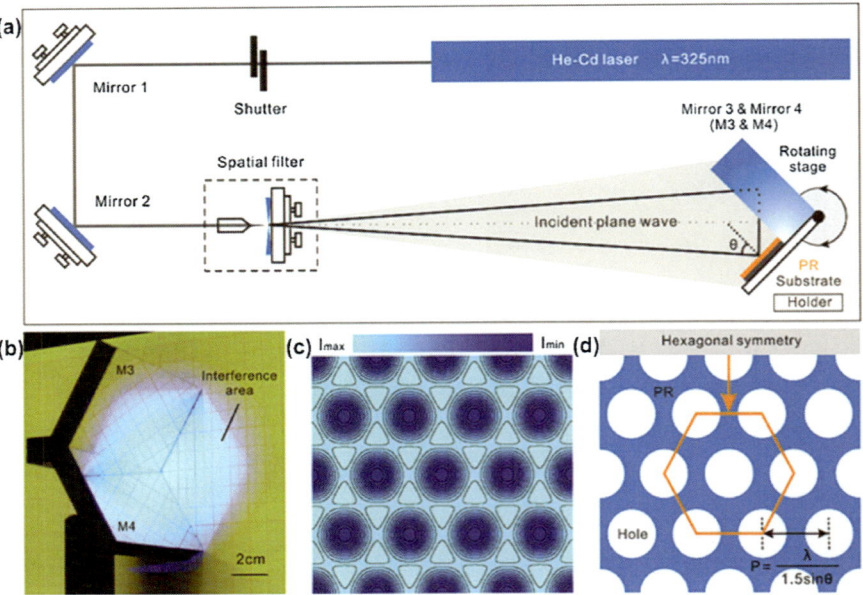

Figure 2.26 Optical setup and working principle of 3BIL. (a) Schematic illustration of the 3BIL system; (b) optical image of the three-beam Lloyd's mirror interferometer; (c) simulated intensity distribution of the hexagonal interference pattern in the interference area; (d) schematic of the corresponding hexagonal PR hole template fabricated by 3BIL with a single exposure.
Reproduced from ref. 174 with permission from The Royal Society of Chemistry.

resulting in the creation of a hexagonal interference intensity pattern (Figure 2.26(c)), which was simulated by MATLAB (The Math Works). The exposure area ($\sim$22 cm^2) allowed for the large-scale fabrication of a hexagonal PR hole template in a single exposure (Figure 2.26(d)). The hole period P is given by:[169]

$$P = \frac{\lambda}{1.5 \sin \theta} \tag{2.10}$$

(different from the 2BIL's), where λ is the laser wavelength and θ is the incident angle. An electronic shutter offers precise control over the exposure time (T_E). Optical components were all covered by a closed plexi-glass cabinet to avoid thermal changes and air movements, which can affect the stability of the interference pattern.

The morphologies of PR hole template and resulting ZnO NRAs were both investigated by SEM. Figure 2.27(a) is the optical image of a *p*-GaN substrate (22 mm$\times$22 mm) covered with a hexagonal PR hole template, which was fabricated *via* 3BIL in a single exposure ($T_E = 30$ s, five times faster than previously due to the use of highly sensitive, chemically amplified PR AR-N 4340).[169] The colour caused by light diffraction indicates a uniform hole template over the whole surface. The top view SEM images of the hole template at different magnifications are shown in Figure 2.27(b). It has a period of 707 nm ($\theta = 18°$), hole diameter of $\sim$500 nm, and vertical sidewalls. With the help of the hole templates, large-scale highly ordered ZnO NRAs were fabricated on *p*-GaN *via* HTS at 95 °C. Figure 2.27(c) is the 45° tilted and top view SEM images of the ZnO NRAs at different magnifications. Obviously, the perfectly hexagonal-faceted ZnO NRAs have almost the same orientation (vertically aligned), diameter ($\sim$550 nm) and length ($\sim$3.8 µm). The nanorod density is $\sim$2.4$\times$10^8 rods cm^{-2} (equal to the hole's). It should be noted that the nanorod diameter is a little larger than the hole's due to there being no template confinement after the nanorod grew out of the hole.[160] Even so, the diameters of the ZnO NRAs could still be controlled by the PR hole template.

Moreover, an important aspect of the 3BIL + HTS approach is that the periods of hole templates and ZnO NRAs could be easily and precisely controlled by rotating the three-beam Lloyd's mirror interferometer to change the incident angle θ. Ref. 30 presents the relationship between the period P and the incident angle θ in 3BIL. For a laser wavelength of 325 nm, the theoretical minimum period is 216 nm ($\theta = 90°$). Here, $\theta = 38°$ was used to bring the period of the ZnO NRAs down to 353 nm. Figure 2.27(d) presents the top view SEM images of the highly ordered ZnO NRAs in this period at different magnifications. The ZnO NRAs have an average diameter of 280 nm and height of 2 µm. The nanorod density is $\sim$9.5$\times$10^8 rods cm^{-2}, which is four times as much as that of $P = 707$ nm. Based on this precise period control method, it will easily realise large-scale highly ordered ZnO NRAs with customised density and diameter. Figure 2.27(e) shows a TEM image of

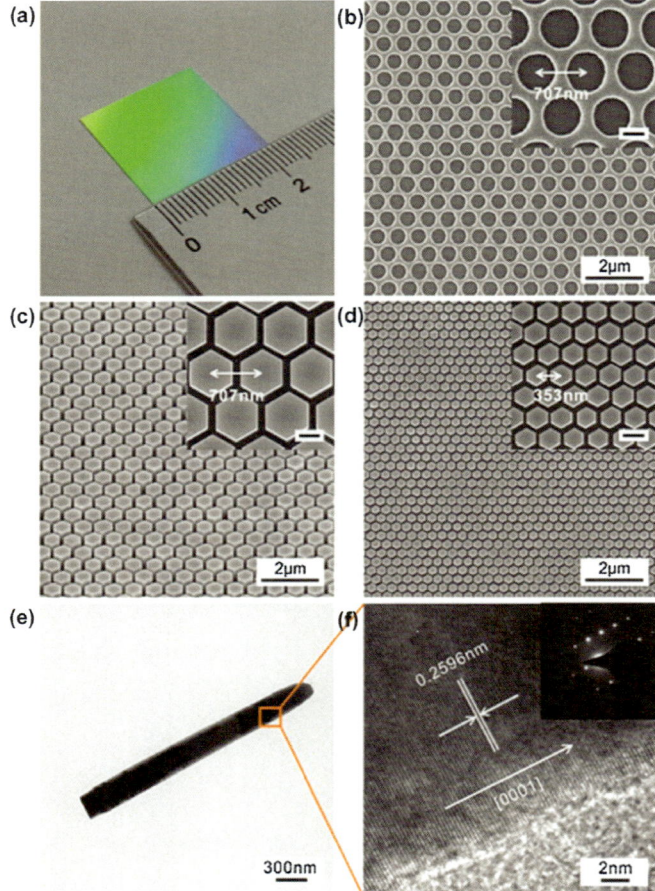

Figure 2.27 Large-scale highly ordered ZnO NRAs with precise period control based on 3BIL and HTS. (a) Optical image of a *p*-GaN substrate with uniform PR hole template; (b) top view SEM image of the PR hole template; (c) 45° tilted view SEM image of the resulting highly ordered ZnO NRAs; (d) top view SEM image of the highly ordered ZnO NRAs; (e–f) TEM (e) and HRTEM (f) images of a *c* axis oriented single crystal ZnO nanorod. Reproduced from ref. 174 with permission from The Royal Society of Chemistry.

a single ZnO nanorod with a diameter of ∼280 nm. It has a relatively small diameter at the root due to the confinement of the PR hole template.[160,168,171]

To achieve large-scale highly ordered ZnO NRAs with uniform distribution, the PR hole template requires the following conditions to be satisfied: circular holes, vertical hole sidewalls, good water affinity, and no residual PR layer or particles at the bottom of the holes. However, there are several negative factors impeding the attainment of these conditions, which are necessary to pay attention to and study. First, holes will not be circular if the Lloyd's mirrors are low in reflectivity, or without aligning with the laser beam centre. Thus, the intensities of the three beams are not equal any more, and the PR hole patterns become deformed accordingly. This kind of problem could be solved by using dielectric mirrors (offering almost full reflectivity) and aligning with the laser beam centre.[169] Second, hole edges will zig–zag due to the vertical interference of incident light and reflection light from the substrate surface, and the roots of the ZnO NRAs will zig–zag accordingly. This effect could be reduced by using chemically amplified AR-N 4340, because its own post-exposure baking (PEB) process promotes the thermally activated diffusion of the photoactive compound from the exposed areas to the unexposed areas in PR, and makes the hole patterns smoother. Third, low contrast PR hole templates will be caused by thermal changes and air movements in the laboratory. These disturbances lead to relative motions between optical components, which translate into a random phase shift between the two interfering fields, degrading the contrast in the dose profile. Undesired ZnO NRAs are fabricated by using the low contrast PR hole templates. These difficulties could be overcome by putting all optical components into a closed plexi-glass cabinet (avoiding thermal changes and air movements). Finally, a hydrophobic PR surface and residual PR layer or particles at the bottom of the holes will make it difficult to achieve uniform distribution of hole templates and ZnO NRAs.[168] Fortunately, adding water-soluble TARC Aquatar is proven to be an easy and powerful method to solve this problem.

In summary, application-oriented tuneable fabrication of large-scale patterned ZnO NRAs has been easily and successfully achieved based on 2BLIL and HTS. By adjusting the substrate, PR template and growth condition, which are the three most important facets among all influencing factors, large-scale patterned ZnO NRAs with tunable arrangement, period and morphology were designed and fabricated. The morphology and symmetry of patterned ZnO NRAs will further be influenced by the orientation of the PR template, especially on single crystalline substrate. Moreover, the resulting ZnO NRAs exhibit excellent crystallisation and PL characteristics. Besides, large-scale highly ordered ZnO NRAs with precise period control and uniform distribution have been easily and quickly fabricated *via* 3BIL, TARC and HTS for the first time. The ZnO NRAs have a highly crystalline structure and high optical quality. This combined approach is cost-effective and available for most laboratories and industrial applications, and compatible with various substrates including polymer due to the maximum temperature being 95 °C. Moreover, wafer-scale could be easily achieved by using larger expanded laser beams and bigger Lloyd's mirrors, and various morphologies could be obtained by changing the growth conditions, including pH, temperature and time. We foresee a bright future for these simple but effective approaches in designing and fabricating large-scale highly ordered semiconductor NRAs, paving the way for low cost and high-performance nanodevices and nanosystems, which are critically important for smart micro/nano-electro-mechanical systems, nanorobot, sensing, personal electronics, medical science, environmental monitoring, and even defence technology.

References

1. Y. Dai, Y. Zhang, Q. Li and C. Nan, Synthesis and optical properties of tetrapod-like zinc oxide nanorods, *Chem. Phys. Lett.*, 2002, **358**(1), 83–86.
2. Y. Dai, Y. Zhang and Z. Wang, The octa-twin tetraleg ZnO nanostructures, *Solid State Commun.*, 2003, **126**(11), 629–633.
3. Y. Huang, Y. Zhang, L. Liu, S. Fan, Y. Wei and J. He, Controlled synthesis and field emission properties of ZnO nanostructures with different morphologies, *J. Nanosci. Nanotechnol.*, 2006, **6**(3), 787–790.
4. Y. Zhang, Y. Huang, J. He, Y. Dai, X. Zhang, J. Liu and Q. Liao, Quasi one-dimensional ZnO nanostructures fabricated without catalyst at lower temperature, *Front. Phys. China*, 2006, **1**(1), 72–84.
5. Y. Huang, Y. Zhang, X. Bai and Y. Zhang, In situ mechanical properties of individual ZnO nanowires and the mass measurement of nanoparticles, *J. Phys.: Condens. Matter.*, 2006, **18**(15), L179–L184.
6. Y. Huang, Y. Zhang, Y. Gu, X. Bai, J. Qi, Q. Liao and J. Liu, Field emission of a single in-doped ZnO nanowire, *J. Phys. Chem. C*, 2007, **111**(26), 9039–9043.
7. Y. Chen, L. Hu, H. Song, H. Jiang, D. Li, G. Miao, Z. Li, X. Sun, Z. Zhang and T. Guo, Optimized performances of tetrapod-like ZnO nanostructures for a triode structure field emission planar light source, *Nanoscale*, 2014, **6**(22), 13544–13549.
8. V. M. Diep, Andrea and M. Armani, Flexible Light-Emitting Nanocomposite Based on ZnO Nanotetrapods, *Nano Lett.*, 2016, **16**(12), 7389–7393.
9. Y. Huang, J. He, Y. Zhang, Y. Dai, Y. Gu, S. Wang and C. Zhou, Morphology, structures and properties of ZnO nanobelts fabricated by Zn-powder evaporation without catalyst at lower temperature, *J. Mater. Sci.*, 2006, **41**(10), 3057–3062.
10. Y. Huang, Y. Zhang, J. He, Y. Dai, Y. Gu, Z. Ji and Z. Cheng, Fabrication and characterization of ZnO comb-like nanostructures, *Ceram. Int.*, 2006, **32**(5), 561–566.
11. X. Wang, K. Chen, Y. Zhang, J. Wan, O. L. Warren, J. Oh, J. Li, E. Ma and Z. I. Shan, Growth conditions control the elastic and electrical properties of ZnO nanowires, *Nano Lett.*, 2015, **15**(12), 7886–7892.
12. X. Yin and X. D. Wang, Kinetics-driven crystal facets evolution at the tip of nanowires: a new implementation of the Ostwald-Lussac law, *Nano Lett.*, 2016, **9**(11), 7078–7084.
13. Y. Li, Y. Bando, T. Sato and K. Kurashima, Zno nanobelts grown on Si substrate, *Appl. Phys. Lett.*, 2002, **81**(1), 144–146.
14. B. Geng, T. Xie, X. Peng, Y. Lin, X. Yuan, G. Meng and L. Zhang, Large-scale synthesis of ZnO nanowires using a low-temperature chemical

route and their photoluminescence properties, *Appl. Phys. A*, 2003, 77(3), 363–366.

15. C. Xu, G. Xu, Y. Liu and G. Wang, A simple and novel route for the preparation of zno nanorods, *Solid State Commun.*, 2002, **122**(3–4), 175–179.

16. J. He, Y. Huang, Y. Zhang, Y. Gu, Z. Ji and C. Zhou, Large-scale synthesis, microstructure and growth mechanism of self-assembled core–shell zno/sio x, nanowires, *Mater. Lett.*, 2006, **60**(2), 150–153.

17. Y. Zhang, H. Jia, R. Wang, C. Chen, X. Luo, D. Yu and C. Lee, Low-temperature growth and Raman scattering study of vertically aligned ZnO nanowires on Si substrate, *Appl. Phys. Lett.*, 2003, **83**(22), 4631–4633.

18. H. Lu, S. Chu and S. Cheng, The vibration and photoluminescence properties of one-dimensional ZnO nanowires, *J. Cryst. Growth*, 2005, **274**(3), 506–511.

19. C. Chang and C. Chang, Site-specific growth to control ZnO nanorods density and related field emission properties, *Solid State Commun.* 2005, **135**(s11–12), 765–768.

20. X. Meng, D. Shen, J. Zhang, D. Zhao, Y. Lu, L. Dong, Z. Zhang, Y. Liu and X. Fan, The structural and optical properties of ZnO nanorod arrays, *Solid State Commun.*, 2005, **135**(3), 179–182.

21. J. Liu, P. Yan, G. Yue, J. Chang, R. Zhuo and D. Qu, Controllable synthesis of undoped/cd-doped ZnO nanostructures, *Mater. Lett.*, 2006, **60**(25), 3122–3125.

22. P. Chang, Z. Fan, D. Wang, W. Tseng, W. Chiou, J. Hong and J. Lu, ZnO nanowires synthesized by vapor trapping CVD method, *Chem. Mater.*, 2004, **16**(24), 5133–5137.

23. Z. Fan, D. Wang, P. Chang, W. Tseng and J. Lu, ZnO nanowire field-effect transistor and oxygen sensing property, *Appl. Phys. Lett.*, 2004, **85**(24), 5923–5925.

24. Y. Wang, L. Zhang, G. Wang, X. Peng, Z. Chu and C. Liang, Catalytic growth of semiconducting zinc oxide nanowires and their photoluminescence properties, *J. Cryst. Growth*, 2002, **234**(1), 171–175.

25. H. Yuan, S. Xie, D. Liu, X. Yan, Z. Zhou and L. Ci, Characterization of zinc oxide crystal nanowires grown by thermal evaporation of ZnS powders, *Chem. Phys. Lett.*, 2003, **371**(3–4), 337–341.

26. C. Lee, T. Lee, S. Lyu, Y. Zhang, H. Ruh and H. Lee, Field emission from well-aligned zinc oxide nanowires grown at low temperature, *Appl. Phys. Lett.*, 2002, **81**(19), 3648–3650.

27. S. Lyu, Y. Zhang, H. Ruh, H. Lee and H. Shim, Low temperature growth and photoluminescence of well-aligned zinc oxide nanowires, *Chem. Phys. Lett.*, 2002, **363**(1–2), 134–138.

28. S. Lyu, Y. Zhang, C. Lee and H. Lee, Low-Temperature Growth of ZnO Nanowire Array by a Simple Physical Vapor-Deposition Method, *Chem. Mater.*, 2003, **15**(17), 3294–3299.

29. Y. Dai, Y. Zhang, Q. Bai and Z. Wang, Bicrystalline zinc oxide nano-wires, *Chem. Phys. Lett.*, 2003, **375**(1), 96–101.

30. X. Wang, Q. Li, Z. Liu, Z. Zhang and Z. liu, Low-temperature growth and properties of ZnO nanowires, *Appl. Phys. Lett.*, 2004, **84**(24), 4941–4943.

31. X. Xu, H. Zhang and Q. Zhao, patterned growth of ZnO nanorad arrays on a large-area stainless steel grid, *Phys. Chem. B.*, 2005, **109**(5), 1699–1702.

32. Y. Kong, D. Yu, B. Zhang and W. Fang, Ultraviolet-emitting ZnO nanowires synthesized by a physical vapor deposition approach, *Appl. Phys. Lett.*, 2001, **78**(4), 407–409.

33. M. Huang, S. Mao, H. Feick, H. Yan and Y. Wu, Room-temperature ultraviolet nanowire nanolasers, *Cheminformatics*, 2001, **292**(40), 1897–1899.

34. M. Huang, Y. Wu, H. Feick, N. Tran and E. Weber, Catalytic Growth of Zinc Oxide Nanowires by Vapor Transport, *Adv. Mater.*, 2001, **13**(2), 113–116.

35. P. Yang, H. Yan, S. Mao, R. Russo and J. Johnson, Controlled Growth of ZnO Nanowires and Their Optical Properties, *Adv. Funct. Mater.*, 2002, **12**(5), 323–331.

36. X. Wang, C. Summers and Z. Wang, Large-scale hexagonal-patterned growth of aligned ZnO nanorods for nano-optoelectronics and nano-sensor arrays, *Nano Lett.*, 2004, **4**(3), 423–426.

37. H. Hou, B. Chen, J. Li and J. Han, Optical properties of single-crystalline ZnO nanowires on m-sapphire, *Appl. Phys. Lett.*, 2003, **82**(82), 2023–2025.

38. Y. Chen, M. Lewis and W. Zhou, Zno nanostructures fabricated through a double-tube vapor-phase transport synthesis, *J. Cryst. Growth*, 2005, **282**(1), 85–93.

39. M. Huang, Y. Wu, H. Feick, N. Tran and E. Wber, Catalytic growth of Zinc Oxide nanowires by vapor transport, *Adv. Mater.*, 2001, **13**(2), 113–116.

40. S. Li, P. Lin and C. Lee, Field emission and photofiuorescent charac-teristics of zinc oxide nanostructure synthesized by a metal catalyzed vapor-liquid-solid process, *J. Appl. Phys.*, 2004, **95**(7), 3711–3716.

41. P. Gao, A. Ding and Z. Wang, Crystallographic orientation-aligned ZnO nanorods grown by a Tin catalys, *Nano Lett.*, 2003, **3**(9), 1315–1320.

42. P. Gao and L. Zhong, Mesoporous polyhedral cages and shells formed by textured self-assembly of ZnO nanocrystals, *J. Am. Chem. Soc.*, 2003, **125**(37), 11299–11305.

43. C. Xu and X. Sun, Field emission from zinc oxide nanopins, *Appl. Phys. Lett.*, 2003, **83**(18), 3806–3808.

44. S. Y. Li, C. Y. Lee and T. Y. Tseng, Copper-catalyzed ZnO nanowires on silicon (1 0 0) grown by vapor–liquid–solid process, *J. Cryst. Growth*, 2003, **247**(3–4), 357–362.

45. J. Qi, Y. Zhang, Y. Huang, Q. Liao and J. Liu, Doping and defects in the formation of single-crystal ZnO nanodisks, *Appl. Phys. Lett.*, 2006, **89**(25), 252115.

46. J. Liu, Y. Zhang, J. Qi, J. He, Y. Huang and X. Zhang, Fabrication and characterization of In-doped zinc oxide nanodisks, *Solid State Phenom.*, 2007, **121–123**(1), 127–130; D. Dummit and J. Labute, Fabrication and Characterization of In-doped Zinc Oxide Nanodisks, *Solid State Phenom.*, 2006, **121–123**(1), 127–130.

47. J. Liu, Y. Zhang, J. Qi, Y. Huang and X. Zhang, In-doped zinc oxide dodecagonal nanometer thick disks, *Mater. Lett.*, 2006, **60**(21), 2623–2626.

48. B. D. Yao, Y. F. Chan and N. Wang, Formation of ZnO nanostructures by a simple way of thermal evaporation, *Appl. Phys. Lett.*, 2002, **81**(4), 757–759; H. Yuan, S. Xie, D. Liu, X. Yan and Z. Zhou, Formation of ZnS nanostructures by a simple way of thermal evaporation, *J. Cryst. Growth*, 2003, **258**(3–4), 225–231.

49. G. Deng, A. Ding, W. Cheng, X. Zheng and P. Qiu, Two-dimensional zinc oxide nanostructure, *Solid State Commun.*, 2005, **134**(4), 283–286.

50. L. Dong, J. Jiao, D. Tuggle, J. Petty and S. Elliff, ZnO nanowires formed on tungsten substrates and their electron field emission properties, *Appl. Phys. Lett.*, 2003, **82**(7), 1096–1098.

51. Z. Zhu, T. Chen, Y. Gu, A. Warren and M. Richard, Zinc oxide nanowires grown by vapor-phase transport using selected metal catalysts: acomparative study, *Chem. Mater.*, 2005, **17**(16), 4227–4234.

52. X. Wang, J. Song, C. Summers, J. Ryou and P. Li, Density-controlled growth of aligned ZnO nanowires sharing a common contact: a simple, low-cost, and mask-free technique for large-scale applications, *J. Phys. Chem. B.*, 2006, **110**(15), 7720–7724.

53. J. Song, X. Wang, E. Riedo and Z. Wang, Systematic study on experimental conditions for large-scale growth of aligned ZnO nanowires on nitrides, *J. Phys. Chem. B.*, 2005, **109**(20), 9869–9872.

54. J. Lao, J. Huang, D. Z. Wang and Z. Ren, ZnO Nanobridges and Nanonails, *Nano Lett.*, 2002, **3**(2), 235–238.

55. J. Lao, J. Wen and Z. Ren, Hierarchical ZnO nanostructures, *Nano Lett.*, 2002, **2**(11), 1287–1291.

56. J. Wu and S. Liu, Low-Temperature Growth of Well-Aligned ZnO Nanorods by Chemical Vapor Deposition, *Adv. Mater.*, 2002, **14**(3), 215–218.

57. B. Zhang, N. Binh, Y. Segawa, Y. Kashiwaba and K. Haga, Photoluminescence study of ZnO nanorods epitaxially grown on sapphire (1120) substrates, *Appl. Phys. Lett.*, 2004, **84**(4), 586–588.

58. H. Yuan and Y. Zhang, Preparation of well-aligned ZnO whiskers on glass substrate by atmospheric MOCVD, *J. Crys. Growth.*, 2004, **263**(1–4), 119–124.

59. W. Park, D. Kim, S. Jung and G. Yi, Metalorganic vapor-phase epitaxial growth of vertically well-aligned ZnO nanorods, *Appl. Phys. Lett.*, 2002, **80**(22), 4232–4234.

60. W. Park, G. Yi, M. Kim and S. Pennycook, Quantum Confinement Observed in ZnO/ZnMgO Nanorod Heterostructures, *Adv. Mater.*, 2003, **15**(6), 526–529.

61. K. Kim and H. Kim, Synthesis of ZnO nanorod on bare Si substrate using metal organic chemical vapor deposition, *Phys. B*, 2003, **328**(3–4), 368–371.

62. W. Lee, M. Jeong and J. Myoung, Catalyst-free growth of ZnO nanowires by metal-organic chemical vapour deposition (MOCVD) and thermal evaporation, *Acta Mater.*, 2004, **52**(13), 3949–3957.

63. Y. Kim, H. Yoo, C. Lee, J. B. Park, H. Baek, M. Kim and G. Yi, Position- and Morphology-controlled ZnO nanostructures grown on graphene layers, *Adv. Mater.*, 2012, **24**(11), 5565–5569.

64. S. Kim, S. Fujita and S. Fujita, ZnO nanowires with high aspect ratios grown by metalorganic chemical vapor deposition using gold nanoparticles, *Appl. Phys. Lett.*, 2005, **86**(86), 153119.

65. S. Kim, S. Fujita, H. Park, B. Yong and H. Kim, Growth of ZnO nanostructures in a chemical vapor deposition process, *J. Cryst. Growth.*, 2006, **292**(2), 306–310.

66. M. Jeong, B. Oh, W. Lee and J. M. Young, Comparative study on the growth characteristics of ZnO nanowires and thin films by metalorganic chemical vapor deposition (MOCVD), *J. Cryst. Growth.*, 2004, **268**(1–2), 149–154.

67. A. Umar, S. Lee, Y. Lee, K. Nahm and Y. Hahn, Star-shaped ZnO nanostructures on silicon by cyclic feeding chemical vapor deposition, *J. Cryst. Growth.*, 2005, **277**(1–4), 479–484.

68. J. Park, H. Oh, J. Kim and S. Sang, Growth of ZnO nanorods via metalorganic chemical vapor deposition and their electrical properties, *J. Cryst. Growth*, 2006, **87**(1), 145–148.

69. J. Baxter and E. Aydil, Epitaxial growth of ZnO nanowires on a- and c-plane sapphire, *J. Cryst. Growth*, 2005, **274**(3–4), 407–411.

70. B. Zhang, K. Wakatsuki, N. Binh, Y. Segawa and N. Usami, Low-temperature growth of ZnO nanostructure networks, *J. Appl. Phys.*, 2004, **96**(1), 340–343.

71. W. Xu, Z. Ye, L. Zhu, Y. Zeng and L. Jiang, ZnO nanostructure networks grown on silicon substrates, *J. Cryst. Growth*, 2005, **277**(1–4), 490–495.

72. Z. Pan, Z. Dai and Z. Wang, Nanobelts of Semiconducting Oxides, *Science*, 2001, **291**(5510), 1947–1949.

73. X. Kong, Y. Ding, R. Yang and Z. Wang, Single-crystal nanorings formed by epitaxial self-coiling of polar nanobelts, *Science*, 2004, **303**(5662), 1348–1351.

74. X. Han, G. Wang, Q. Wang, L. Cao, R. Zou and J. Liu, Ultraviolet lasing and time-resolved photoluminescence of well-aligned ZnO nanorod arrays, *Appl. Phys. Lett.*, 2005, **86**(22), 223106.

75. K. Keem, H. Kim, G. Kim, J. Lee, B. Min, K. Cho and M. Sung, Photocurrent in ZnO nanowires grown from Au electrodes, *Appl. Phys. Lett.*, 2004, **84**(22), 4376–4378.

76. Q. Zhao, M. Willander, R. Morjan, Q. Hu and E. Campbell, Optical recombination of ZnO nanowires grown on sapphire and Si substrates, *Appl. Phys. Lett.*, 2003, **83**(1), 165.

77. S. Bae, H. Choi, C. Na and J. Park, Influence of In incorporation on the electronic structure of ZnO nanowires, *Appl. Phys. Lett.*, 2005, **86**(3), 033102.

78. J. Jie, G. Wang, X. Han, Q. Yu, L. Yuan, G. Li and J. Hou, Indium-doped zinc oxide nanobelts, *Chem. Phys. Lett.*, 2004, **387**(4–6), 466–470.

79. M. Yan, H. Zhang, E. Widjaja and R. Chang, Self-assembly of well-aligned gallium-doped zinc oxide nanorods, *J. Appl. Phys.*, 2003, **94**(8), 5240.

80. A. Hartanto, X. Ning, Y. Nakata and T. Okada, Growth mechanism of ZnO nanorods from nanoparticles formed in a laser ablation plume, *Appl. Phys., A.*, 2004, **78**(3), 299–301.

81. Y. Zhang, R. Russo and S. Mao, Femtosecond laser assisted growth of ZnO nanowires, *Appl. Phys. Lett.*, 2005, **87**(13), 133115.

82. T. Premkumar, Y. S. Zhou, Y. F. Lu and K. Baskar, Optical and field-emission properties of ZnO nanostructures deposited using high-pressure pulsed laser deposition, *ACS Appl. Mater. Interfaces*, 2010, **2**(10), 2863–2869.

83. Y. Sun, G. Fuge and M. Ashfold, Growth of aligned ZnO nanorod arrays by catalyst-free pulsed laser deposition methods, *Chem. Phys. Lett.*, 2004, **396**(1–3), 21–26.

84. H. Hsu, Y. Tseng, H. Cheng, J. Kuo and W. Hsieh, Selective growth of ZnO nanorods on pre-coated ZnO buffer layer, *J. Cryst. Growth*, 2004, **261**(4), 520–525.

85. Y. Zhang, R. Russo and S. Mao, Quantum efficiency of ZnO nanowire nanolasers, *Appl. Phys. Lett.*, 2005, **87**(4), 043106.

86. V. Gupta, P. Bhattacharya, Y. Yuzuk, K. Sreenivas and R. Katiyar, Optical phonon modes in ZnO nanorods on Si prepared by pulsed laser deposition, *J. Cryst. Growth*, 2006, **287**(1), 39–43.

87. S. Choopun, H. Tabata and T. Kawai, Self-assembly ZnO nanorods by pulsed laser deposition under argon atmosphere, *J. Cryst. Growth*, 2005, **274**(1–2), 167–172.

88. Y. Heo, V. Varadarajan, M. Kaufman and D. Fleming, Site-specific growth of Zno nanorods using catalysis-driven molecular-beam epitaxy, *Appl. Phys. Lett.*, 2002, **81**(16), 3046.

89. X. Zhang, S. Xie, Z. Jiang, X. Zheng, Z. Tian, Z. Xie, R. Huang and L. Zheng, Rational design and fabrication of ZnO nanotubes from nanowire templates in a microwave plasma system, *J. Phys. Chem. B*, 2003, **107**(37), 10114–10118.

90. J. Baxter, F. Wu and E. Aydil, Growth mechanism and characterization of zinc oxide hexagonal columns, *Appl. Phys. Lett.*, 2003, **83**(18), 3797.

91. S. Choopun, N. Hongsith, S. Tanunchai, T. Chairuangsri, C. Krua-in, S. Singkarat, T. Vilaithong, P. Mangkorntong and N. Mangkorntong, Single-crystalline ZnO nanobelts by RF sputtering, *J. Cryst. Growth*, 2005, **282**(3–4), 365–369.

92. M. Chen and J. Ting, Sputter deposition of ZnO nanorods/thin-film structures on Si, *Thin Solid Films*, 2006, **494**(1–2), 250–254.

93. T. Chou and J. Ting, Deposition and characterization of a novel integrated ZnO nanorods/thin film structure, *Thin Solid Films*, 2006, **494**(1–2), 291–295.

94. L. Vayssieres, K. Keis, S. Lindquist and A. Hagfeldt, Purpose-built anisotropic metal oxide material: 3D highly oriented microrod array of ZnO, *J. Phys. Chem. B.*, 2001, **105**(17), 3350–3352.

95. L. Vayssieres, Growth of arrayed nanorods and nanowires of ZnO from aqueous solutions, *Adv. Mater.*, 2003, **15**(5), 464–466.

96. C. Lin, S. Chen and S. Cheng, Retracted: Nucleation and growth behavior of well-aligned ZnO nanorods on organic substrates in aqueous solutions, *J. Cryst. Growth*, 2005, **283**(1–2), 141–146.

97. H. Hu, K. Yu, J. Zhu and Z. Zhu, ZnO nanostructures with different morphologies and their field emission properties, *Appl. Surf. Sci.*, 2006, **252**(24), 8410–8413.

98. F. Li, Z. Li and F. Jin, Structural and luminescent properties of ZnO nanorods prepared from aqueous solution, *Mater. Lett.*, 2007, **61**(8–9), 1876–1880.

99. Q. Liao, Y. Yang, L. Xia, J. Qi, Y. Zhang, H. Huang and Z. Qin, High intensity, plasma-induced emission from large area ZnO nanorod array cathodes, *Phys. Plasmas*, 2008, **15**(11), 114505.

100. Z. Qin, Y. Huang, Q. Liao, Z. Zhang, X. Zhang and Y. Zhang, Stability improvement of the ZnO nanowire array electrode modified with Al_2O_3 and SiO_2 for dye-sensitized solar cells, *Mater. Lett.*, 2012, **70**, 177–180.

101. Z. Qin, G. Zhang, Q. Liao, Y. Qiu, Y. Huang and Y. Zhang, Influences of low temperature thermal treatment on ZnO nanowire arrays and nanoparticles based flexible dye-sensitized solar cells, *Colloid Surf., A*, 2012, **402**, 127–131.

102. W. Wang, Z. Zhang, Q. Liao, T. Yu, Y. Shen, P. Li, Y. Huang and Y. Zhang, Two-step epitaxial synthesis and layered growth mechanism of bisectional ZnO nanowire arrays, *J. Cryst. Growth*, 2013, **363**, 247–252.

103. Y. Liu, Z. Lin, W. Lin, K. Moon and C. Wong, Reversible superhydrophobic-superhydrophilic transition of ZnO nanorod/epoxy composite films, *ACS Appl. Mater. Interface*, 2012, **4**(8), 3959–3964.

104. L. Greene, M. Law, J. Goldberger, F. Kim, J. Johnson, Y. Zhang, R. Saykally and P. Yang, Low-temperature wafer-scale production of ZnO nanowire arrays, *Angew. Chem., Int. Ed.*, 2003, **42**(26), 3031–3034.

105. Y. Lin, S. Yang, S. Tsai, H. Hsu, S. Wu and I. Chen, Visible photoluminescence of ultrathin ZnO nanowire at room temperature, *Cryst. Growth Des.*, 2006, **6**(8), 1951–1955.

106. J. Cui, C. Daghlian, U. Gibson, R. Püsche, P. Geithner and L. Ley, Low-temperature growth and field emission of ZnO nanowire arrays, *J. Appl. Phys.*, 2005, **97**(4), 044315.

107. M. Guo, P. Diao and S. Cai, Hydrothermal growth of perpendicularly oriented ZnO nanorod array film and its photoelectrochemical properties, *Appl. Surf. Sci.*, 2005, **249**(1–4), 71–75.

108. M. Guo, P. Diao, X. Wang and S. Cai, The effect of hydrothermal growth temperature on preparation and photoelectrochemical performance of ZnO nanorod array films, *J. Solid State Chem.*, 2005, **178**(10), 3210–3215.

109. M. Guo, P. Diao and S. Cai, Hydrothermal growth of well-aligned ZnO nanorod arrays: Dependence of morphology and alignment ordering upon preparing conditions, *J. Solid State Chem.*, 2005, **178**(6), 1864–1873.

110. S. Liou, C. Hsiao and S. Chen, RETRACTED: Growth behavior and microstructure evolution of ZnO nanorods grown on Si in aqueous solution, *J. Cryst. Growth*, 2005, **274**(3–4), 438–446.

111. L. Greene, B. Yuhas, M. Law, D. Zitoun and P. Yang, Solution-grown zinc oxide nanowires, *Inorg. Chem.*, 2006, **45**(19), 7535–7543.

112. X. Liu and Y. Zhou, Seed-mediated synthesis of uniform ZnO nanorods in the presence of polyethylene glycol, *J. Cryst. Growth*, 2004, **270**(3–4), 527–534.

113. X. Liu, Z. Jin, S. Bu, J. Zhao and K. Yu, Preparation of ZnO nanorods and special lath-like crystals by aqueous chemical growth method, *Mater. Sci. Eng. B.*, 2006, **129**(1–3), 139–143.

114. Y. Tak and K. Yong, Controlled growth of well-aligned ZnO nanorod array using a novel solution method, *J. Phys. Chem. B.*, 2005, **109**(41), 19263–19269.

115. Y. Sun, N.-A. N. George, D. Jason Riley and N. R. Ashfold Michael, Synthesis and photoluminescence of ultra-thin ZnO nanowire/nanotube arrays formed by hydrothermal growth, *Chem. Phys. Lett.*, 2006, **431**(4–6), 352–357.

116. S. Joon, P. Jae-Hwan and P. Jae-Gwan, *Patterned growth of ZnO nanorods by micromolding of sol-gel-derived seed layer. Appl. Phys. Lett.*, 2005, **87**(13), 133112.

117. A. Kargar, K. Sun, Y. Jing, C. Choi, H. Jeong, Y. Zhou, K. Madsen, P. Naughton, S. Jin, G. Y. Jung and D. Wang, Tailoring n-ZnO/p-Si Branched nanowire heterostructures for selective photoelectrochemical water oxidation or reduction, *Nano Lett.*, 2013, **13**(7), 3017–3022.

118. E. Della Gaspera, D. F. Kennedy, J. van Embden, A. S. R. Chesman, T. R. Gengenbach, K. Weber and J. J. Jasieniak, Flash-assisted processing of highly conductive Zinc Oxide electrodes from water, *Adv. Funct. Mater.*, 2015, **25**(47), 7263–7271.

119. Z. Feng, R. Jia, B. Dou, H. Li, Z. Jin, X. Liu, F. Li, W. Zhang and C. Wu, Fabrication and properties of ZnO nanorods within silicon nanostructures for solar cell application, *Appl. Phys. Lett.*, 2015, **106**(5), 053118.

120. J. J. Cheng, S. M. Nicaise, K. K. Berggren and S. Gradečak, Dimensional tailoring of hydrothermally grown Zinc Oxide nanowire arrays, *Nano Lett.*, 2016, **16**(1), 753–759.

121. L. Chen, F. Xue, X. Li, X. Huang, L. Wang, J. Kou and Z. L. Wang, Strain-gated field effect transistor of a MoS_2 – ZnO 2D – 1D hybrid structure, *ACS Nano*, 2016, **10**(1), 1546–1551.

122. D. I. Son, B. W. Kwon, D. H. Park, W. Seo, Y. Yi, B. Angadi, C. Lee and W. K. Choi, Emissive ZnO-graphene quantum dots for white-light-emitting diodes, *Nat. Nanotechnol.*, 2012, **7**(7), 465–471.

123. H. Shingo, T. Nobuo, S. Shu, M. Kyosuke, I. Katsuhiko, T. Hideo and K. Makoto, Room-temperature nanowire ultraviolet lasers: An aqueous pathway for zinc oxide nanowires with low defect density, *J. Appl. Phys.*, 2005, **98**(9), 094305.

124. B. Liu and H. Zeng, Room temperature solution synthesis of mono-dispersed single-crystalline ZnO nanorods and derived hierarchical nanostructures, *Langmuir*, 2004, **20**(10), 4196–4204.

125. U. Pal, S. Garcia, P. Santiago, G. Xiong, K. Ucer and R. Williams, Synthesis and optical properties of ZnO nanostructures with different morphologies, *Opt. Mater.*, 2006, **29**(1), 65–69.

126. M. Yin, Y. Gu, I. Kuskovsky, T. Andelman, Y. Zhu, G. Neumark and S. O'Brien, Zinc Oxide quantum rods, *J. Am. Chem. Soc.*, 2004, **126**(20), 6206–6207.

127. X. Zhang and Y. Kang, Large-scale synthesis of perpendicular side-faceted one-dimensional ZnO nanocrystals, *Inorg. Chem.*, 2006, **45**(10), 4186–4190.

128. X. Zhang, Y. Kim and Y. Kang, Low-temperature synthesis and shape control of ZnO nanorods, *Curr. Appl. Phys.*, 2006, **6**(4), 796–800.

129. H. Hou, Y. Xiong, Y. Xie, Q. Li, J. Zhang and X. Tian, Structure-direct assembly of hexagonal pencil-like ZnO group whiskers, *J. Solid State Chem.*, 2004, **177**(1), 176–180.

130. X. Zhang, H. Zhao, X. Tao, Y. Zhao and Z. Zhang, Sonochemical method for the preparation of ZnO nanorods and trigonal-shaped ultrafine particles, *Mater. Lett.*, 2005, **59**(14–15), 1745–1747.

131. X. Hou, F. Zhou, Y. Sun and W. Liu, Ultrasound-assisted synthesis of dentritic ZnO nanostructure in ionic liquid, *Mater. Lett.*, 2007, **61**(8–9), 1789–1792.

132. Y. Ji, L. Guo, H. Xu, P. Simon and Z. Wu, Regularly shaped, single-crystalline ZnO nanorods with wurtzite structure, *J. Am. Chem. Soc.*, 2002, **124**(50), 14864–14865.

133. Y. Lv, Y. Zhang, C. Li, L. Ren, L. Guo, H. Xu, L. Ding, C. Yang, W. Ge and S. Yang, Temperature-dependent photoluminescence of ZnO nanorods prepared by a simple solution route, *J. Lumin.*, 2007, (122–123), 816–818.

134. Y. Liu, Z. Liu and G. Wang, Synthesis and characterization of ZnO nanorods, *J. Cryst. Growth*, 2003, **252**(1–3), 213–218.

135. M. Izaki and T. Omi, Transparent zinc oxide films prepared by electrochemical reaction, *Appl. Phys. Lett.*, 1996, **68**(17), 2439–2441.

136. M. Izaki, S. Watase and H. Takahashi, Room-temperature ultraviolet light-emitting zinc oxide micropatterns prepared by low-temperature electrodeposition and photoresist, *Appl. Phys. Lett.*, 2003, **83**(24), 4930–4932.

137. B. Cao, Y. Li, G. Duan and W. Cai, Growth of ZnO nanoneedle arrays with strong ultraviolet emissions by an electrochemical deposition method, *Cryst. Growth Des.*, 2006, **6**(5), 1091–1095.
138. Y. Wang, A. Ouslim and G. Wang, Structure study of electrodeposited ZnO nanowires, *Microchem. J.*, 2005, **36**(7), 625–628.
139. Y. Wang, G. Wang, X. Zhang and D. Yu, Study on the microstructure and growth mechanism of electrochemical deposited ZnO nanowires, *J. Crys. Growth*, 2006, **287**(1), 89–93.
140. J. Cui and U. Gibson, Electrodeposition and room temperature ferromagnetic anisotropy of Co and Ni-doped ZnO nanowire arrays, *Appl. Phys. Lett.*, 2005, **87**(13), 133108.
141. S. Kawai and I. Ishiguro, Recording characteristics of anodic oxide films on aluminium containing electrodeposited ferromagnetic metals and alloys, *J. Electrochem. Soc.*, 1976, **123**(7), 1047–1051.
142. Y. Li, G. Meng, L. Zhang and F. Phillipp, Ordered semiconductor ZnO nanowire arrays and their photoluminescence properties, *Appl. Surf. Sci.*, 2000, **76**(15), 2011–2013.
143. M. Zheng, L. Zhang, G. Li and W. Shen, Fabrication and optical properties of large-scale uniform zinc oxide nanowire arrays by one-step electrochemical deposition technique, *Chem. Phys. Lett.*, 2002, **363**(1–2), 123–128.
144. G. Wu, Y. Zhuang, Z. Lin, Y. Yuan, T. Xie and L. Zhang, Synthesis and photoluminescence of Dy-doped ZnO nanowires, *Phys. E*, 2006, **31**(1), 5–8.
145. M. Lai and D. Riley, Templated electrosynthesis of Zinc Oxide nanorods, *Chem. Mater.*, 2006, **18**(9), 2233–2237.
146. Z. Chen, Q. Gao, M. Ruan and J. Shi, Zinc oxide nanoarrays in nanoporous nickel phosphate with a huge blueshift ultraviolet-visible exciton absorption peak, *Appl. Phys. Lett.*, 2005, **87**(9), 093113.
147. H. Wu, X. Wei, M. Shao and J. Gu, Synthesis of zinc oxide nanorods using carbon nanotubes as templates, *J. Crys. Growth*, 2004, **265**(1–2), 184–189.
148. G. Wu, T. Xie, X. Yuan, Y. Li, L. Yang, Y. Xiao and L. Zhang, Controlled synthesis of ZnO nanowires or nanotubes via sol–gel template process, *Solid State Commun.*, 2005, **134**(7), 485–489.
149. S. Ozturk, N. Kılınc, N. Tas-altın and Z. Z. Ozturk, Fabrication of ZnO nanowires and nanorods, *Phys. E*, 2012, **44**(6), 1062–1065.
150. X. Yang, C. Shao, H. Guan, X. Li and J. Gong, Preparation and characterization of ZnO nanofibers by using electrospun PVA zinc acetate composite fiber as precursor, *Inorg. Chem. Commun.*, 2004, 7(2), 176–178.
151. R. Kaur, A. Singh, K. Sehrawat, N. Mehra and R. Mehra, Sol-gel derived yttrium doped ZnO nanostructures, *J. Non-Cryst. Solids*, 2006, **352**(23–25), 2565–2568.
152. Y. J. Hong, S. J. An, H. S. Jung, C. H. Lee and G. C. Yi, Position-controlled selective growth of ZnO nanorods on Si substrates using facet-controlled GaN micropatterns, *Adv. Mater.*, 2007, **19**(24), 4416–4419.

153. C. Cheng, M. Lei, L. Feng, T. L. Wong, K. M. Ho, K. K. Fung, M. M. Loy, D. Yu and N. Wang, High-quality ZnO nanowire arrays directly fabricated from photoresists, *ACS Nano*, 2009, **3**(1), 53–58.

154. D. F. Liu, Y. J. Xiang, X. C. Wu, Z. X. Zhang, L. F. Liu, L. Song, X. W. Zhao, S. D. Luo, W. J. Ma, J. Shen, W. Y. Zhou, G. Wang, C. Y. Wang and S. S. Xie, Periodic ZnO nanorod arrays defined by polystyrene microsphere self-assembled monolayers, *Nano Lett.*, 2006, **6**(10), 2375–2378.

155. C. Li, G. Hong, P. Wang, D. Yu and L. Qi, Wet chemical approaches to patterned arrays of well-aligned ZnO nanopillars assisted by monolayer colloidal crystals, *Chem. Mater.*, 2009, **21**(5), 891–897.

156. J. J. Dong, X. W. Zhang, Z. G. Yin, S. G. Zhang, J. X. Wang, H. R. Tan, Y. Gao, F. T. Si and H. L. Gao, Controllable growth of highly ordered ZnO nanorod arrays via inverted self-assembled monolayer template, *ACS Appl. Mater. Interfaces*, 2011, **3**(11), 4388–4395.

157. J. W. Hsu, Z. R. Tian, N. C. Simmons, C. M. Matzke, J. A. Voigt and J. Liu, Directed spatial organization of zinc oxide nanorods, *Nano Lett.*, 2005, **5**(1), 83–86.

158. J. K. Hwang, S. Cho, E. K. Seo, J. M. Myoung and M. M. Sung, Large-area fabrication of patterned ZnO-nanowire arrays using light stamping lithography, *ACS Appl. Mater. Interfaces*, 2009, **1**(12), 2843–2847.

159. H. H. Park, X. Zhang, K. W. Lee, K. H. Kim, S. H. Jung, D. S. Park, Y. S. Choi, H.-B. Shin, H. K. Sung, K. H. Park, H. K. Kang, H.-H. Park and C. K. Ko, Position-controlled hydrothermal growth of ZnO nanorods on arbitrary substrates with a patterned seed layer via ultraviolet-assisted nanoimprint lithography, *CrystEngComm*, 2013, **15**(17), 3463–3469.

160. S. Xu, Y. Wei, M. Kirkham, J. Liu, W. Mai, D. Davidovic, R. L. Snyder and Z. L. Wang, Patterned growth of vertically aligned ZnO nanowire arrays on inorganic substrates at low temperature without catalyst, *J. Am. Chem. Soc.*, 2008, **130**(45), 14958–14959.

161. R. B. Erdélyi, T. Nagata, D. J. Rogers, F. H. Teherani, Z. E. Horváth, Z. N. Lábadi, Z. F. Baji, Y. Wakayama and J. N. Volk, Investigations into the impact of the template layer on ZnO nanowire arrays made using low temperature wet chemical growth, *Cryst. Growth Des.*, 2011, **11**(6), 2515–2519.

162. D. B. Zhang, S. J. Wang, K. Cheng, S. X. Dai, B. B. Hu, X. Han, Q. Shi and Z. L. Du, Controllable fabrication of patterned ZnO nanorod arrays: investigations into the impacts on their morphology, *ACS Appl. Mater. Interfaces*, 2012, **4**(6), 2969–2977.

163. S. J. Wang, C. S. Song, K. Cheng, S. X. Dai, Y. Y. Zhang and Z. L. Du, Controllable growth of ZnO nanorod arrays with different densities and their photoelectric properties, *Nanoscale Res. Lett.*, 2012, **7**(1), 246–252.

164. D. S. Kim, R. Ji, H. J. Fan, F. Bertram, R. Scholz, A. Dadgar, K. Nielsch, A. Krost, J. Christen, U. Gosele and M. Zacharias, Laser-interference

lithography tailored for highly symmetrically arranged ZnO nanowire arrays, *Small*, 2007, **3**(1), 76–80.

165. T.-U. Kim, J.-A. Kim, S. M. Pawar, J.-H. Moon and J. H. Kim, Creation of nanoscale two-dimensional patterns of ZnO nanorods using laser interference lithography followed by hydrothermal synthesis at 90 °C, *Cryst. Growth Des.*, 2010, **10**(10), 4256–4261.

166. Y. Wei, W. Wu, R. Guo, D. Yuan, S. Das and Z. L. Wang, Wafer-scale high-throughput ordered growth of vertically aligned ZnO nanowire arrays, *Nano Lett.*, 2010, **10**(9), 3414–3419.

167. D. Yuan, R. Guo, Y. Wei, W. Wu, Y. Ding, Z. L. Wang and S. Das, Heteroepitaxial patterned growth of vertically aligned and periodically distributed ZnO nanowires on GaN using laser interference ablation, *Adv. Funct. Mater.*, 2010, **20**(20), 3484–3489.

168. K. S. Kim, H. Jeong, M. S. Jeong and G. Y. Jung, Polymer-templated hydrothermal growth of vertically aligned single-crystal ZnO nanorods and morphological transformations using structural polarity, *Adv. Funct. Mater.*, 2010, **20**(18), 3055–3063.

169. J. D. Boor, N. Geyer, U. Gösele and V. Schmidt, Three-beam interference lithography: upgrading a Lloyd's interferometer for single-exposure hexagonal patterning, *Opt. Lett.*, 2009, **34**(12), 1783–1785.

170. Y. G. Wei, W. Z. Wu, R. Guo, D. J. Yuan, S. Das and Z. L. Wang, Wafer-scale high-throughput ordered growth of vertically aligned ZnO nano-wire arrays, *Nano Lett.*, 2010, **10**(9), 3414–3419.

171. X. Chen, X. Q. Yan, Z. M. Bai, P. Lin, Y. W. Shen, X. Zheng, Y. Y. Feng and Y. Zhang, Facile fabrication of large-scale patterned ZnO nanorod arrays with tunable arrangement, period and morphology, *CrystEngComm*, 2013, **15**(39), 8022–8028.

172. Y. Bai, H. Yu, Z. Li, R. Amal, G. Q. Lu and L. Wang, In situ growth of a ZnO nanowire network within a TiO_2 nanoparticle film for enhanced dye-sensitized solar cell performance, *Adv. Mater.*, 2012, **24**(43), 5850–5856.

173. H. Tang, G. Meng, Q. Huang, Z. Zhang, Z. Huang and C. Zhu, Arrays of cone-shaped ZnO nanorods decorated with Ag nanoparticles as 3D surface-enhanced raman scattering substrates for rapid detection of trace polychlorinated biphenyls, *Adv. Funct. Mater.*, 2012, **22**(1), 218–224.

174. X. Chen, X. Q. Yan, Z. M. Bai, Y. W. Shen, Z. Z. Wang, X. Z. Dong, X. M. Duan and Y. Zhang, High-throughput fabrication of large-scale highly ordered ZnO nanorod arrays via three-beam interference lithography, *CrystEngComm*, 2013, **15**(42), 8416–8421.

CHAPTER 3

Property Characterisation and Optimisation

XU SUN, YANWEI SHEN, PEI LIN, ZHUO KANG AND
YUE ZHANG*

University of Science and Technology Beijing, Beijing, China
*Email: yuezhang@ustb.edu.cn

3.1 Electronic Properties

3.1.1 Electronic Structure

As the band structures of materials and their nanostructures are vital in determining their properties and potential uses, enormous efforts have been paid to accurately describing them. ZnO is a direct bandgap semiconductor (both the valence band maxima (VBM) and the conduction band minima (CBM) at the Γ point), and the bandgap of ZnO is 3.44 eV at low temperatures, but 3.37 eV at room temperature.[1] However, there are large differences among different theoretical approaches for the calculated band structures of ZnO and their nanostructures.[2,3] The band structures of ZnO are mostly calculated by traditional DFT methods involving the LDA or GGA functional. However, the results of standard LDA or GGA calculations show obvious drawbacks; for example, the LDA calculations underestimate the ZnO bandgap as smaller than 1.0 eV, in comparison with $E_{g(exp)} = 3.37$ eV. Moreover, the interactions with the anion p valence bands are artificially enlarged, leading them to shift unphysically towards the conduction bands and their overestimated dispersion. Consequently, band correction *via* plus Hubbard potentials or self-interaction correction (SIC) shown in Figure 3.1,

Nanoscience & Nanotechnology Series No. 43
ZnO Nanostructures: Fabrication and Applications
By Yue Zhang
© Yue Zhang 2017
Published by the Royal Society of Chemistry, www.rsc.org

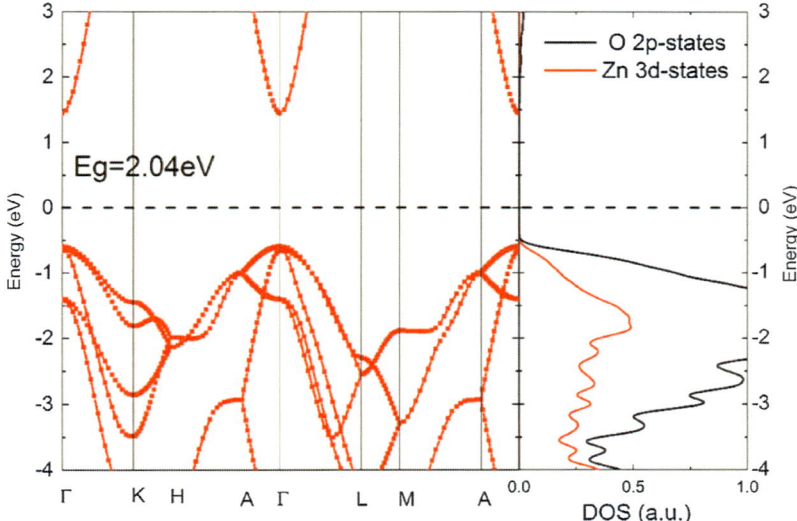

Figure 3.1 The calculated band structure of ZnO by considering the self-interaction correction (SIC) pseudopotentials. The energy of the valence-band maximum was set to zero.

as well as more accurate and time-consuming methods such as HSE hybrid functional method or G_0W_0 method,[4] has been used to overcome this drawback to obtain more justifiable bandgap and energy levels of the occupied cationic d band. By using the HSE hybrid functional method, the bandgap is determined to be 3.77 eV.[5] At the same time, the Zn 3d levels correspond to energy levels of about −9 eV, and the O 2p levels range from −5 to 0 eV.[6] The conduction bands consist of unoccupied Zn 3s states, while the O 2s states are located at deep energy levels around −20 eV with respect to the core-like energy levels. Meanwhile, photoelectron spectroscopy (PES) and angle-resolved photoelectron spectroscopy (ARPES) have been used to investigate the energy region of ZnO. The more accurate band structure found with the G_0W_0 method is in excellent agreement with the ARPES data both in terms of the band energies and band curvatures.[7]

3.1.2 Electronic Structure Modulation

Generally, the effects of strains on the band structures could be described by deformation potentials of ZnO,[8,9] which are the linear coefficients for the band structures to external strain responses. However, these strain parameters are not easily determined from experiments. The state-of-the-art first principles methods in combination with the k·p approach[10] usually aims to derive the effective-mass Hamiltonians for the deformation potentials of semiconductors in terms of Kane's model.[11] This method provides a theoretical approach for calculating the strain-induced electronic band structures and optical transitions of bulk and quantum-well wurtzite semiconductors.

Energy dispersion around the valence band maximum at Γ point can be obtained by calculating eigenvalues of the Bir-Pikus Hamiltonian matrix H for WZ semiconductors as a function of k:[12]

$$H = \begin{bmatrix} F & -K^* & -H^* & 0 & 0 & 0 \\ -K & G & H & 0 & 0 & \Delta \\ -H & H^* & \lambda & 0 & \Delta & 0 \\ 0 & 0 & 0 & F & -K & -H \\ 0 & 0 & \Delta & -K^* & G & -H^* \\ 0 & \Delta & 0 & H^* & -H & \lambda \end{bmatrix} \begin{matrix} |u_1\rangle \\ |u_2\rangle \\ |u_3\rangle \\ |u_4\rangle \\ |u_5\rangle \\ |u_6\rangle \end{matrix} \tag{3.1}$$

$$F = \Delta_1 + \Delta_2 + \lambda + \theta$$

$$G = \Delta_1 - \Delta_2 + \lambda + \theta$$

$$\lambda = \frac{\hbar^2}{2m_0}[A_1 k_z^2 + A_1(k_x^2 + k_y^2)] + \lambda_\varepsilon$$

$$\lambda_\varepsilon = D_1 \varepsilon_{zz} + D_2(\varepsilon_{xx} + \varepsilon_{yy})$$

$$\theta = \frac{\hbar^2}{2m_0}[A_3 k_z^2 + A_4(k_x^2 + k_y^2)] + \theta_\varepsilon \tag{3.2}$$

$$\theta_\varepsilon = D_2 \varepsilon_{zz} + D_4(\varepsilon_{xx} + \varepsilon_{yy})$$

$$K = \frac{\hbar^2}{2m_0}A_5(k_x + ik_y)^2 + D_5 \varepsilon_+$$

$$H = \frac{\hbar^2}{2m_0}A_6 k_z(k_x + ik_y) + D_6 \varepsilon_{z+}$$

$$\Delta = \sqrt{2}\Delta_3$$

where $\varepsilon_\pm = \varepsilon_{xx} \pm 2i\,\varepsilon_{xx} - \varepsilon_{yy}$, $\varepsilon_{z,\pm} = \varepsilon_{zx} \pm 2i\,\varepsilon_{yz}$. It should be pointed out that under the cubic approximation,[13,14] the following relationship holds for the parameters A_i:

$$A_1 - A_2 = -A_3 = 2A_4, \quad A_3 + 4A_5 = \sqrt{2}A_6, \quad \Delta_2 = \Delta_3$$

$$D_1 - D_2 = -D_3 = 2D_4, \quad D_3 + 4D_5 = \sqrt{2}D_6 \tag{3.3}$$

Therefore, only five band-structure parameters, *i.e.* A_1, A_2, A_5, Δ_1 and Δ_2, and the deformation potentials are necessary for the calculation of the valence-band structures. The band-edge energies with the corresponding basis functions from the eigenvalues and eigenvectors of the Hamiltonian can be written as:[12]

$$E_1 = E_v + \Delta_1 + \Delta_2$$

$$E_2 = E_v + \frac{\Delta_1 - \Delta_2}{2} + \sqrt{\left(\frac{\Delta_1 - \Delta_2}{2}\right)^2 + 2\Delta_3^2}$$

$$E_3 = E_v + \frac{\Delta_1 - \Delta_2}{2} - \sqrt{\left(\frac{\Delta_1 - \Delta_2}{2}\right)^2 + 2\Delta_3^2}$$

(3.4)

The band-edge energies can also be described by considering the following simplified case. If, without the spin–orbit interaction effects $\Delta_2 = \Delta_3 = 0$, we have $E_1 = E_2 = \Delta_1 > 0$, and $E_3 = 0$. When the spin–orbit interaction is taken into account, the top valence band energy is E_1, and the conduction-band edge energy is $E_c = E_g + \Delta_1 + \Delta_2$, if the energy of valence band is set to zero. If the common case is $\Delta_1 > \Delta_2 > 0$, the three bands from top to bottom can be labelled as heavy-hole (HH), light-hole (LH), and crystal-field split-off hole (CH) bands, respectively. The energies Δ_1 and Δ_2 are obtained from the crystal-field split energy Δ_{cr} and the spin–orbit split-off energy Δ_{so} by:

$$\Delta_1 = \Delta_{cr}, \quad \Delta_2 = \Delta_3 = \frac{1}{3}\Delta_{so}$$

(3.5)

The conduction-band edge has a hydrostatic energy shift $P_{c\varepsilon}$:[6]

$$E_c = E_v + \Delta_1 + \Delta_2 + E_g + P_{c\varepsilon}$$

$$P_{c\varepsilon} = a_{cz}\varepsilon_{zz} + a_{ct}(\varepsilon_{xx} + \varepsilon_{yy})$$

(3.6)

The net band-gap shift is determined by:[6]

$$E_c - E_1 = E_g + P_{c\varepsilon}$$

(3.7)

Considering that four different types of strain may be present in binary wurtzite systems like ZnO, biaxial strain in the c plane $\varepsilon_{xx} = \varepsilon_{yy}$, $\varepsilon_\perp = \varepsilon_{xx} + \varepsilon_{yy}$, uniaxial strain along the c axis ε_{zz}, anisotropic strain in the c plane $|\varepsilon_{xx} - \varepsilon_{yy}|$ and shear strain ε_{xy} and ε_{yz}, the expressions for the three highest valence-to-conduction-band transitions (E_1, E_2 and E_3) can be obtained in terms of strain deformation potentials. For biaxial strain in the c plane and uniaxial strain out of the c plane, the transition energies are given by:[6]

$$E_{1/2} = E_{1/2}(0) + (a_{cz} - D_1)\varepsilon_{zz} + (a_{ct} - D_2)\varepsilon_\perp - (D_3\varepsilon_{zz} + D_4\varepsilon_\perp)$$

$$E_3 = E_3(0) + (a_{cz} - D_1)\varepsilon_{zz} + (a_{ct} - D_2)\varepsilon_\perp$$

(3.8)

In the case of anisotropic strain in the c plane $|\varepsilon_{xx} - \varepsilon_{yy}| \neq 0$, the crystal symmetry lowers from C_{6v} to C_{2v}, which will shift the degeneracy of the Γ_6 state and lead to a splitting between the heavy-hole (HH) and light-hole (LH) bands. The deformation potential D_5 is expressed by following:[6]

$$\Delta E = |E_{HH} - E_{LH}| = 2|D_5(\varepsilon_{xx} - \varepsilon_{yy})| \tag{3.9}$$

The effect of shear strain on the band structure of wurtzite materials is given by the deformation potential D_6. The crystal-field split-off (CH) band energy for the unstrained system is set as zero and the topmost three valence band energies are as follows:[6]

$$E_1 = \Delta_{cr}$$

$$E_{2,3} = \frac{\Delta_{cr}}{2} \pm \frac{\sqrt{\Delta_{cr}^2 + 8D_6^2\varepsilon_{xz}^2}}{2} \tag{3.10}$$

These various expressions can now be fitted from the band structures in first-principle calculations for different strain conditions to determine the deformation potentials. A comparison of the available experimental deformation potentials and theoretical calculations from HSE, G_0W_0 and GGA-PBE calculations is presented in Table 3.1.

The uniaxial strain-dependent band structure and exciton-polariton transitions of wurtzite ZnO have been thoroughly investigated using modern first-principle calculations based on the HSE + G_0W_0 approach, k · p modelling using the deformation potential framework and polarised photoluminescence measurements.[17] The uniaxial deformation potentials and their optical transitions are obtained with or without the effects of the spin–orbit interaction. The excitonic deformation potentials and the exciton polaritons are investigated. These effects provide a comprehensive understanding of the electronic structures of ZnO by considering the crystal–field interaction, spin–orbit coupling, exchange interaction, and exciton–polariton splitting. In addition, the HSE hybrid functional and the exact exchange-based quasi-particle energy calculations in the G_0W_0@OEPx(cLAD) approach have been applied to investigate deformation potentials of wurtzite II–VI semiconductors under different realistic strain conditions.[18]

Table 3.1 The bandgap E_g, crystal splitting Δ_{cr} and the deformation potentials (eV) of wurtzite ZnO.

	Method	E_g (eV)	Δ_{cr} (meV)	$a_{cz} - D_1$ (eV)	$a_{ct} - D_2$ (eV)	D_3 (eV)	D_4 (eV)	D_5 (eV)	D_6 (eV)
ZnO	Exp.[15]	—	—	−3.80	−3.80	0.80	−1.40	−1.20	1.0
	Exp.[16]	—	—	−3.90	−4.13	1.15	−1.22	−1.53	−0.92
	Exp.[17]			−3.41	−4.33	−2.26	1.49		
	HSE06[18]	2.48	66	−3.06	−2.46	0.47	−0.84	−1.21	−1.77
	RPBE-SIC	2.04	55	−3.016	—	—	−1.260	—	—

However, the band alignments among group IV, II–VI and III–V of semi-conductors can be investigated by taking into account the deformation potentials.[19,20] Furthermore, the valence and conduction band offsets from heterointerface calculations, ionisation potential (IPs) and electron affinity (EAs) from surface calculations, and the valence band maximum (VBMs) and conduction band minimum (CBMs) relative to branch point energies or charge neutrality level, from bulk calculations, have been considered to determine the band alignments of semiconductors.[21]

Based on these band parameters and deformation potentials, including effective masses, valence-band (Luttinger or Luttinger-like) parameters and strain deformation potentials, they are useful for understanding the band structure of ZnO. These investigations on strained ZnO also pave the way for considering ZnO based device structures, such as the band offsets, optical properties, and the novel behaviours among ZnO and ZnO-based hetero-junctions by extracting from their deformation potential parameters.

Although the $k \cdot p$ method is good at evaluating the strained band structures and strain parameters of bulk semiconductors, the exciton deformation potentials of GaN in the wurtzite phase have been found to deviate from the quasicubic approximation assumed, as mentioned above, to calculate the deformation potentials parameters.[22] Meanwhile, it is observed that there are many non-linear relationships between strain and strain-induced electronic properties on the nanostructures of semiconductors. As shown in Figure 3.2(a), the ZnO nanowires were investigated under uniaxial and bent strains,[23] the sizes of which varied from $d_1 = 7.52$ Å, $d_2 = 10.07$ Å and $d_3 = 14.12$ Å, and it was found that the nanowires 1 and 2 of respective cross sections d_1 and d_2 show a non-linear behaviour of bandgap *vs.* uniaxial strain that is caused by surface atom ratios and their sizes. Similar non-linear behaviour has also been found in bent nanowires, shown in Figure 3.2(c). In the case of thicker nanowire d_3, the non-monotonic behaviour of ZnO nanowires disappears and there is a tendency to recover the bulk-like behaviour, and it is proposed that a very thick nanowire is necessary to have bulk-like behaviour,[24] shown in Figure 3.2(b). Different to the non-linear bandgap variations of ZnO nanowires, the uniaxial strain can result in a significant reduction of the bandgap of zigzag ZnO nanotubes due to a quantum-confined Stark effect induced by the built-in electric polarisation, as shown in Figure 3.2(d).[25]

3.1.3 Influence of Defects

As mentioned in the previous section, most growth techniques will introduce impurities through their sources or contaminants, whereas these dopant impurities and defects can strongly affect the physical properties of semiconductors. The intrinsic defects like oxygen vacancies have often been invoked as sources of *n*-type conductivity in ZnO. However, most of these arguments are based on indirect evidence, *e.g.* that the electrical conductivity increases as the oxygen partial pressure decreases.[26,27] Here, we address

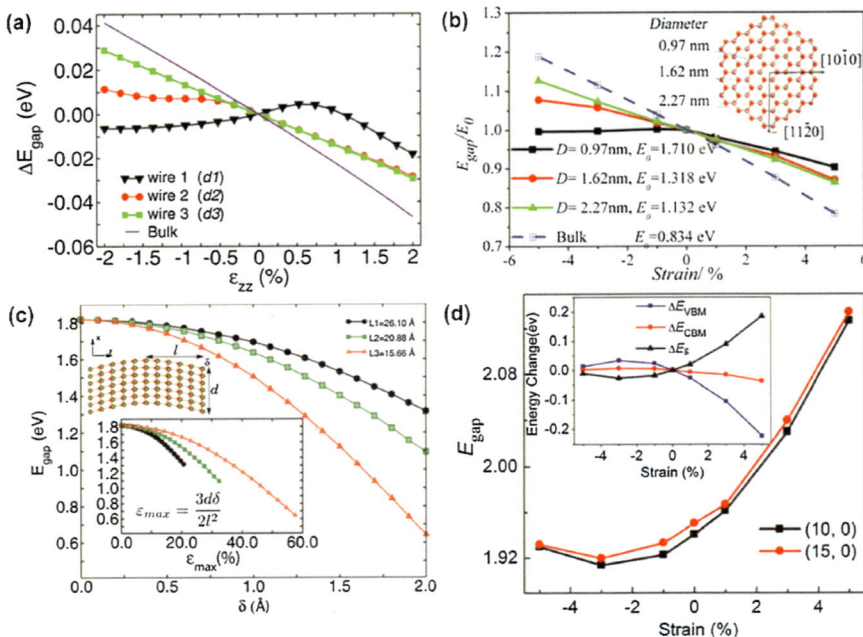

Figure 3.2 (a) The bandgap of the nanowires varied with their bending amplitude calculated *via* GGA + U method ($U_{eff} = 6.5$ eV); (b) the *c* axial strain-induced bandgap of ZnO NWs containing 48, 108, and 192 atoms in the cross section with diameters of 0.97, 1.62, and 2.27 nm, respectively. Inset plot shows a sketch of the NW section; (c) the bandgap of the NWs *vs.* uniaxial strain calculated through the GGA functional; (d) variations of bandgap of zigzag ZnO NTs under axial strain. The inset plot is the energy shifts of the band edge states. (a) and (c) Reprinted with permission from W. A. Adeagbo, *Phys. Rev. B*, **89**, 195135, 2014. Copyright (2014) by the American Physical Society.[23] (b) Reprinted from ref. 24 with permission from John Wiley and Sons, Copyright © 2009 WILEY-VCH Verlag GmbH & Co. KGaA, Weinheim. (d) Reprinted with permission from ref. 25. Copyright (2011) American Chemical Society.

the first-principle calculations to provide some insights into defect characteristics and corresponding characteristics of ZnO nanostructures.

Assuming thermodynamic equilibrium and neglecting defect–defect interactions (*i.e.* in the dilute regime), the concentration of defects in a crystal is given as:[28]

$$c = N_{sites} \exp(-E_f/kT) \tag{3.11}$$

where E_f is the formation energy, N_{sites} is the number of sites where the defect can be incorporated, k is the Boltzmann constant, and T is the temperature. Eqn (3.12) shows that defects with high formation energies will occur in low concentrations. From this equation, the formation energy of a defect and its concentration can be computed from first-principle calculations without resorting to experimental data, but actually it still depends on the growth or annealing conditions.[29] The formation energy of a defect can be described as follows:

$$\Delta H_f(\alpha,q) = E(a,q) - E(\text{host}) - \sum_i n_i \mu_i + q(E_F + E_V) \tag{3.12}$$

where $E(a,q)$ and $E(\text{host})$ represent the total energy of supercell containing a defect in charge state q or that in a perfect supercell, respectively. n_i and μ_i are the differences in the number of constituent atoms of type i in the supercells or that of the chemical potentials in constituent atoms of type i, respectively. E_F and E_V are the Fermi level and the energy on the valence band maximum in the supercells, respectively.

Due to the chemical potentials of ZnO varying within limits determined by phase equilibria, the correlation and range of μ_{Zn} and μ_O are given by:

$$\mu_{Zn} + \mu_O = \mu_{ZnO} \tag{3.13}$$

In order to describe μ_{Zn} and μ_O under various circumstances, growth conditions are described by the stoichiometric parameter λ,[30] which varies between 0 and 1 under extreme Zn- and O-rich conditions, respectively.

The terms μ_O and μ_{Zn} are given as:

$$\mu_{Zn} = \mu_{ZnO} + \lambda\Delta H_f \quad \text{and} \quad \mu_O = \frac{1}{2}\mu_{O_2} + (1-\lambda)\Delta H_f \tag{3.14}$$

where ΔH_f represents the heat of formation of ZnO. For the reference chemical potentials (μ_{ZnO}, μ_{Zn} and μ_O), the calculated total energies of the ZnO, Zn crystals and the O_2 molecule were used.

In addition, the thermodynamic transition level of a defect between charge states q and q' corresponding to the energy level $\varepsilon(q/q')$ can be derived from the formation energies of charge states q and q'[31] as follows:

$$\varepsilon(q/q') = \frac{\Delta E^q_{f,\text{VBM}} - \Delta E^{q'}_{f,\text{VBM}}}{q' - q} \tag{3.15}$$

where $\Delta E^q_{f,\text{VBM}}$ represents the defect formation energy of charge state q when the Fermi level is located at the VBM.

3.1.3.1 Native Defects of ZnO and ZnO Nanostructures

By means of the methods mentioned above, native defects in ZnO have been discussed and reviewed in studies based on density functional theory with various approaches.[32-35] Vacancies, interstitials, and anti-sites of Zn and of O are considered to be the native defects in ZnO. The O vacancy, Zn interstitial and Zn anti-site, which are often associated with O deficiency or Zn excess, are donor-type defects, whereas the Zn vacancy, O interstitial, and O anti-site are acceptor-type defects associated with Zn deficiency or O excess.

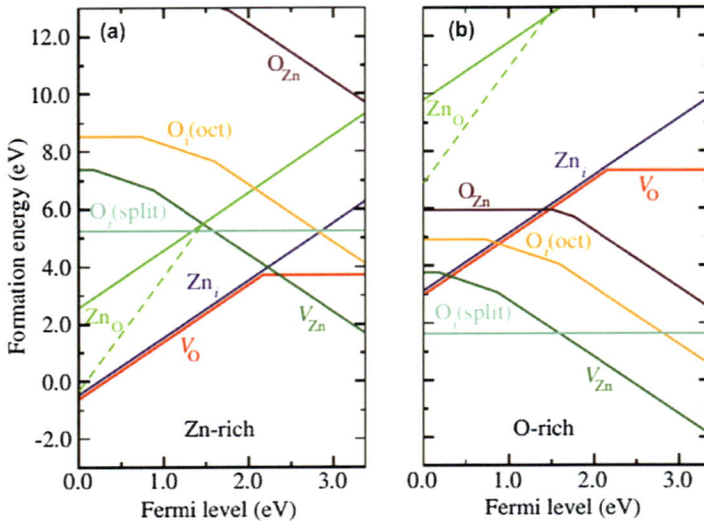

Figure 3.3 Formation energies of native point defects in ZnO *vs.* Fermi-level position on (a) Zn-rich and (b) O-rich conditions. The valence-band maximum corresponds to the zero point of Fermi level. The slope of these defects indicates their charge state. The knee points in the curves indicate transitions between different charge states.
Reprinted with permission from A. Janotti and C. G. Van de Walle, *Phys. Rev. B*, **76**, 165202, 2007. Copyright (2007) by the American Physical Society.[33]

In the case of O vacancy, the neutral O vacancy induces a deep and localised one-electron state in the bandgap shown in Figure 3.3.[33] The O vacancy is considered to be the dominant donor-type defect associated with non-stoichiometry toward the O-deficient side, but is unlikely to provide carrier electrons. The Zn interstitial and anti-site are both shallow donors. However, they have high formation energies in *n*-type ZnO and, hence, do not form at high concentration under thermal equilibrium. With lowering of the Fermi level, the O vacancy, Zn interstitial and Zn anti-site become energetically favourable. These donor-type defects can compensate for acceptor-type defects in *p*-type ZnO.

3.1.3.2   The p-type Defects of ZnO and their Nanostructures

In order to achieve *p*-type ZnO to meet the demands for *p–n* homojunctions and functionalised nanodevices, many dopants have been attempted by experimentalists and theorists in previous decades. Many theoretical predictions and designs are proposed to achieve effective *p*-type conducting.[33,36] Despite many reports in the literature, reliable *p*-type doping of ZnO remains very difficult.[37–39] The low solubility of *p*-type dopants and the compensation by abundant donor defects and impurities are often raised as the main issues.

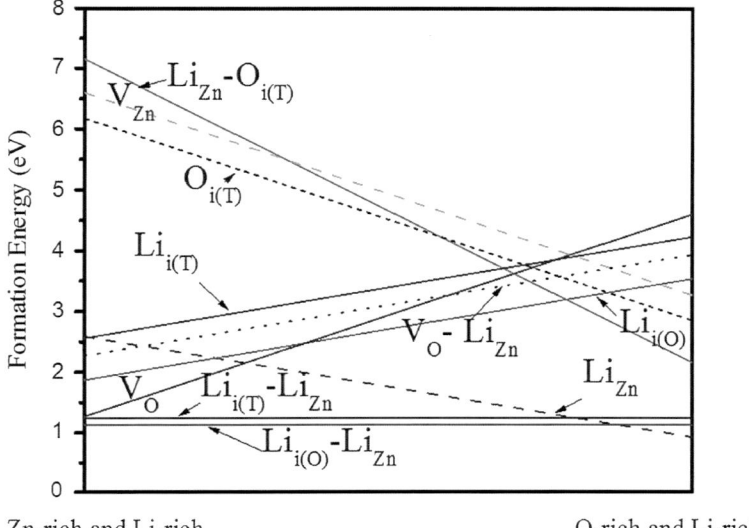

Figure 3.4 Formation energies of the native point defects and the defect complexes in Li-doped ZnO under Zn-rich and O-rich ambient conditions.
Reprinted with permission from X. Sun, Y. Gu, X. Wang and Y. Zhang, Defects Energetics and Electronic Properties of Li Doped ZnO: A Hybrid Hartree-Fock and Density Functional Study, *Chin. J. Chem. Phys.*, 2012, **25**(3), 261–268. Chinese Physical Society.[40]

Known acceptor impurities include group-I elements Li, Na, K, Cu, and Ag, and group-V elements N, P, As and Sb. Lithium doped into ZnO may behave both as a donor and as an acceptor in ZnO.[40,41] As shown in Figure 3.4, the donor behaviour is exhibited when lithium locates at an interstitial site; the acceptor behaviour is exhibited when lithium substitutes on a Zn site.[40] From the consideration of possible defects existing in Li-doped ZnO under a Zn-rich atmosphere, *p*-type conduction is impossible, since all the defects will result in either an intrinsic or *n*-type semiconductor. Efficient *p*-type conduction can be obtained only in an O-rich environment, as Li_{Zn} will exist in high concentration in the view of formation energies and overcome the thermodynamic diffusions and compensation effects under the O-rich environment. Notably, the defect complex made up of Li_{Zn} and $Li_{i(O)}$ exhibit the lowest energy formation. This is similar to the Li defect pair that was observed as distorted Li_2O in ZnO reported by Wardle *et al.*[42] They considered the neutral Li_{Zn}–Li_i as an inactive complex defect and relatively stable. From this viewpoint, the formation of Li pair complexes is the main obstacle to realising *p*-type conductance in Li-doped ZnO. Additionally, V_{Zn} and P_{Zn}–$2V_{Zn}$ are dominant acceptors under O-rich conditions when P_2O_5 and Zn_3P_2 are used as P dopant sources, respectively.[30] Likewise, EPR studies of Li in ZnO were measured by Kasai[37] and by Schirmer.[38] For the axial configuration, the measured g values were $g_{||} = 2.0028$ and $g_{\perp} = 2.0253$.[38] For group-V impurities, Limpijumnong *et al.*[39] argued that N_O is relatively deep

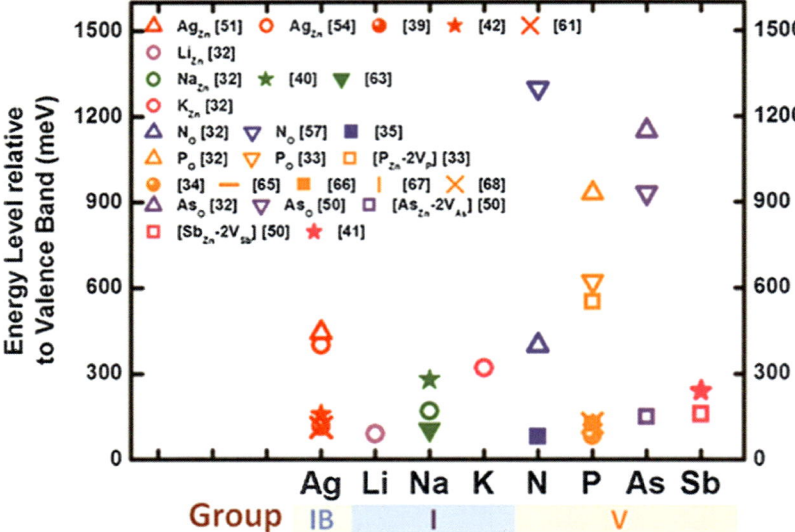

Figure 3.5 Experimental and theoretical acceptor energy levels for various element-doped ZnO. The acceptor defects adopted in those theoretical works are indicated using the empty symbols. The experimental acceptor energy levels were extracted approximately from either low-temperature PL or cathodoluminescence data measured from *p*-type ZnO NWs by using solid symbols. The references in the plot can be found there.
Reprinted from Nano Energy, 1(2), M. Lu, M. Lu and L. Chen, p-Type ZnO nanowires: From synthesis to nanoenergy, 247–258, Copyright (2012), with permission from Elsevier.[44]

and As_{Zn}–$2V_{Zn}$ and Sb_{Zn}–$2V_{Zn}$ defects are responsible for the experimentally obtained As- and Sb-doped *p*-type ZnO under O-rich conditions, respectively.

For P-doped ZnO NWs, the P_{Zn} defects have low formation energies as a donor to exhibit *n*-type conducting under the Zn- and P-rich conditions, and V_{Zn} and P_{Zn}–$2V_{Zn}$ defects can lead to *p*-type ZnO nanowires under the O- and P-rich conditions.[43] For the other dopants, some significant discrepancies are evident between the experimental and calculated data, as illustrated in Figure 3.5.[44] However, the mechanism to form stable acceptor defects in the experiments is very complex, and many growth or post-annealing conditions, such as the growth temperature, dopant element, dopant concentration, and growth environment, should be taken into account. Achieving effective manipulation of donor-type and acceptor-type defects in the conducting behaviour of ZnO and their nanostructures is required for further research in the future.

3.2 Electrical Properties

As mentioned above, undoped ZnO usually exhibits *n*-type conductivity, whereas the *p*-type doping of ZnO has been considered to be the main obstacle to achieving high-quality ZnO-based homojunctions. Here, the electrical properties of ZnO nanostructures are discussed below and the contacts with typical metals are presented as well.

3.2.1 Electrical Transport Properties

The resistivity ρ is expressed by:

$$\rho = 1/(ne\mu_n + pe\mu_p) \qquad (3.16)$$

where n and p are the concentrations of electron and hole, e is the unit electronic charge, and μ_n and μ_p are the mobilities of electron and hole. Due to the unintentional defects and self-compensation effects, the *n*-type conductivity with high electron concentration and low electron mobility dominates the electrical properties of ZnO nanostructures. Thus, the resistivity ρ of *n*-type ZnO is inversely proportional to the free-electron concentration and electron mobility. Furthermore, the mobility is related to the scattering time by $\mu = q\langle\tau\rangle/m^*$, where m^* is the electron effective mass, and $\langle\tau\rangle$ is the relaxation time averaged over energy distribution of electrons. According to the Matthiessen's rule,[45,46] the lattice scattering mechanisms consist of optical–polar phonon, piezoelectric potential, and deformation potential terms, and then the total relaxation time (τ_{total}) is given by:[2]

$$\frac{1}{\tau_{\text{total}}} = \frac{1}{\tau_{\text{opt}}} + \frac{1}{\tau_{\text{piezo}}} + \frac{1}{\tau_{\text{deform}}} + \frac{1}{\tau_{\text{ion}}} \qquad (3.17)$$

where τ_i is the relaxation time controlled by various mechanisms. The lattice vibration of an optical mode contributes to the optical–polar phonon scattering (τ_{opt}) in the crystal, which is caused by the interactions of ionic charges with an electric field. The lattice vibration of an acoustic mode induces the second electric fields produced by strains in a crystal without inversion symmetry, and results in the piezoelectric potential scattering (τ_{piezo}). The lattice vibration of an acoustic mode contributes to the deformation potential scattering (τ_{deform}), and leads to the change of lattice spacing, and induces the energy change of the band edges. The lattice scattering, which corresponds to impurities and defects, is also proportional to temperature. The defect scattering mechanisms are dominated by ionised impurities. Ionised impurity scattering (τ_{ion}) arises from the carriers deflected by the long-range coulomb potential of charged centres induced by intrinsic defects or external impurities. If a semiconductor system possesses a high density of dislocations or native defects, the dislocation scattering and scattering through defects should also be considered as possible scattering mechanisms. Dislocation scattering arises from the fact that acceptor centres are introduced along the dislocation line, which will trap electrons from the conduction band in an *n*-type semiconductor. The electrons are also scattered by these negatively charged dislocations to reduce the mobility. Notably, among these scattering mechanisms, only ionised impurity scattering is dependent on the crystal qualities of the samples, whereas the others are dependent on the characteristics of the materials. All these scattering mechanisms are simultaneously involved in the electron transport of semiconductors. Therefore, over several decades, investigation of the carrier mobility in *n*-type ZnO has been performed with a temperature range from 4 K to room temperature.[47–49]

The mobilities of the ZnO thin film are achieved at low temperature (up to 5000 cm^2 V^{-1} s^{-1} for the maximum mobility[50]), compared to 2500 cm^2 V^{-1} s^{-1} in bulk material.[51] The electrical characteristics of ZnO NR based field-effect transistors (FETs) were fabricated by coating a polyimide thin layer on the NR surface.[52] ZnO NR FETs exhibit a large turn-on/off ratio of 10^4–10^5, a high transconductance of 1.9 mS, and high electron mobility above 1000 cm^2 V^{-1} s^{-1}, while ZnO NR FETs without surface coating show electrical characteristics with a transconductance of ~140 nS and a mobility of 75 cm^2 V^{-1} s^{-1}.

In situ electrical transport measurements on individual bent ZnO nanowires have been performed inside a high-resolution transmission electron microscope, and it was found that the conductance of the ZnO nanowires dropped significantly with an increase of the bending deformation.[53] The current–voltage (*I–V*) characteristics of single ZnO nanowires have been investigated in the humid air, dry air, vacuum and under ultraviolet (UV) irradiation.[54] A model of a single ZnO nanowire connected with two opposite diodes was proposed to explain the observed *I–V* behaviours. These electrical characteristics are dominated by the reverse barrier height. The barrier height can be adjusted by surface adsorption, which is ascribed to the effect of surface states on surface band bending and Fermi level pining. Furthermore, the surface states have a pronounced effect on the electronic transport in single ZnO nanowires, due to an enormous increase of conductance of the nanowire upon UV irradiation and a considerable persistent photocurrent after withdrawal of the UV excitation.

As shown in Figure 3.6, temperature dependent electron transport properties of individual ZnO micro/nanowires from 293–473 K were measured and it was found that the conductivity increased with the increasing temperature.[55] The plot of the conductivity as a function of 1000/T of one ZnO MW sample can be quantitatively described by the equation

$$\sigma = \sigma_1 \exp\left(-\frac{E_1}{k_B T}\right) + \sigma_2 \exp\left(-\frac{E_2}{k_B T}\right),$$ where σ_1 and σ_2 are pre-exponential

factors, and E_1 and E_2 are the activation energies associated with two kinds of thermal activation conduction processes. The calculated values of E_1 and E_2 for this selected sample are 406 meV and 29 meV, respectively. The activation energy E_2 related to shallow donors is effectively independent of the radius, suggesting that it is associated with the surface conduction channel of ZnO wire, while the size dependence of the activation energy E_1 related to deep donors on the wire diameter may result from the dielectric confinement effect.

Besides the temperature dependent transport properties, the strain modulated electrical properties at room temperature have been investigated

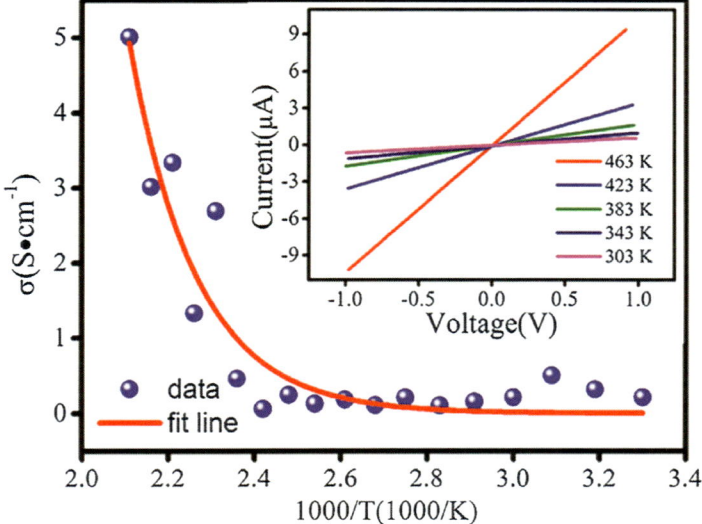

Figure 3.6 The conductivities and the double exponential fit of the ZnO microwires as a function of temperature. The inset plot is the temperature dependent *I–V* characteristics of a four-terminal structure. Reprinted from ref. 55 with permission from AIP Publishing.

using *in situ* scanning tunnelling microscope transmission electron microscope.[56] Figure 3.7(a)–(g) shows a typical strain-free ZnO NW with a length of 822 nm and a width of 136 nm. By controlling the piezo-motor movement (each fine step corresponding to a movement of 0.6 nm), the ZnO NW can be either stretched or compressed, and then the corresponding *I–V* measurements can be carried out simultaneously by applying an external voltage under a given strain to explore the intrinsic piezo-resistance of strained *n*-type [0001] ZnO NWs. The resistance of ZnO nanowires reduces with an increase of compressive strain, but enlarges with the increase of tensile strain, indicating that piezo-resistive characteristics dominate the transport properties, as shown in Figure 3.8. With the decrease of E_g (induced by tensile strain), the conductance increases monotonically, indicating that ZnO NWs are indeed very sensitive to strain induced variation of E_g and thus conductivities.

Additionally, a negative differential resistance (NDR) was observed in the PtIr/ZnO ribbon/sexithiophen hybrid double diodes consisting of the back-to-back Schottky and *p–n* junction diodes.[57] The NDR can be explained by current-induced breakdown of sexithiophen (6T) film rather than electron resonant tunnelling through the double diodes. The transport properties of a superlattice-structured ZnO nanohelix superlattice have been fabricated, which is a helical structure made by coiling a nanobelt, as illustrated in Figure 3.9.[58] This nanohelix superlattice is about 300–800 nm in diameter, and the nanobelt is ~10–30 nm in thickness and ~100–600 nm in width.

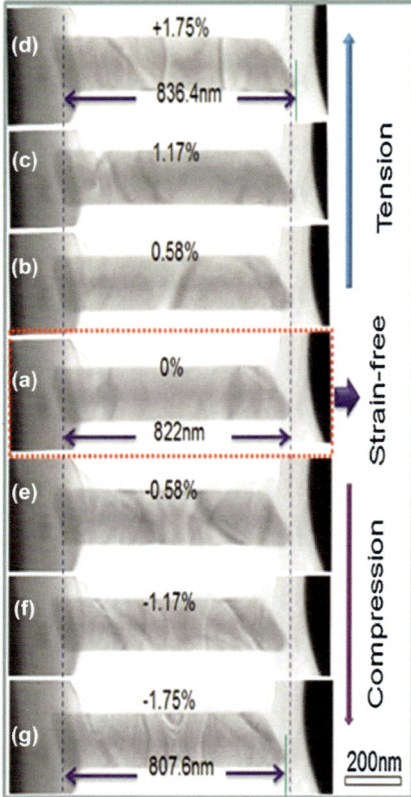

Figure 3.7   TEM images of the ZnO NWs under (a) unstrained, (b)–(d) tensile deformed and (e)–(g) compressive deformed conditions.
Reprinted from ref. 56 with permission from The Royal Society of Chemistry.

The observed superlattice lengths are in the range of ~1–500 μm. The nanohelix was expected to form another new type of band structure modulated superlattice, which is in parallel to the traditional heterostructured superlattices obtained by chemical fabrication or doping modulation. The non-linear electronic transport properties of ZnO nanohelix and temperature dependent electrical characteristics were investigated using a four-probe characterisation method. With a current sweeping from −4 to 4 μA across the nanohelix, it exhibited a series of non-linear *I–V* curves at 293, 240, 190, and 90 K, shown in Figure 3.10. The non-linear electronic transport behaviour of the nanohelix might be due to a major contribution from nanostripe boundaries and surfaces, where a built-in periodic back-to-back energy barrier might occur across the nanostripe interfaces as a result of polar charges and interface-strain-induced piezoelectric effect.

Doped ZnO nanostructures have also been investigated by others.[59–65] Gallium-doped zinc oxide fibres have been fabricated by a vapour-phase transport method,[59] and it was found that with a gallium doping concentration of 0.73 at.%, the carrier concentration and resistivity were

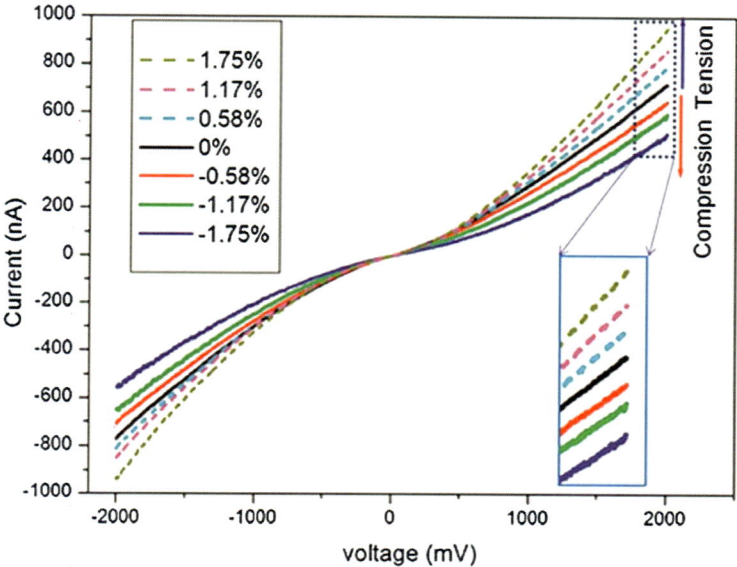

Figure 3.8 The relationship of the resistivity as a function of the applied strain. The *I–V* enlarged characteristics of ZnO NWs are inset in the plot. Reprinted from ref. 56 with permission from The Royal Society of Chemistry.

3.77×10^{20} cm^{-3} and 8.9×10^{-4} Ω cm, respectively. For undoped ZnO, the carrier concentration and resistivity of undoped ZnO were 6.66×10^{18} cm^{-3} and 0.13 Ω cm, respectively. In particular the resistivity of Ga-doped ZnO is more than two orders of magnitude smaller than that of the undoped ZnO. Zhou *et al.*[60] investigated the electrical properties of Ga-doped individual ZnO nanowires, and confirmed that the resistivities of Ga-doped ZnO nanowires (when randomly choosing ten nanowires) is on the order of 10^{-3} Ω cm, while the resistivities of undoped ZnO nanowires are in the range of 10^{-1} Ω cm.

The transport properties of ZnO films with different Ga dopants are listed in Table 3.2. It was found that room-temperature resistivities of the films were observed to decrease with an increase in Ga concentration up to 5% Ga, which displayed the lowest value of resistivity.[61] Additionally, the temperature dependent resistivity of these films exhibited a metal-semiconductor transition, which was rationalised by localisation of degenerate electrons. Below the transition temperature, a linear variation of conductivity with $\sqrt{T}$ suggests that the degenerate electrons are in a weak-localisation regime. Moreover, the transition temperature is dependent on the Ga concentration and is related to the increase in disorder induced by dopant addition. Besides this, the Cd-doped ZnO[62] has been investigated in terms of temperature dependence conductance. And P- and As-doping have also been reported to achieve *p*-type conducting in the ZnO nanostructures.[63–65]

On the theoretical framework, the density functional theories are well-established and widely used for simulating electronic conductance of

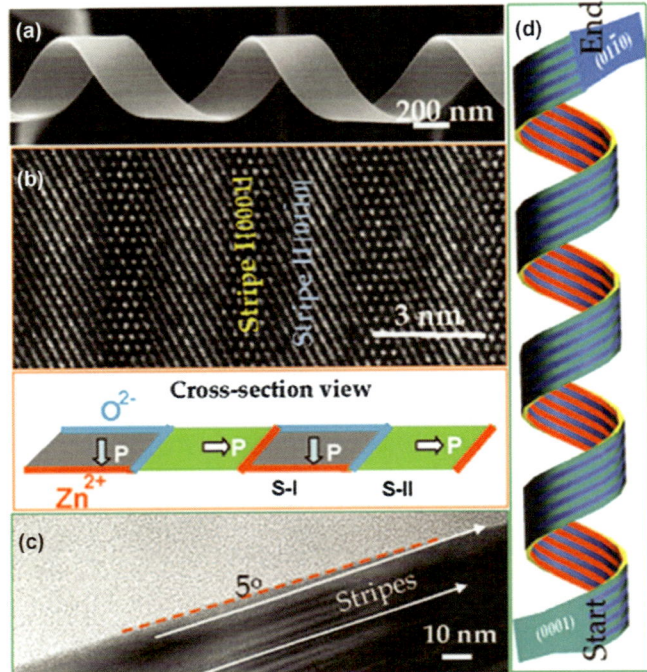

Figure 3.9 (a) SEM image of a nanohelix superlattice; (b) high-resolution transmission electron microscopy image of the nanohelix with the superlattice structures. The nanostripes I and II are dominated by (0001) polar surfaces and {01$\bar{1}$0} non-polar surfaces, respectively. The schematic diagram of the cross section of the nanostripes is shown beneath HRTEM imaging; (c) about ∼5° angle offset is observed between the running directions of nanobelt and the nanostripes; (d) a schematic model of the structure of the nanohelix.
Reprinted with permission from ref. 58. Copyright (2009) American Chemical Society.

Table 3.2 List of the values of resistivity and the metal–semiconductor transition temperature of ZnO films with varying Ga concentrations.

%Ga	Resistivity (Ω cm)	$L_i = d_{Ga}$ (nm)	T_c (measured) (K)	T_c (calculated) (K)
0	2.59×10^{-1}	—	—	—
2	4.71×10^{-4}	1.33	93	92
3.65	4.29×10^{-4}	1.10	135	134
5	1.57×10^{-4}	0.98	170	171
7	2.57×10^{-3}	—	—	—

nanomaterials,[66–72] based on a anon-equilibrium Green function (NEGF) method.[73,74] The current–voltage curves could be obtained *via* the Landauer-Büttiker formula:

$$I = \frac{2e}{h} \int dE (f_L - f_R) T(E) \tag{3.18}$$

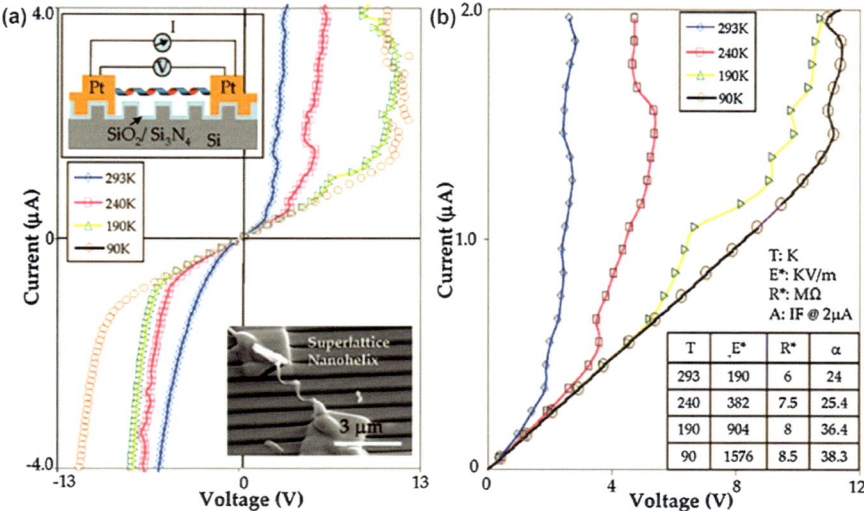

Figure 3.10 (a) The non-linear *I–V* curves of superlattice nanohelix (the length of nanohelix about ~4 µm) measured in the range from 90–293 K. The inset plot is the schematic diagram of the device; (b) a 0 to 2 µA forward current sweeping *I–V* curves of the device at different temperatures, showing different breakdown strengths, non-linear ideal factors, and low electric field resistances. The inset table shows the evaluated parameters.
Reprinted with permission from ref. 58. Copyright (2009) American Chemical Society.

where $T(E)$ is the transmission matrix and could be calculated from Green's functions, and $f_{L/R}$ are the Fermi-Dirac distribution function of left or right electrodes.

Since the contact behaviour has crucial influence on the device property and is very sensitive to the fabrication processes, the transport properties of ZnO need to be investigated by contact with metal electrodes. The transport properties of ZnO bulk were calculated by coupling with Au and Mg electrodes, respectively.[66] It was found that the Au–ZnO–Au interface shows Schottky contact behaviour while the Mg–ZnO–Mg interface shows Ohmic contact, as shown in Figure 3.11. The transport properties of ZnO nanowires have been studied in contact with Al electrodes,[67] and it was found that the contact interface played important roles in electron transport. Single-walled ZnO nanotubes were also studied to show the length-dependence transport characteristics under lithium electrodes.[68] The transport properties of ZnO nanobelts under a Cu electrode were affected by interfacial spacing and nanobelt widths,[69] and it was found that the conductance decreased exponentially with the widths of nanobelts and the interfacial spacing. Additionally, the transport properties of Cu/ZnO-nanobelt/Cu nanostructures under uniaxial strains were studied using first-principles calculation.[70] The conductivity G is very sensitive to strain, and it decreases linearly when strain varies from compression to tension,

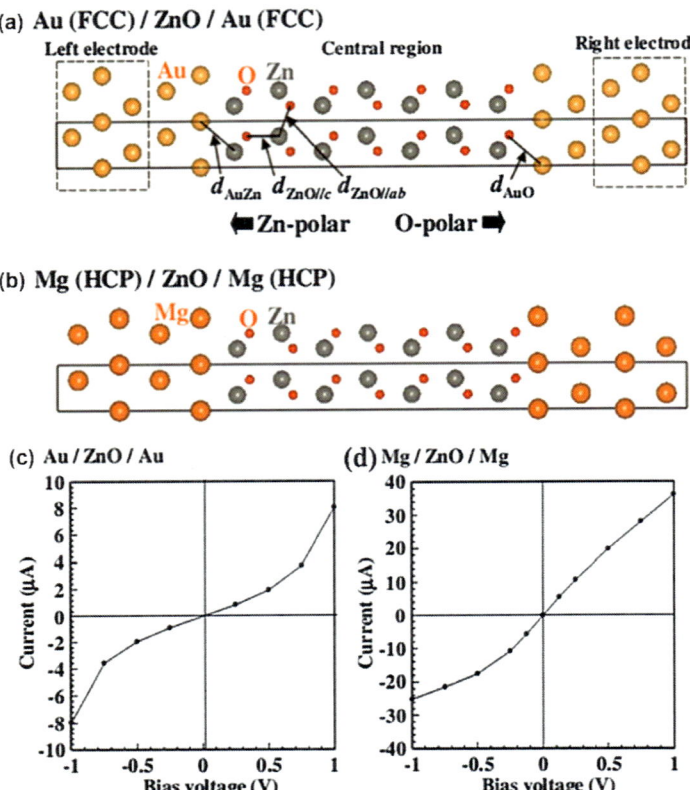

Figure 3.11 Two-probe models of (a) Au/ZnO/Au and (b) Mg/ZnO/Mg nanojunctions. The *I–V* characteristics of the relaxed (a) Au/ZnO/Au and (b) Mg/ZnO/Mg nanojunctions.
Reprinted with permission from John Wiley and Sons.[66] Copyright © 2008 WILEY-VCH Verlag GmbH & Co. KGaA, Weinheim.

as shown in Figure 3.12. The number of carriers nearby the Fermi level decreased obviously when the strains changed from compression to tension, which could be used to explain the linearly decreased conductance. The strain induced transport characteristics and the relative physical properties of nanoscale zinc oxide (ZnO) tunnel junctions were investigated[71] and an intrinsic giant piezoelectric resistance (GPR) effect was found in this nanojunction. The increase of up to 2000% of the electro-resistance ratio with respect to the zero strain state can be reached at the same bias voltage and with a 5% compression.

3.2.2 Ohmic and Schottky Contacts

It is known that the interface and contact is one of the critical technologies before the fabrication and modulation of nanoscale devices. Here, we

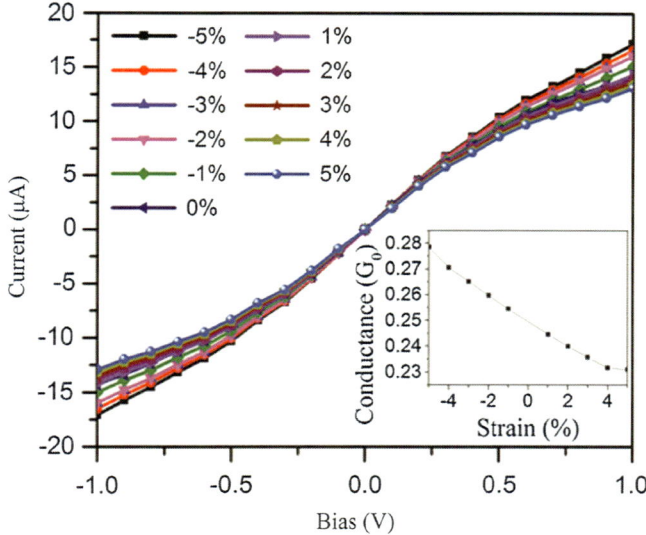

Figure 3.12 The *I–V* characteristics of Cu/ZnO/Cu nanojunctions under different strains and the inset plot shows the conductance *G* as a function of strain.
Reprinted with permission from John Wiley and Sons.[70] Copyright © 2015 WILEY-VCH Verlag GmbH & Co. KGaA, Weinheim.

briefly discuss the Schottky barriers and ohmic contacts in ZnO and their nanostructures. Figure 3.13 shows the energy band diagrams in the metal/semiconductor interface before and after junction formation. In the Ohmic contact, electron transport is always dominated by the field emission that allows electrons to tunnel through the Schottky barrier, whereas the Schottky contact is determined by the thermionic emission that allows electrons to jump over the Schottky barriers. In order to form ohmic contact to ZnO nanostructures, the barriers should be sufficiently narrow at or near the bottom of the conduction bands to allow thermally excited electrons to be tunnelled directly. When putting ZnO and their nanostructures on metal substrates or connecting with metal electrodes, the current–voltage features appear as ohmic behaviours, and the contact resistance between the two materials is dependent on doping concentration in ZnO nanostructures and unintentionally formed interface states.

To distinguish the ohmic and Schottky contacts, the Schottky barrier can be justified by the difference of the metal work function and the semiconductor electron affinity. It means that the Schottky barrier height is independent of interface structures and states, surface and bulk conductivities. However, the actually obtained Schottky barrier height, often called the effective Schottky barrier height, is different from the theoretical value because various factors, such as interface states, residual electrons, and edge

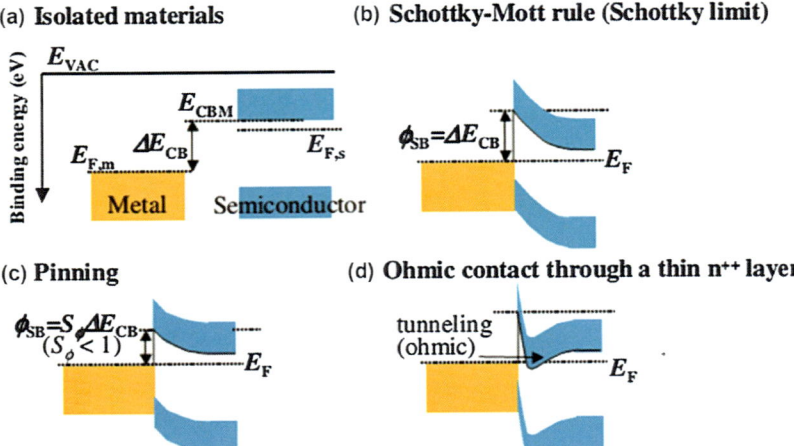

Figure 3.13 Energy band diagrams at metal-semiconductor interfaces. (a) Energy alignments in metal and semiconductor before contact; (b) the ideal Schottky contact between metal and semiconductor; (c) the effective contact between metal and semiconductor. The pinning factor Sφ is smaller than unity; (d) ohmic contact.
Reprinted with permission from John Wiley and Sons.[66] Copyright © 2008 WILEY-VCH Verlag GmbH & Co. KGaA, Weinheim.

and screw dislocations generate leakage current paths, which can sharply affect the Schottky barrier.

According to a classical Schottky model, $\Phi_{SB} = \Phi_M - \chi_{SC}$, χ_{SC} and Φ_M are the semiconductor electronic affinity and metal work function. Considering the interface states and the dipoles that screen part of the potential difference between metal and semiconductor, Φ_{SB} can be written to $\Phi_{SB} = \Phi_M - \chi_{SC} - \Delta\chi$, and $\Delta\chi$ represents the dipole at the interface.

To evaluate the Schottky contact is to estimate the Schottky diode characteristics using I–V measurements. This is derived by theoretical calculation and fitting using the diode equation, which is expressed by:[75]

$$I = I_0 \exp\left(\frac{q(V - IR)}{nkT}\right)$$

$$I_0 = A^*AT^2 \exp\left(-\frac{\Phi_{SB}}{kT}\right)$$

(3.19)

where I is the current, I_0 is the reverse saturation current, q is the unit charge of an electron, V is the applied voltage, R is the series resistance, n is the ideality factor, k is the Boltzman constant, T is the temperature, A^* is the Richardson constant ($32\,\mathrm{A\,cm^{-2}\,K^{-2}}$), A is the contact area, and Φ_{SB} is the Schottky barrier height. In the experiments, the Schottky barrier height is determined by the built-in potential and the donor density, which is expressed by:

$$\Phi_{SB} = qV_{bi}^{real} + kT \ln\left(\frac{N_c}{N_d}\right) \tag{3.20}$$

where N_c and N_d are the effective density of states and the donor density in the conduction band, and V_{bi}^{real} is the real built-in potential.

In terms of first-principle calculations, the interfacial Schottky barriers between metal and ZnO nanostructures can also be evaluated *via* several approaches, like the potential,[77] potential profile line up[78] and chemical bond polarisation methods.[79] The Schottky barriers could be directly predicted from the differences between the potentials of ZnO nanostructures and metal electrodes in the potential method. The SBHs using the potential profile line up method can be expressed as ΔE_n *i.e.* the difference between the Fermi level of combined system and conduction band edge of the semiconductor:

$$\begin{aligned}
\Delta E_n &= E_c - E_F = E_g - (E_c - E_V) \\
&= E_g - \{(E_F - \langle V_1 \rangle) - (E_F - \langle V_2 \rangle)\}
\end{aligned} \tag{3.21}$$

The atomic core potentials are introduced to describe the average potentials. E_F and $\langle V_1 \rangle$ come from the junction systems and $\langle V_1 \rangle$ can be evaluated from the sum of atomic core potentials of the ZnO atomic layer furthest away from the interface. E_V and $\langle V_2 \rangle$ are derived from the calculations of ZnO nanostructures, and E_g is the band gap of ZnO nanostructures.

The SBH Φ_{SB} in the chemical bond polarisation method (CBP) can be expressed as:[79]

$$\Phi_{SB} = \gamma_B(\Phi_M - \chi_s) + (1 - \gamma_B)\frac{E_g}{2} \tag{3.22}$$

where γ_B is the chemical bond polarisation parameter:

$$\gamma_B = 1 - \frac{e^2 N_B d_{MS}}{\varepsilon_{int}\varepsilon_0(E_g + \kappa)} \tag{3.23}$$

d_{MS} is the distance between the metal and ZnO nanostructures at the interface, and N_B is the density of chemical bonds at the interface. κ is the sum of all the hopping interactions, which could be written as $\kappa = 4J_{SS} + 4J_{MM} - 2J_{MS}$, where $J_{SS} = e^2/4\pi\varepsilon_{it}d_{SS}$ and $J_{NM} = e^2/4\pi\varepsilon_{it}d_{MM}$ are the interactions for the bonded metal–ZnO pair, ZnO–ZnO pair and the metal–metal interactions along the interface plane, respectively. ε_{it} is the dielectric constant of ZnO and $d_{SS/MM}$ is the plane spacing of ZnO or metal electrode.

More calculated and measured SBHs Φ_B for various ZnO–metal interface configurations are listed in Table 3.3. In order to form ohmic

Table 3.3 The n-type Schottky barriers Φ_B for the ZnO-metal interfaces.

	ZnO(0001) O-terminated		ZnO(0001) Zn-terminated		ZnO (10$\bar{1}$0)	
	Exp.	Theo.	Exp.	Theo.	Exp.	Theo.
Ag	0.69[76] 0.80[82]	0.60 1.0[84] 1.11[85]	0.77[82] 0.99[86]	0.03 −0.2[84] −0.02[85]	0.68[87]	−0.03
Au	1.2[88] 0.81[88]	1.14 0.7[84] 0.69[85]	0.43[89] 0.77[88] 1.07[88]	0.62 −0.2[84] 0.1[84] 0.30[85]	0.65[94]	0.41
Al	Ohmic[95]	−0.12 −0.60[84] −0.87[85]	Ohmic	−0.28 −0.3[84] −0.1[84] −0.35[85]	Ohmic[95]	−0.34
Pt	0.61[90] 0.85[91] 0.93[92]	0.88 0.74[85]	0.42[93] 0.51[82]	1.13 1.36[85]	0.75[94]	0.25
Ti	Ohmic[95]					
Pd	1.14[94] 0.73[8]	0.73[85]	0.68[83]	0.55[85]	<0.3[96]	—
Ir	—	0.16[85]	0.54[93] 0.69[89]	0.68[85]	—	—
Zr	—	−0.99[85]	—	0.63[85]	—	—

contact to *n*-type ZnO, the work functions of the metals should be less than that of ZnO's electron affinity ($\sim$4.2–4.35 eV).[80,81] Therefore, Al and Ti are good candidates to form ohmic contacts with *n*-type ZnO due to their work functions of approximately 4.28 and 4.33 eV, respectively. In contrast, Au, Pd, Pt and *etc.*, acting as high work function metals, are good candidates to form Schottky contacts, although ambient gas diffusion through Au and Pd may introduce interface dipoles that can alter the barriers.[76] Ag provides relatively high barriers once oxidised, while Ir can yield relatively high barriers owing to its high work function, weak reactivity with ZnO, and stability in air.

3.3 Magnetic Properties

Previously, we discussed many *n*-type and *p*-type defects or impurities in ZnO and their nanostructures, but the incorporation of transition metals (TM) that have partially filled d states (Sc, Ti, V, Cr, Mn, Fe, Co, Ni and Cu, *etc.*) or rare earth elements that have partially filled f states (Eu, Gd, Er, *etc.*) into ZnO will form diluted magnetic semiconductors (DMS), or semi-magnetic semiconductors. In 1999, the BeZnMnSe system was the first successful realisation by using semi-magnetic semiconductors instead of a ferromagnetic (FM) metal.[97] Consequently, the extensive experimental and computational searches for multifunctional materials have resulted in the development of semiconductors such as CdMnTe, HgMnSe, (Ga, Mn) N, or (Zn, Cr) Te, which exhibit surprisingly stable ferromagnetic signatures despite having a small or nominally zero concentration of magnetic elements.[98] For the transition-metal doped III–V and II–VI semiconductors, ZnO and GaN became the most extensively studied materials as promising candidates to realise DMS with Curie temperatures above room temperature, which were predicted based on mean field theory by Dietl *et al.*[99] The Zener model described a hole-mediated exchange coupling, with the degree of exchange coupling depending on free hole concentration. This prediction prompted extensive research focusing on diluted magnetic oxide semiconductors to understand the mechanisms involved, and attempts to design and fabricate relative materials suitable for practical spintronic applications.

The first investigation of room temperature ferromagnetism in 3d TM-doped ZnO films was performed by Ueda *et al.*[100] in Co-doped ZnO films, and the magnetic property depends on the concentration of Co ions and carriers. Additionally, anisotropic ferromagnetism is observed in Co-doped ZnO, with the largest moment of 2.6 μ_B/dopant atom under parallel magnetic field.[101] With increasing Co concentration, the moment/Co decreases due to the increase of Co–Co antiferromagnetic coupling between neighboring dopants. The Mn-doped ZnO with room temperature FM was first observed by Sharma *et al.*,[102] where an average magnetic moment of $\sim$0.16 μ_B was obtained. A further study demonstrated that room temperature FM in Mn-doped ZnO does not originate from carrier induced exchange coupling between Mn dopants, but rather from a new metastable ferromagnetic phase.[103] As shown in Figure 3.14, an abnormal peak located at

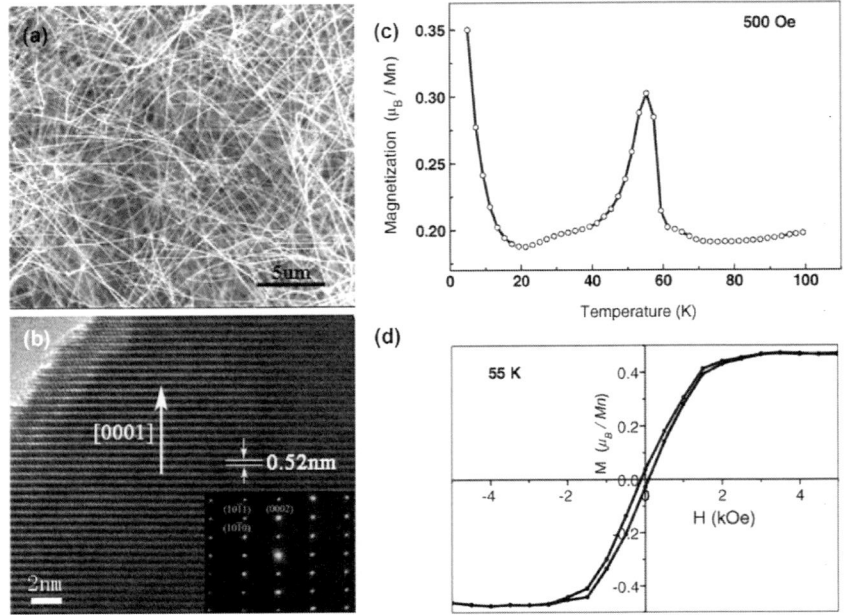

Figure 3.14 The SEM image (a) and HRTEM image (b) of $Zn_{1-x}Mn_xO$ nanowires. The inset plot is the corresponding electron diffraction pattern; (c) the M–T curves in an external field of 500 Oe; (d) magnetic hysteresis curve measured at 55 K.
Reproduced with permission from Y. Gu, X. Zhang, X. Wang, Y. Huang, J. Qi and Y. Zhang, A quantum explanation of the abnormal magnetic behaviour in Mn-doped ZnO nanowires, *J. Phys.: Condens. Matter*, **19**(23), 236223, 2007 IOP Publishing Ltd.[105]

55 K was observed in the M–T curves of $Zn_{1-x}Mn_xO$ nanowire under an external magnetic field of 500 Oe.[104,105] The ferromagnetic hysteresis loop measured at 55 K further ruled out the possibility of an abrupt change of susceptibility due to phase transition in residual O_2 condensation.

Besides, various 3d TM elements, such as Ti,[106,107] V,[101,107] Fe,[108,109] Cr,[110,111] and Ni,[112,113] have also been incorporated into ZnO to explore the magnetic properties of ZnO. It was concluded that these materials are magnetic semiconductors with strong sd- and pd-interaction. Moreover, large amounts of literature reported that ZnO without any magnetic ions also shows ferromagnetic behaviour. Ferromagnetism has also been observed in undoped ZnO.[114,115] Zn vacancies and other p-type defects can stabilise ferromagnetism in $Zn_{1-x}Cr_xO$, $Zn_{1-x}Mn_xO$ and $Zn_{1-x}Fe_xO$ thin films, whereas O vacancies or Zn interstitials can greatly enhance the ferromagnetic coupling in $Zn_{1-x}Co_xO$ thin films. There are many magnetic properties reported in nonmagnetic ions doped ZnO, like carbon,[116] Sc,[106,107] Cu,[117–119] Al,[120,121] doped ZnO and even Ar-implanted ZnO.[122] Thus, such a complicated behaviour may be the reason for the numerous conflicting experimental results.[123] All these results demonstrate that either the magnetic properties of non-magnetic doped ZnO, or ZnO contaminated by magnetic impurities, include complex interactions between the defects and host materials. These mechanisms are not yet understood in detail. More efforts should be made to illustrate the complex ferromagnetism interplay occurring in ZnO to eventually lead to emerging spintronics technologies.

3.4 Mechanical Properties

3.4.1 Elastic Properties

For wurtzite ZnO, non-zero and independent elastic stiffness constants are $c_{11}, c_{12}, c_{13}, c_{33}, c_{44}$ and $c_{66} = (c_{11} - c_{12})/2$. The stress components $\tau_{\alpha\beta}$ represent linear function of strain components $\varepsilon_{\alpha\beta}$ in given as:[124]

$$
\begin{bmatrix} \tau_{xx} \\ \tau_{yy} \\ \tau_{zz} \\ \tau_{yz} \\ \tau_{zx} \\ \tau_{xy} \end{bmatrix} = \begin{bmatrix} c_{11} & c_{12} & c_{13} & 0 & 0 & 0 \\ c_{12} & c_{11} & c_{13} & 0 & 0 & 0 \\ c_{13} & c_{13} & c_{33} & 0 & 0 & 0 \\ 0 & 0 & 0 & c_{44} & 0 & 0 \\ 0 & 0 & 0 & 0 & c_{44} & 0 \\ 0 & 0 & 0 & 0 & 0 & c_{66} \end{bmatrix} \begin{bmatrix} \varepsilon_{xx} \\ \varepsilon_{yy} \\ \varepsilon_{zz} \\ \varepsilon_{yz} \\ \varepsilon_{zx} \\ \varepsilon_{xy} \end{bmatrix} \tag{3.24}
$$

where x, y and z direction denote the orthogonal $[2\bar{1}\bar{1}0]$, $[01\bar{1}0]$, and $[0001]$ direction, and the subscripts 1, 2, 3, 4, 5 and 6 denote xx, yy, zz, yz, zx and xy, respectively. The elastic stiffness constants c_{11} and c_{33} represent the longitudinal coefficients along the $[1000]$ and $[0001]$ directions, while the c_{44} and c_{66} dominate the transverse modes along the $[0001]$ and $[1000]$ directions, respectively. Additionally, c_{13} describes the velocity of modes in low symmetrical directions like $[0011]$ in combination with the other four moduli. These elastic constants can be obtained by using ultrasonic measurements[125,126] electric-field-induced resonant excitation[127–130] or contact resonance atomic force microscopy measurement.[131]

Electric-field induced resonance can be used with *in situ* transmission electron microscopy (TEM) to measure the axial Young's modulus of pure and doped ZnO nanostructures.[127–129] As shown in Figure 3.15, the amplitude of vibration of the ZnO nanowire increases or decreases, changing with the frequency of the applied voltage. The harmonic resonance of the nanowire (269 KHz) is illustrated in Figure 3.15(d). In terms of the fundamental resonance frequency derived from the classical elasticity theory for a rod or a wire, the elastic bending modulus is given by:

$$
E = \rho \left[\frac{8\pi \nu_i L^2}{\beta_i^2 D} \right]^2 \tag{3.25}
$$

where β_i is a constant for the ith harmonic, $\beta_1 = 1.875$ and $\beta_2 = 4.694$, E is the elastic bending modulus, L is length of the nanowire, and ρ is the mass density. Then, the average of elastic bending modulus of the ZnO nanowires (the diameter changing from 43–110 nm) is ~58 GPa. This value is in

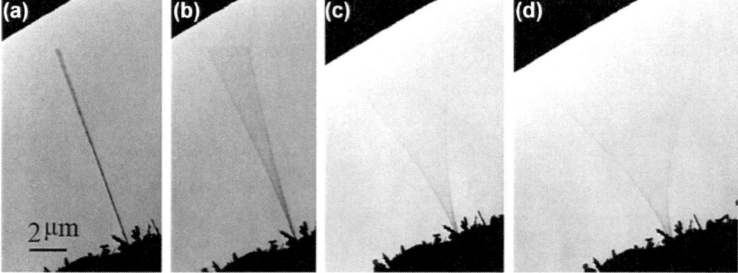

Figure 3.15   The mechanical resonances of ZnO nanowire.
Reprinted with permission from Y. Huang, X. Bai and Y. Zhang, *In-situ* mechanical properties of individual ZnO nanowires and the mass measurement of nanoparticles, *J. Phys.: Condens. Matter*, 2006, **18**(15), L179, IOP publishing Ltd.[127]

agreement with the elastic modulus measured by a nanoindenter for ZnO nanobelts,[132] and it is slightly higher than that of ZnO nanobelts (~52 GPa) measured by the same *in situ* TEM technique.[128] For In-doped (17.2 at.%) ZnO nanowires, the average of the elastic modulus is ~99 GPa with a diameter from 38–86 nm. Therefore, the modulus of experimental In-doped ZnO nanowires along the [1̄010] direction is 120% higher than that of pure ZnO nanowires along the same direction, comparing 99 GPa with 44.8 GPa. This tendency is in good agreement with the first-principle calculations that there is about a 42% increase in the Young's module of ZnO along the [1̄010] direction due to the 12.5% In doping, up to 151.1 GPa in comparison with that of undoped ZnO of 106.3 GPa. As shown in Figure 3.16, the dependence of the Young's modulus has also been reported with diameters of ZnO NWs smaller than 150 nm, and it was found that an approximate core–shell composite NW model in terms of the surface stiffening effect can effectively explain the size dependence of Young's modulus.[130]

However, the radial elastic moduli of [0001] ZnO nanowires with diameters smaller than 150 nm can be accurately measured using contact resonance atomic force microscopy (CR-AFM).[131] Both the radial indentation moduli and tangential shear moduli show a pronounced increase when the wire diameter is reduced below 80 nm. The Young's modulus of [0001] oriented ZnO nanowires as a function of wire diameter were also measured by using the micromechanical system (MEMS) based nanoscale material testing system in situ with a transmission electron microscope.[133] The moduli E enlarges from ~140 to 160 GPa as the nanowire diameter reduces from 80 to 20 nm.

Besides that, the strain tensor and the moduli of ZnO nanostructures have been explored by first-principle calculations.[134–141] For a solid subject to a finite deformation, elastic energy per unit volume can be written as:

$$E = \frac{c_{11}}{2}\left(\varepsilon_{xx}^2 + \varepsilon_{yy}^2\right) + \frac{c_{33}}{2}\varepsilon_{zz}^2 + c_{12}\varepsilon_{xx}\varepsilon_{yy} + c_{13}\left(\varepsilon_{xx} + \varepsilon_{yy}\right)\varepsilon_{zz}$$
$$+ 2c_{44}\left(\varepsilon_{yz}^2 + \varepsilon_{zx}^2\right) + 2c_{66}\varepsilon_{xy}^2$$

(3.26)

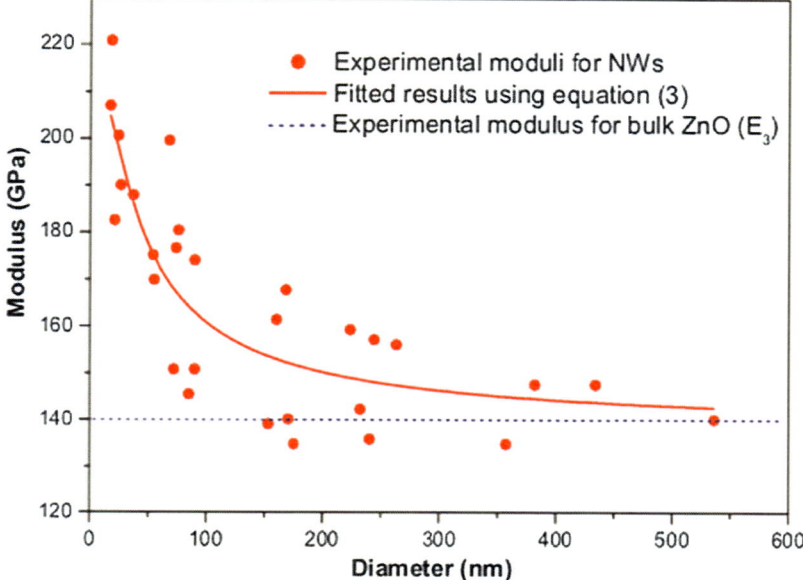

Figure 3.16 Diameter dependence of effective Young's modulus of ZnO nanowires along with [0001] direction. The bending experimental results (red dot), the fitted results from the core–shell composite NW model (solid line), and the modulus for bulk ZnO (E_3) calculated using the experimental data (blue dashed line) come from ref. 138.
Reprinted with permission from C. Q. Chen *et al.*, *Phys. Rev. Lett.*, 2006, **96**, 075505, 2007. Copyright (2007) by the American Physical Society.[130]

If hydrostatic pressure $\tau_{xx} = \tau_{yy} = \tau_{zz}$ is applied:

$$\varepsilon_{xx} = \varepsilon_{yy}, \quad \varepsilon_{zz} = \frac{c_{11} + c_{12} - 2c_{13}}{c_{33} - c_{13}} \varepsilon_{xx} \tag{3.27}$$

and the strains can be obtained from eqn (3.24), whereas all the shear strains are zero. If a biaxial strain along the hexagonal (0001) plane is applied, the strains obey the relations of

$$\varepsilon_{xx} = \varepsilon_{yy}, \quad \varepsilon_{zz} = -\frac{2c_{13}}{c_{33}} \varepsilon_{xx} \tag{3.28}$$

because $\tau_{xx} = \tau_{yy}$ and $\tau_{zz} = 0$.
When considering the non-linear elasticity, the elastic energy ΔE is needed to define the energy change caused by elastic deformation, in terms of the Lagrangian strain through Taylor series expansion. Then the elastic energy is expressed as follows:

$$\Delta E(X, \eta_{ij}) = \frac{V}{2} \sum_{ijkl} C_{ijkl} \eta_{ij} \eta_{kl} + \frac{V}{6} \sum_{ijklmn} C_{ijklmn} \eta_{ij} \eta_{kl} \eta_{mn} + \cdots, \tag{3.29}$$

where C_{ijkl} and C_{ijklmn} are the second- and third-order elastic constants. Likewise, the elastic modulus of the ZnO nanostructures can still be calculated from $Y = \partial^2 E / \partial \varepsilon^2$, and the second partial derivatives of the elastic energy density with respect to strain ζ is given by:

$$Y = \frac{\partial^2}{\partial \zeta^2} \Delta \rho_E(X, \eta) = Y^{(0)} + Y^{(1)} \zeta + o(\zeta^2) \tag{3.30}$$

$$Y^{(0)} = C_{33} - C_{13} \cdot \lambda$$

$$Y^{(1)} = C_{333} - 6C_{133} \cdot \lambda + (C_{112} + 5C_{113} + 6C_{123})\lambda^2$$
$$+ (-4C_{111} - 6C_{112} + 2C_{222})\lambda^3 \qquad . \tag{3.31}$$

$$\lambda = \frac{C_{13}}{C_{11} + C_{12}}$$

where $Y^{(0)}$ and λ are the constant terms, and $Y^{(1)}$ is the linear terms with respect to ζ.

Wang *et al.*[136] obtained the third-order elastic constants of ZnO bulk *via* first principles calculations, and the strain dependent Young's moduli of ZnO NWs and its size effect from molecular mechanical simulations as shown in Figure 3.17, and found that $Y^{(0)}$ decreases as the diameter decreases, while $Y^{(1)}$ increases rapidly as the diameter decreases. The strain dependent Young's moduli were evaluated for the [0001] direction ($Y = 142 - 173.4\zeta$ GPa). If the non-linear elasticity is not considered, $Y^{(1)}$ becomes the zero, and then the Young's moduli can be calculated from the second-order elastic constants.

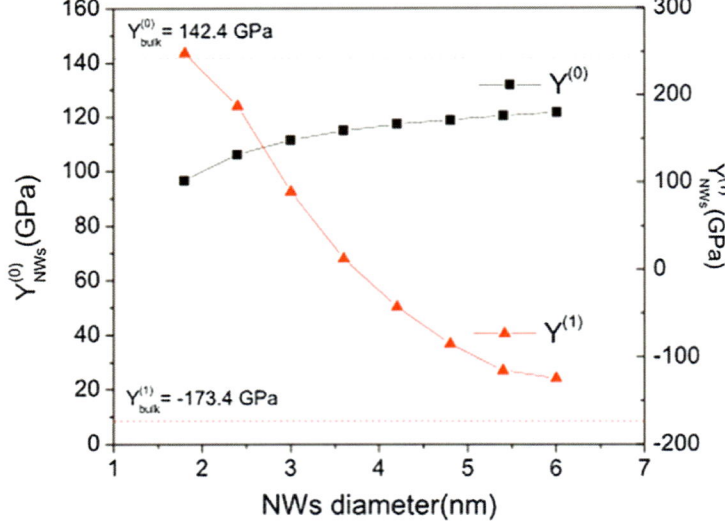

Figure 3.17   The calculated Young's moduli of ZnO NWs. $Y^{(0)}$ and $Y^{(1)}$ are the constant terms and the linear term of Young's modulus, respectively. Reprinted from ref. 136 with the permission of AIP Publishing.

Table 3.4 Some experimental results and theoretical predictions of the elastic moduli of one-dimensional ZnO nanomaterials.

Objects	Size (nm)	Modulus (GPa)	Method
ZnO NWs[138]	~45	29 ± 8	AFM
ZnO NWs[130]	17–550	140–220	E-field Resonance
ZnO NWs[139]	18–303	133 ± 15	AFM bending
ZnO NWs[140]	25–100	50–130	Surface tension
ZnO NWs[141]	0.7–1.6	165–210	DFT calculation
ZnO NWs[136]	1.2–1.8	136–138	DFT calculation
In-ZnO NWs[127]	43–110	49–70	*In situ* TEM
Mn-doped NWs	13–18	129–136	DFT calculation
H-terminated ZnO NWs[141]	1.1–2.0	60–90	DFT calculation
ZnO NBs[128]	55×33	52	*In situ* TEM

Some experimental results and theoretical predictions of the elastic moduli of one dimensional ZnO nanomaterials are listed in Table 3.4.

3.4.2 High-pressure Induced Phase Transition

Aside from the strain induced properties of ZnO nanostructures, the structural transitions from the wurtzite (B4) structure (space group *P63mc*) to the rocksalt (B1) structure (space group *Fm3m*) at high pressure have been intensively investigated. This is due to the WZ-ZnO phase showing high-quality *n*-type conductivity in ambient conditions, whereas RS-cubic phase shows an electronic structure with an indirect bandgap that may be compatible for high *p*-type doping.[142,143]

Experimentally, *in situ* observations of ZnO were performed in terms of energy dispersive X-ray diffraction,[144–146] Raman scattering,[147] and X-ray absorption near-edge structure (EXAFS).[148] The phase transformation of ZnO nanomaterials, for instance ZnO nanoparticles,[149,150] nanorods,[151] nanotubes,[152] and nanocrystalline materials,[153] have been investigated comprehensively. As shown in Figure 3.18, the WZ-ZnO to RS-ZnO transformation in the pressure range (5–7 GPa) occurs in the temperature range 550–1000 K,[154] which is far below the Tamman temperature of ZnO (~1200 K).[155] However, the WZ to RS transition is reversible under ambient conditions. At room temperature, ~9 GPa applied pressure leads to a WZ to RS phase transition but transforms back to WZ phase on release of pressure. Recently, it was reported that cubic ZnO nanocrystals[156] that are sub-20 nm can exist *via* use of defect-induced stabilisation conditions under ambient conditions, which was achieved by ball milling fabrication. Meanwhile, theoretical works on the phase transformation in ZnO nanowires under tensile load[157,158] and the critical dimension for phase transition of nanowires[159,161] have been investigated. Gao *et al.*[160] calculated the WS to RS phase transition of pure-ZnO and Mn-doped ZnO nanowires, including the size effect and doping effect. The critical pressure of structural transition for nanowires (6.35 GPa) is lower than that of the bulk (8.92 GPa), and the critical pressure decreases

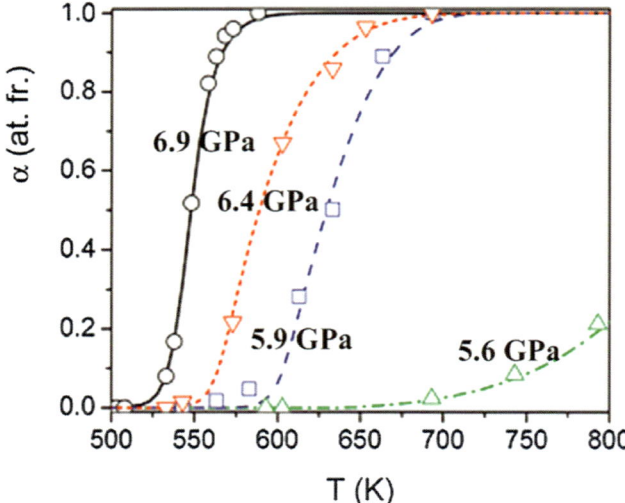

Figure 3.18 Non-isothermal kinetic curves of the wurtzite-to-rock-salt phase transformation for ZnO bulk with different pressures as a function of temperature.
Reprinted with permission from ref. 154. Copyright (2011) American Chemical Society.

from 7.95 GPa to 6.35 GPa as the diameter of the nanowire decreases from 0.975 nm to 1.62 nm. Additionally, Mn dopants can also reduce the critical pressure of structural transition (5.51 GPa) in comparison with the bulk transition points. Moreover, the surface reconstruction can affect the structural transition of ZnO.[160]

3.5 Optical Properties

Direct wide bandgap and high exciton binding energy are the two most important features of ZnO's optical properties, and lay the foundation for promising applications in short-wavelength optoelectronics, such as ultraviolet light emitting devices and ultraviolet detectors. Here, we address the optical properties of ZnO nanostructures in detail from three aspects. First, we make a brief introduction on the photoluminescence properties of ZnO nanostructures and detail several important factors affecting their photoluminescence properties. Then, the stimulated emission and waveguide properties of ZnO nanostructures are presented. Finally, an overview of their nonlinear optical properties is provided.

3.5.1 Photoluminescence

The optical properties of low-dimensional ZnO nanostructures have been investigated by photoluminescence (PL) spectroscopy. The photoluminescence spectrum of ZnO nanomaterials, excited by He–Cd laser with a

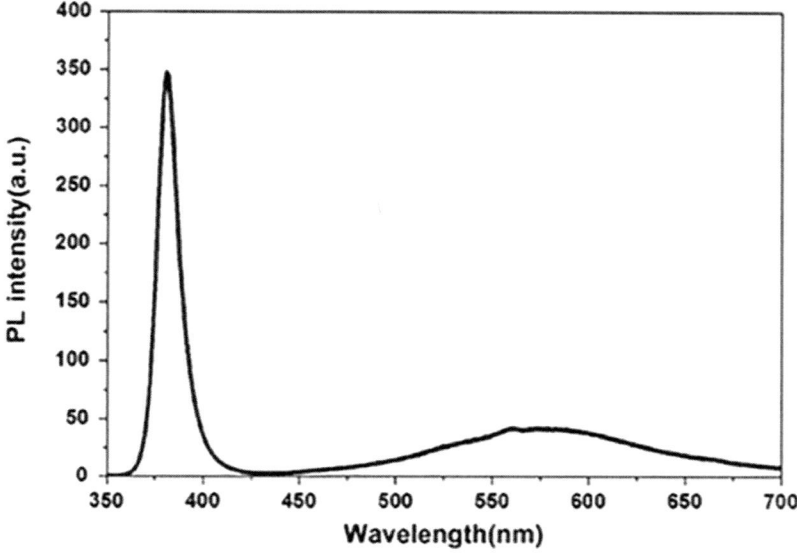

Figure 3.19   The photoluminescence spectrum of ZnO nanowire arrays.
Reproduced from Curr. Appl. Phys., 13(1), ZnO nanowire array ultraviolet photodetectors with self-powered properties, Z. Bai, X. Yan, X.
Chen, H. Liu, Y. Shen and Y. Zhang, 165–169, Copyright (2013), with
permission from Elsevier.[166]

wavelength of 325 nm at room temperature, can be recorded with a fluo-
rescense spectrophotometer and usually has two emission bands, namely,
the UV emission and visible emission. It is generally acknowledged that the
UV emission is attributed to the near-band-edge excitonic transition processes and the visible emission presumably stems from the electron–hole
recombination related to some deep-level deffects in the bandgap, *e.g.*, zinc
interstitials, oxygen vacancies and some hydroxyl groups.[162–165] A typical
room-temperature PL spectrum for ZnO nanowires is shown in Figure 3.19,
where a dominant UV emission peak centered at around 380 nm and a weak
visible emisson peak at around 575 nm can be observed.[166] The PL properties of ZnO nanomaterials can be influenced by the size, annealing treatment, and element doping, as well as the excitation light source and
environment, which have been elaborately studied by researchers in rencent
years as follows.

3.5.1.1   Size-dependent Photoluminescence Properties

The PL property of the ultra-narrow ZnO nanobelts can be affected obviously
by size.[167] Comparing the PL spectra acquired from the 6 nm wide ZnO
nanobelts with the ∼200 nm wide ZnO nanobelts, a 120 meV blue shift took
place in the PL curves of the 6 nm wide ZnO nanobelts. For ZnO NRs, with
increasing diameter of ZnO NRs, the band edge emission increases while the

defect-related visible emission decreases, which is attributed to the difference in surface to volume ratio of NRs with different diameters.[168] Furthermore, the intensity ratio of the band edge and surface defect emissions can be described by:[169]

$$\frac{I_{BE}}{I_{SD}} = C\left(\frac{r^2}{2rt - t^2} - 1\right) \tag{3.32}$$

where I_{BE} and I_{SD} are the PL intensities of band edge emission and surface-defect emission, respectively. r is the radius of NRs, and C and t are constants that can be determined by experiments.

3.5.1.2 Effect of Annealing Treatment

The annealing treatment to purify ZnO NRs at different temperatures can affect the relative integrated photoluminescence intensity ratio between the UV emission (I_{UV}) and deep level emission (I_{DLE}).[170] The intensity ratio (I_{UV}/I_{DLE}) is related to the sample's crystallisation, and a larger ratio indicates better crystallisation and thus fewer deep level defects.[171] The I_{UV}/I_{DLE} value first increased and then decreased with the increase of annealing temperature. A maximum value emerged at 500 °C. The as-grown ZnO nanopillar demonstrated a moderate UV-to-visible emission ratio of ~4.2, which could be substantially enhanced by annealing.[172] A considerable enhancement in the UV-to-visible emission ratio (16.7–28.6) could be achieved by annealing at 300 °C in all three atmospheres studied (air, 5% H_2/95% N_2, and vacuum), and annealing at temperatures above 500 °C would create a considerable number of defects.

3.5.1.3 Effect of Doping

The influence of different doping concentrations of Co on the room temperature photoluminescence properties of hydrothermally grown ZnO nanowires had been systematically studied.[173] As the Co concentration increased, the band edge emission peak shifted slightly towards longer wavelengths and its full-width at half-maximum (FWHM) increased, which may be attributed to the doping induced band gap renormalisation effect.[174,175] Meanwhile, the relative visible emission intensity increased significantly at higher Co concentration due to the doping-induced additional oxygen vacancies in ZnO. The photoluminescence behaviour of Ga-, In-, and Sn-doped ZnO nanowires was also considered.[176] As the doping elements entered the ZnO crystal lattices, the defect energy levels would be introduced into ZnO, thus causing the broadening down to lower energy regions of the photoluminescence emission peak, especially for the Sn-doping. In particular, the Sn-doped nanowires had a green emission band at around 2.5 eV, which may originate from more oxygen vacancies related to the largest charge density of Sn.

3.5.1.4   Effect of Excitation Intensity

The density of excitation light had a considerable influence on the intensity ratio of the near-bandgap emission to the deep-level emission from ZnO. The photoluminescence spectra of ZnO nanoparticles, NRs, microparticles, and single crystals under various conditions of ultraviolet excitation had been comprehensively investigated.[177] Notably, a different non-linear dependence relationship existed both in the intensities of emissions and excitation density. Therefore, the value of intensity ratio could not be simply employed to unequivocally evaluate the quality of the ZnO. Moreover, when measuring the number of defects in the sample using the value of intensity ratio, the same excitation intensity should be used with caution.

3.5.1.5   Effect of Test Temperature

At low temperatures, the photoluminescence spectrum of ZnO has two near-band edge emission peaks; one is the free exciton peak at the high-energy side, the other one is the bond exciton peak with relative lower energy, and some single or multiple phonon replicas of bound exciton may also exist. The relative intensity ratio between bound exciton peak and free exciton peak indicates the sample's crystallinity, *i.e.* the weaker the bound exciton peak, the fewer the defects and the higher the crystal quality of materials. As the temperature increases, the width of the bandgap narrows, so that both the bound exciton peak and free exciton peak shift toward the longer wavelength side. Meanwhile, with the increase of thermal vibration energy, the bound states disappear and only the free exciton emission peaks exist.[178,179]

A typical low-temperature photoluminescence spectrum of ZnO nanowires is shown in Figure 3.20.[180] When the temperature decreased from room temperature to 4.2 K, the dominant UV emission peak was blue-shifted to 368 nm due to the temperature-dependent bandgap increase following the Varshni relationship.[181] At the same time, it can be clearly observed that the photoluminescence spectrum of ZnO nanowires had the longitudinal and transversal optical (LO and TO) phonon replicas.

3.5.2   Stimulated Emission and Waveguide Properties

For wide bandgap semiconductors, a high carrier concentration is generally required to achieve an electron–hole plasmonic state to produce stimulated emission, and thus their threshold of stimulated emission is very high.[182] Specifically, the exciton binding energy of 60 meV for ZnO at room temperature can produce a high concentration of excitons. Importantly, the efficiency of exciton recombination is much higher than that of electron–hole plasma. The low threshold stimulated emission can be achieved in a ZnO material system.[183] However, the nanowire structure is a sub-wavelength waveguide, which can laterally confine the light traveling inside due to the

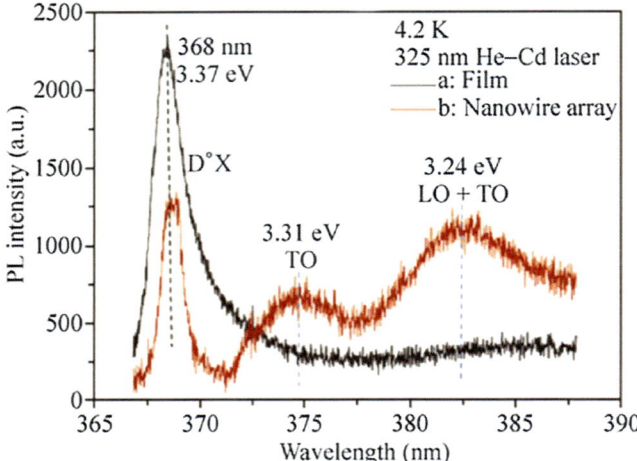

Figure 3.20 Photoluminescence spectra of ZnO nanowires and ZnO film at 4.2 K. Reproduced with permission from ref. 180. Copyright (2007) American Chemical Society.

high refractive index of ZnO ($n = 2.45$). At the same time, each ZnO nanowire has two parallel highly reflective top and bottom end facets, which can form a Fabry-Perot cavity.[184] The size effect of ZnO nanowires also can produce some confinement effects on the carriers, which further improves the efficiency of radiation recombination and reduces the threshold of stimulated emission.

The ZnO nanowire arrays grown on Al_2O_3 (110) substrates could produce room-temperature lasing.[185] It was found that single or multiple sharp peaks at wavelengths from 370–400 nm emerged from ZnO nanowire arrays under an excitation threshold of 40 kW cm^{-2}. The line widths of these stimulated emission peaks were less than 0.3 nm and at least 50 times smaller than the linewidth of the spontaneous emission peak. The lasing threshold was much lower than the random stimulated emission threshold for ZnO disordered particles or thin films (about 300 kW cm^{-2}),[186] indicating that ZnO nanowire arrays have excellent lasing properties. In addition, the stimulated emission properties were also studied for solution processed ZnO nanowire arrays[184] and polycrystalline ZnO nanowire arrays filled in the anodic aluminium oxide (AAO) template.[187] Since there were more defects in the ZnO nanowires prepared by these two methods, the stimulated emission threshold was higher (70–100 kW cm^{-2}).

More importantly, electrically pumped lasing from a p-polymer/n-ZnO micro/nanowire heterostructure has been demonstrated under an injection current of $\sim$1 A cm^{-2}, which could be explained using a whispering-gallery mode resulting from ZnO-structure-related cavity properties.[188] The electrically pumped waveguide lasing was also achieved in ZnO homojunction laser diodes based on Sb-doped p-type ZnO nanowires and n-type ZnO thin

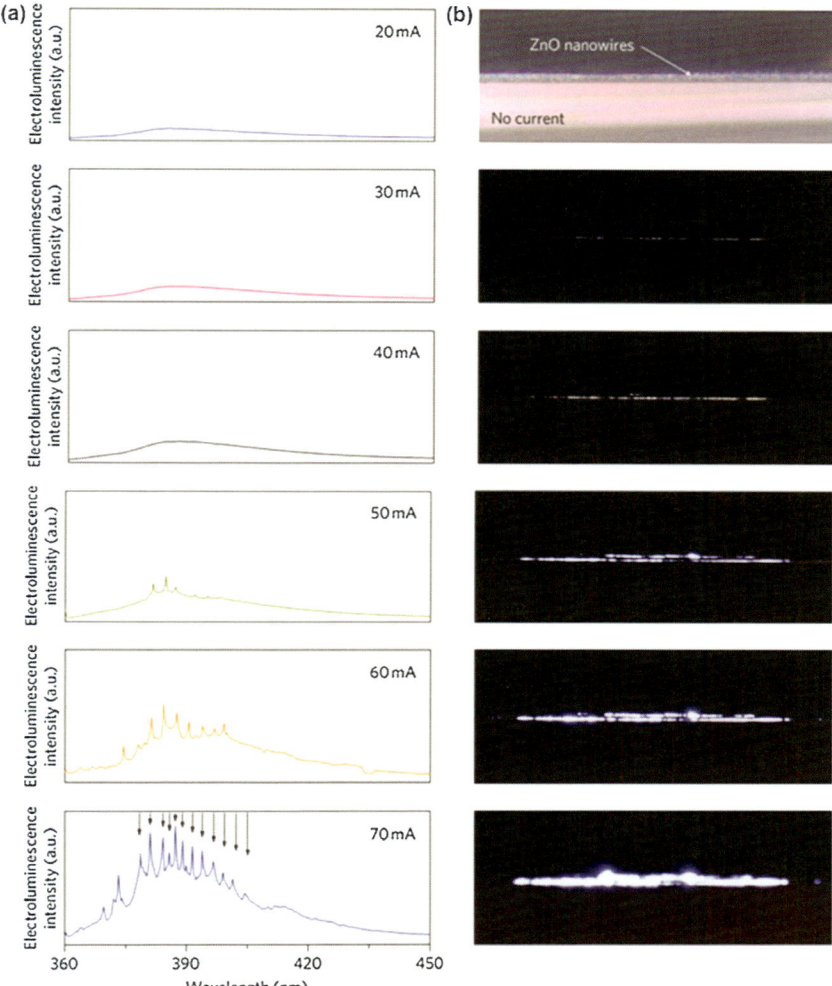

Figure 3.21 Electroluminescence spectra and corresponding side-view optical microscope images of the laser device operated between 20–70 mA. Reproduced by permission from Macmillan Publishers Ltd: *Nat. Nanotechnol.* (ref. 189), copyright (2011).

films.[189] As seen from Figure 3.21, only spontaneous emission assigned as the free exciton recombination could be observed at ∼385 nm under low injection currents (20–40 mA). When the injection current increased to the pumping threshold of ∼50 mA, multiple sharp emission peaks emerged and their line-widths were less than 0.5 nm, indicating that the lasing action took place. As the injection current increased further, more lasing peaks appeared. Although some good results have been achieved in the study of electrically pumped lasing from ZnO nanowire lighting devices, further research is still needed to improve laser performance.

3.5.3 Non-linear Optical Properties

As an important optical character, the non-linear optical property of oxide nanowires has good prospects in frequency converters and logic components.[190] Linear optical properties of nanowires, *e.g.* photoluminescence polarised light, are based on the traditional electromagnetic properties, while the coherent non-linear optical phenomena are completely determined by the materials' crystal structure. Due to the non-centrosymmetric crystal structure, the non-linear optical properties of a single ZnO nanowire had been investigated using near-field scanning optical microscopy.[190] It was found that a single ZnO nanowire shows a high non-resonant second-order non-linearity based on a theoretical model of a hexagonal·lattice structure. The second-harmonic generation from ZnO nanowire arrays had also been comprehensively researched under excitation with femtosecond pulses.[191]

In the experiments, femtosecond light pulses were generated by a wavelength-tunable near-infrared (700–900 nm) femtosecond oscillator with an energy of 12 nJ and 50 fs duration. Three different sets of emission peaks exist in the spectra, namely, the resonant second-harmonic generation at a wavelength of 360 nm (wavelength of femtosecond oscillator excitation was 720 nm), the near-band edge photoluminescence and the visible light emission from deep-level defect centres. An obvious red-shift took place during the first 27 s after unblocking the excitation laser, which could be attributed to the band-gap shrinkage due to the temperature increase inside the nanowires. Meanwhile, a phenomenological four-layer model could be used to simulate the angular dependence of the polarisation curves.

3.6 Piezoelectric and Dielectric Properties

Here, we present experimental and theoretical investigations into the piezoelectric property of ZnO nanostructures. First, the basic principle and origin of piezoelectricity in ZnO is described based on crystallography. Then, piezoresponse force microscopy (PFM), a state-of-art characterisation method to identify local piezoelectric property of nanomaterials, is presented. This is followed by examples of measured ZnO piezoelectric coefficient with different morphologies, such as nanowire, nanorod and so forth. Several important influencing factors on piezoelectric property are detailed. Finally, some feasible approaches to enhancing the effective ZnO piezoelectric coefficient are proposed.

3.6.1 Origin of Piezoelectricity in ZnO

The piezoelectric effect represents a linear relationship between mechanical and electrical states. A more accurate description of piezoelectricity could be given based on crystallography, which originates from the lack of center symmetry in a crystal structure. For ZnO, it crystallises preferentially in a hexagonal wurtzite structure in ambient conditions, which can be described

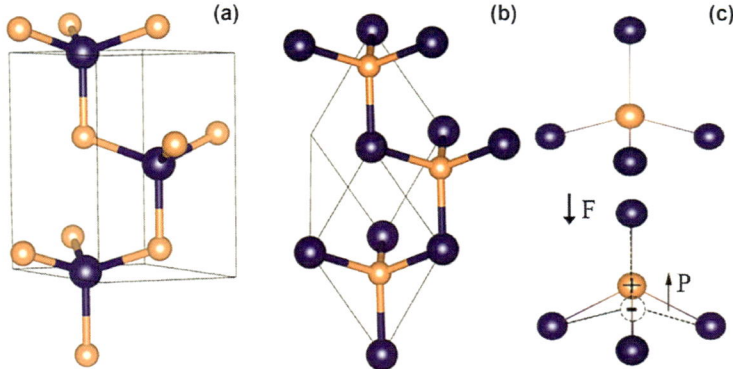

Figure 3.22 Ball-and-stick model of wurtzite ZnO (a) and zinc blende ZnO (b); (c) schematic showing the piezoelectric effect in a tetrahedrally coordinated Zn–O unit.

as a number of alternating planes composed of tetrahedrally coordinated Zn^{2+} and O^{2-} ions.[192–194] Although the Zn^{2+} and O^{2-} planes are electrically charged, the ZnO surface could still remain atomically flat and stable and exhibit no reconstruction.[195] The typical ball-and-stick model of a wurtzite structure is shown in Figure 3.22(a). Therefore, applying pressure along the cornering direction of a tetrahedron will induce the center of Zn^{2+} charges to split up from that of O^{2-} charges, which represents the direct piezoelectric effect, as shown in Figure 3.22(c).[196] Conversely, the internal generation of mechanical strain resulting from an applied external electrical field is called the reverse piezoelectric effect.[197] Among the tetrahedrally bonded semiconductors, ZnO has the highest piezoelectric tensor, which makes it a potential candidate for actuators, sensors and high frequency RF filters.[198] The zinc blende ZnO is metastable and could be only achieved by growing on substrates with a cubic lattice structure.[199] This ZnO structure also exhibits piezoelectric properties because of the tetrahedral coordination, but has different stacking sequences with wurtzite ZnO, as shown in Figure 3.22(b).[200]

Normally, wurtzite ZnO has three fastest growth direction, $\langle 0001 \rangle$, $\langle 01\bar{1}0 \rangle$ and $\langle 2\bar{1}\bar{1}0 \rangle$, resulting in more diverse morphologies than any other nanomaterials, but the piezoelectric effect along the $\langle 0001 \rangle$ direction is the most significant.[193] Meanwhile, the wurtzite ZnO exhibits crystallographic polarity, which indicates the direction of bonds because of the non-central symmetry and polar surface characteristics. The convention is that positive z direction points from the face of the O plane to the Zn plane and the polarity is referred to as Zn polarity. However, it is referred to as O polarity.[199] Piezoresponse of a wurtzite structure depends largely on this polarity direction. For example, exerting compressive strain on a ZnO nanorod growing along the $+c$ direction (Zn polarity) will induce negative piezopotential on the top surface and positive on bottom, while things are opposite for the O polar.[201]

To describe the piezoelectric property, the piezoelectric tensor components e_{ij}, called piezoelectric stress coefficients, are given by the polarisation components P_i with respect to strain ε_{ij}. Considering wurtzite spacing symmetry, the piezoelectric tensor has three independent non-zero components. In Voigt notation, the wurtzite piezoelectric stress moduli are labelled as e_{33}, e_{31}, and e_{15}, yielding the piezoelectric tensor E:[3]

$$E = \begin{pmatrix} 0 & 0 & 0 & 0 & e_{15} & 0 \\ 0 & 0 & 0 & e_{15} & 0 & 0 \\ e_{31} & e_{31} & e_{33} & 0 & 0 & 0 \end{pmatrix} \tag{3.33}$$

The x, y and z components of piezopolarisation are given by:

$$P_x = e_{15}\varepsilon_{zx}, \quad P_y = e_{15}\varepsilon_{yz}, \quad P_z = e_{31}(\varepsilon_{xx} + \varepsilon_{yy}) + e_{33}\varepsilon_{zz} \tag{3.34}$$

where P_x and P_y are induced by shear strains, and the c-directed polarisation P_z is induced by a strain $\varepsilon_{zz} = (c - c_0)/c_0$ along the c-axis, and by a strain $\varepsilon_{xx} = \varepsilon_{yy} = (a - a_0)/a_0$ in one of the basal planes.

Besides, compared with ferroelectricity and ferromagnetism, which may be greatly reduced or even vanish for reduced material size, the piezoelectricity is a crystal-structure-determined effect and this property could be well preserved down to the nanoscale.[196] This unique characteristic established the foundation for nanodevice applications by taking advantage of piezoelectricity in nanomaterials.

3.6.2 Piezoresponse Force Microscopy

Because of its ability to probe the coupling between electrical field and mechanical strain on the nanometre scale, piezoresponse force microscopy (PFM) has become the most powerful tool to investigate local piezoelectric properties of nanomaterials.[202] The determination of the piezoelectric coefficient and spontaneous polarisation of ZnO could be achieved with this method.[203] Typical measurement setup geometries of PFM study of piezoelectricity are shown in Figure 3.23. The principle of this technique is based on the reverse piezoelectric effect. When applying a voltage to the conductive tip in contact with piezoelectric material, the resulting local deformation of samples could be monitored through mechanical displacement of a cantilever. The local piezoelectric coefficient is calculated from the linear slope of deflection amplitude against applied driving voltage. While the concept is straightforward, accurate quantification of the piezoelectric coefficient can be very complicated and challenging due to the complex sample-probe signal transfer mechanism, because the tip motion can be a combination of piezoelectricity, electrostriction and electrostatic interactions.[204] Moreover, the measuring accuracy can be also greatly compromised by the applied voltage frequency, sample conductivity and morphology. Normally, the PFM setup should be calibrated on a standard sample with known size and

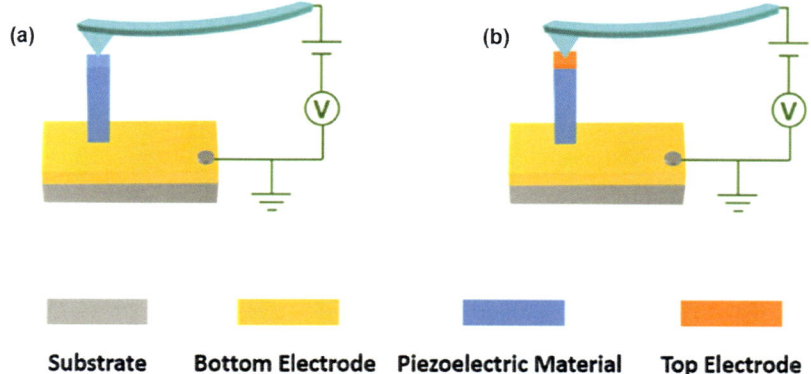

Figure 3.23   Typical measurement setup of piezoresponse force microscopy study of piezoelectric property.

piezoelectric coefficient before measurement, and the measuring parameters should be very carefully chosen.[205–207]

The measurement of the effective piezoelectric coefficient d_{33} for ZnO nanobelts using PFM has been reported. After dispersion of nanobelts on 100 nm Pd coated Si wafers, the top surface of nanobelt was coated with another 5 nm Pd to serve as the top electrode. The measured d_{33} coefficient was found to vary from 14.3–26.7 pm V^{-1} and was frequency dependent. It is considered that pinning of spontaneous polarisation due to surface charge effect and imperfect electrical contact between the bottom of the ZnO and Pd layer might contribute to such unexpected frequency dependence. Even so, those measured coefficients are still considerably larger than that of bulk ZnO (9.9 pm V^{-1}).[208] The effective piezoelectric coefficient d_{33} of ZnO NRs grown by electrochemical deposition approach was characterised by employing PFM.[209] By optimising the growth parameters, a mean effective coefficient of 11.8 pm V^{-1} was measured over several individual NRs, which shows an increase of ~18% in respect to the bulk ZnO.

By using quantitative PFM, the measured average d_{33} value of solution grown ZnO NRs was 4.41 pm V^{-1} with a standard deviation of 1.73 pm V^{-1}.[207] These values are typically smaller than other reported results and several reasons were put forward for this. First, the conductive tip acts as the top electrode, as in the case shown in Figure 3.23(a), so the applied electric field on NRs are inhomogeneous. The clamping effect would lower the measured piezoresponse, whereas a true d_{33} can only be achieved by using a large top electrode to apply a uniform bias to the sample.[210] From this aspect, the measurement setup geometry in Figure 3.23(b) would be more suitable for accurate piezoelectric coefficient determination. But the fabrication of the top electrode is usually difficult, especially for individual nanowires. Second, the conducting nature of solution grown ZnO would also affect the observed piezoresponse. For thermally evaporated ZnO nanopillars, a larger piezoelectric coefficient 7.5 pm V^{-1} is measured, which is expected to be less conducting than solution synthesised ZnO.[211] The above results indicate that accurate determination of ZnO nanomaterials is still challenging, and the measured piezoelectric coefficient could vary considerably because of the different experimental setups, materials growth techniques, and material sizes and morphologies.

Besides the piezoelectric coefficient measurement, PFM could also be used to determine the polarity orientation, which is another important factor that has a profound impact on piezoelectric response.[212] When applying positive tip bias to the (0001) face of ZnO, negative field with respect to the [0001] direction will be induced and produces contraction for Zn-polar ZnO. In contrast, applying negative tip bias to the (0001) face will result in material expansion. This situation is reversed for the [000$\bar{1}$] oriented nanorod. Under applied alternating current tip bias, this opposite response of opposite polar surface is manifested in the phase shift between the applied voltage and the PFM response. With this method, the polarity distribution in ZnO NRs and films have been successfully characterised.[203,207]

An optimised PFM method for measuring the piezoelectric effect is presented. Measurements were performed with the conductive tip in contact with the top electrode and current was applied through an external probe. This configuration offers several advantages like well-defined applied field, large piezoelectrically excited region and minimal tip degradation. The measurement of RF sputtered *c* axis ZnO film using such optimised PFM yielded a piezoelectric constant in the range of 2–13 pm V^{-1}, which never exceeds the single crystal value.[204] It is believed that the polycrystalline nature of the sputtered film may degrade the piezoelectric property. Besides PFM, the nanoindentation technique could also be employed to identify the piezoelectric properties.[213] The effective piezoelectric coefficient of ZnO nanowire grown on paper substrate was measured using such a technique. The calculated result of effective piezoelectric coefficient d_{33} is about 9.2 pm V^{-1}.

3.6.3   Piezoelectric Property Enhancement

3.6.3.1   Size Effect

The size effect is a typical characteristic for nanomaterials and material properties are often enhanced relative to the bulk due to surface effect, low defect density and high surface-to-volume ratio. The size dependence of ZnO piezoelectric coefficient with a diameter ranging from 0.6–2.4 nm was reported by using a first-principles based density functional theory (DFT) method.[214] A giant piezoelectric size effect was identified and approximately two orders of magnitude of improvement in the piezoelectric coefficient can be obtained for nanowires with a diameter of less than 1 nm. This observed size effect was discussed from the aspect of surface charge redistribution. First-principle calculations were carried out to study the piezoelectric and polarisation features of ZnO nanowires and the size effect, as shown in Figure 3.24.[215] The calculated effective piezoelectric coefficient of nanowire

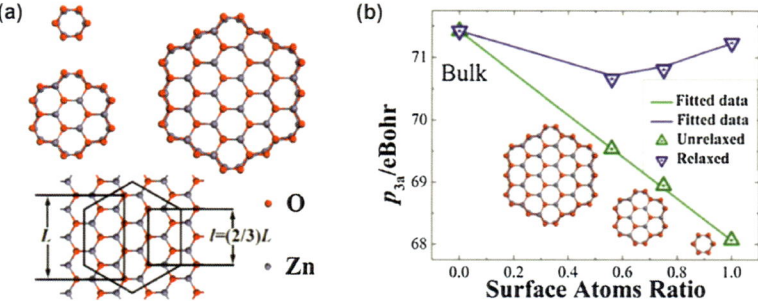

Figure 3.24 (a) Cross sections of ZnO nanowire with different sizes and the method used to estimate sectional area; (b) average electric dipole along *c* axis as a function of surface ratio for relaxed and unrelaxed ZnO nanowires and bulk ZnO.

Reproduced from Nano Res., Structural dependence of piezoelectric size effects and macroscopic polarization in ZnO nanowires: A first-principles study, 8(6), 2015, 2073–2081, C. Qin, Y. Gu, X. Sun, X. Wang and Y. Zhang, (© Tsinghua University Press and Springer-Verlag Berlin Heidelberg 2015) With permission from Springer.[215]

is larger than the bulk value and increases with the decrease in diameter. Furthermore, it is demonstrated that the polarisation behaviour of the nanowire mainly depends on surface atom ratio. Although the diameter of ZnO used in simulation was too small and was not likely to be achieved in experiment, these results provide us some references to enhance the piezoresponse of ZnO by taking advantage of the size effect.

3.6.3.2 Carriers Concentration Modulation

As mentioned previously, the conducting nature of ZnO can affect its observed piezoresponse. Therefore, it can be speculated that better piezoelectric properties could be achieved through appropriate manipulation of the materials'carrier concentration. A correlated relationship between the piezoelectric response and resistivity of individual ZnO nanorods was measured using scanning force microscopy.[216] The PFM and C-AFM techniques are switched to characterise piezoresponse and resistivity of the same NR at the same time without disturbing the alignment of the AFM tip. It demonstrates a variation in resistivity of three orders of magnitude, and in the piezoelectric coefficient which ranged from 0.4–9.5 pm V^{-1} in the same batch of ZnO NRs. The results indicated that NRs with low piezoelectric response display low resistivity and share a similar trend to that observed in CdS.[217]

Many other theoretical calculations and experiments studied the relationship of carrier concentration and piezoelectric property from the output of ZnO piezoelectric nanogenerators. Enhancement of nanogenerators through engineering of ZnO free carriers was reported.[218] As we know, the ZnO was intrinsic *n*-type due to the existence of native defects, and the *p*-type dopant Li could be used to effectively tune carrier concentration in ZnO. Simulation of piezopotential distribution in ZnO nanowires with donor concentration ranging from $10^{12} \sim 5 \times 10^{17}$ cm^{-3} was performed. It is clearly demonstrated that a pronounced screening effect would be induced due to higher carrier concentration and thus there would be a larger reduction in piezopotential. Other simulations give similar results when it comes to the relationship between piezoelectric properties and ZnO carrier concentration.[219,220] Actually, suppressing the carrier concentration through various approaches such as *p*-type polymer coating or surface decoration have been an effective method to enhance the piezoelectric nanogenerator output.[221–223]

3.6.3.3 Transition Element Doping

In the semiconductor industry, doping is the most widely used method to improve material properties. This approach also applies to the piezoelectric property manipulation. It was reported that a giant piezoelectric coefficient d_{33} of 110 pC/N could be obtained in RF sputtered ZnO films with vanadium doping.[224] It was proposed that the switchable spontaneous polarisation induced by the V dopant and the accompanying relatively high permittivity should be responsible for the piezoresponse enhancement. Moreover, an easier rotation of V–O bonds, which are non-collinear with the *c* axis under an electric field, might be another microscopic origin of this anomaly, because it was reported that the dominant effect of applied electric field in piezoelectric ZnO is to rotate the Zn–O bonds that are non-collinear with the *c* axis rather to elongate the polar Zn–O chemical bonds.[225] A piezoelectric coefficient d_{33} of 121 pm V^{-1} was obtained through electrospinning vanadium-doped ZnO piezoelectric nanofibers, which is more than 10 times larger than that of bulk ZnO.[226]

Beside vanadium, copper doping was also adopted to enhance the ZnO piezoelectric coefficient.[227] The Cu-doped ZnO nanowires and films were synthesised through a hydrothermal method with a Cu content ranging from 0–10 at.%. PFM was employed to characterise the piezoelectric coefficient evolution, and it was observed that the d_{33} value increased with an increasing amount of Cu-doping and a d_{33} constant of 79.8 pm V^{-1} was obtained for the 10% Cu–ZnO film.

3.6.4 Dielectric Properties

As mentioned previously, the electrical, mechanical, and electromechanical properties of ZnO nanomaterials have been extensively investigated *via* consideration of the piezoelectric coefficients. Likewise, the dielectric coefficients of ZnO nanostructures have also been investigated to show the size-dependent dielectric constant of the ZnO nanowires.[228] In the presence of a force, F, between the AFM tip and the nanowire, the phase shift $\Delta\Phi$ can be written as:

$$\tan(\Delta\Phi) = \frac{Q}{2k}[C_1''(h) - C_2''(h)](V_{tip} + \varphi)^2 \qquad (3.35)$$

where $Q = \gamma w_0$ is the quality factor of the cantilever (γ, damping coefficient), k is the spring constant, and $C_i''(h)$ is the second derivative of capacitance of the AFM tip-sample system ($i = 1$, AFM tip-SiO$_2$/Si substrate; $i = 2$, AFM tip-ZnO nanowire-SiO$_2$/Si substrate).[229] Over the bare SiO$_2$/Si substrate, $C_1''(h)$ can be given by:

$$C_1''(h) = 2\varepsilon_0(\pi R_{tip}^2)\frac{1}{(h + t/\varepsilon_S)^3}$$

$$\qquad (3.36)$$

$$C_2''(h) = 2\varepsilon_0(\pi R_{tip}^2)\frac{1}{(h + t/\varepsilon_S + D/\varepsilon_f)^3}$$

where R_{tip} is the radius of AFM tip, t is the thickness of the SiO$_2$, and ε_S is the dielectric constant of SiO$_2$. D is the diameter of the nanowire, and ε_f is the dielectric constant of ZnO nanowire. According to the eqn (3.29), the size-dependent dielectric constant $\varepsilon(D)$ can be given by:

$$\varepsilon(D) = 1 + \frac{\varepsilon_{bulk} - 1}{1 + (\Delta E/E_g)^2} \qquad (3.37)$$

where E_g is the bandgap of the material, and ΔE is the change in the bandgap.[230]

According to the eqn (3.37), scanning conductance microscopy (SCM) is used to measure the dielectric constant of a single pencil-like ZnO nanowire with diameters ranging from 85–285 nm. As the diameter decreases, the dielectric constant of ZnO nanowire is found to dramatically decrease from 6.4 to 2.7, which is much smaller than that of the bulk ZnO of 8.66,[231] illustrated in the Figure 3.25. The size dependence of dielectric constant is well explained by an approximate core–shell composite nanowire model in terms of the surface dielectric weakening effect.

3.7 Photocatalytic Properties of ZnO Nanostructures

In 1972, Fujishima and Honda[232] discovered the phenomenon of photocatalytic water splitting with a TiO$_2$ electrode, which started a new area of heterogeneous photocatalytic studies. In the heterogeneous photocatalytic system, the initial electron–hole pairs are generated inside the catalyst molecule, and then the electron transfers to the organic molecules. This finally leads to the oxidisation and decomposition of organic substances. Tetra-needle ZnO is reported to retain higher photocatalytic activity when

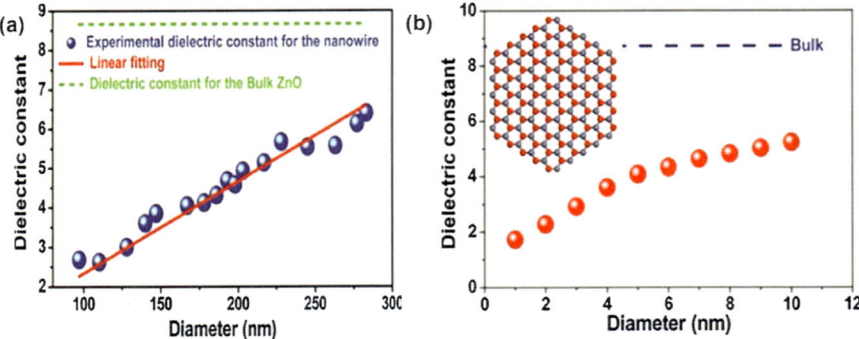

Figure 3.25 (a) Diameter dependence of dielectric constant in the single pencil-like ZnO nanowire: (blue dot) experimental results, (red solid line) fitted results by the core–shell composite nanowire model, and (green dashed line) dielectric constant of bulk ZnO; (b) diameter dependence of dielectric constant ε_{33} in ZnO nanowires by using the theoretical calculation. The inset shows a ZnO nanowire model image along the c direction with a diameter of 3 nm.
Reproduced with permission from ref. 228. Copyright (2012) American Chemical Society.

compared with commercial P_{25} TiO_2 nanoparticles.[233] A large number of oxygen vacancies on the surface like O^-, O_2^- or O^{2-} can capture the photo-generated holes, decreasing the recombination of photo-generated electron–hole pairs.

However, the specific surface area of ZnO nanoparticles is higher than that of its bulk materials, which is favourable for improved catalytic reaction rate and strong adsorption ability towards pollutants. Moreover, the particle size decrease usually reduces the probability of photo-generated charge re-combination, which results in the further improvement of catalytic activity. The bandgap of semiconductors will be broadened with the quantum size effect. Under this condition, the potential of the conduction band is nega-tively shifted while the potential of the valence band is positively shifted, contributing to a stronger redox capability.

The mechanism for photocatalytic pollutant degradation with ZnO nanostructures is shown in Figure 3.26. Such reactions can be divided into the following few processes:[234]

The first step is the excitation of semiconductor photocatalyst in a certain wavelength of light. Electrons can be excited from the valence band to the conduction band, thereby generating electron–hole pairs. The reaction for-mula is represented as follows:

$$semiconductor + h\upsilon \rightarrow e_{CB}^- + h_{VB}^+ \tag{3.38}$$

The second step is the separation of photo-generated electron–hole pairs with the electric field. Electrons and holes move to different surface lo-cations to participate in the redox reactions, so that the substrate molecules adsorbed on the photocatalyst surface can be reduced and oxidised,

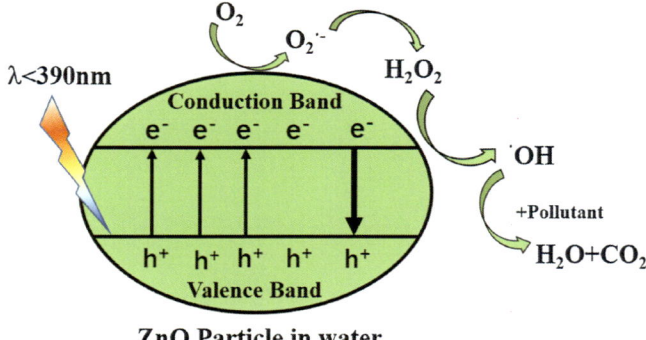

ZnO Particle in water

Figure 3.26 The basic photocatalytic principle of ZnO.

respectively. The photo-generated holes with strong oxidising properties have a strong ability to capture the electrons of the organic compounds on the surface of semiconductor particles. As a result, materials without original light absorption capability can be activated. Moreover, the photo-generated holes react with water to produce the hydroxyl free radical ($^{\bullet}$OH) which is capable of oxidising most of the organic matter. The detailed reaction is as follows:

$$h_{VB}^{+} + H_2O \rightarrow H^{+} + OH \tag{3.39}$$

At the same time, the electron transfers from the conduction band to react with the molecular oxygen from the solution to form hydrogen peroxide and protons. Then the hydrogen peroxide further reacts with electrons to generate $^{\bullet}$OH as follows:

$$e_{CB}^{-} + O_2 \rightarrow O_2^{-} + H^{+} \rightarrow HO_2^{\bullet}$$

$$HO_2^{\bullet} + O_2^{-} + H^{+} \rightarrow H_2O_2 + O_2 \tag{3.40}$$

$$HO_2^{\bullet} + e_{CB}^{-} \rightarrow OH^{-} + {}^{\bullet}OH$$

The photo-induced electrons and holes have the possibility for recombination. The reserved energy would be consumed by recombination as follows in a few nanoseconds if there was no appropriate electron or hole trapping agent:

$$h_{VB}^{+} + e_{CB}^{-} \rightarrow semiconductor + energy \tag{3.41}$$

The third step is the oxidation of organic compounds. The organic matter is oxidised by strong holes and hydroxyl free radical to generate organic free radicals, which then can be oxidised to oxygen free radical in the presence of molecular oxygen. These intermediates can be degraded to water and carbon dioxide *etc.* through thermodynamic reactions as follows:

$$h_{VB}^{+} + organic \rightarrow organic^{\bullet +} \rightarrow oxidation\ of\ oragnic \tag{3.42}$$

$${}^{\bullet}OH + organic \rightarrow organic^{\bullet +} \rightarrow oxidation\ of\ oragnic \tag{3.43}$$

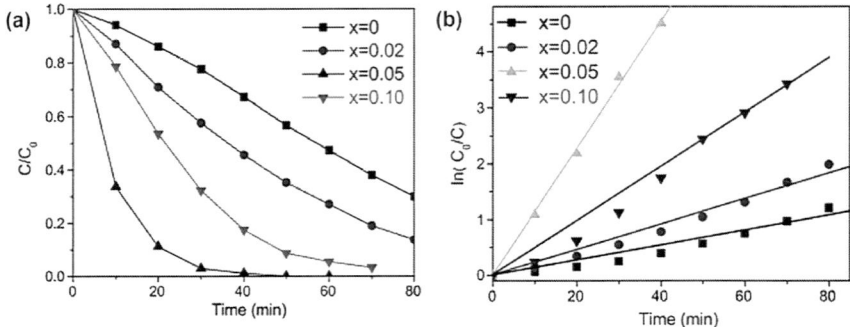

Figure 3.27 Degradation rates of rhodamine B (a) in linear and (b) natural logarithm scales.
Reprinted from Mater. Res. Bull., 46(8), J. Zhao, L. Wang, X. Yan, Y. Yang, Y. Lei, J. Zhou, Y. Huang, Y. Gu and Y. Zhang, Structure and photocatalytic activity of Ni-doped ZnO nanorods, 1207–1210, Copyright (2011), with permission from Elsevier.[235]

Common factors affecting the photocatalytic performance of ZnO are mainly as follows:

ZnO is a wide bandgap semiconductor (3.37 eV) with large exciton binding energy of 60 meV at room temperature. Doping various kinds of ions, such as Mn, Co, Ni, N *etc.*, into ZnO is a useful method to improve their properties. Ni-doped ZnO NRs with different concentration and their photocatalytic degradation rates of rhodamine B were studied.[235] Ni-doped ZnO NRs exhibited higher photocatalytic activity than un-doped ZnO, and the order of photocatalytic activities was $Zn_{0.95}Ni_{0.05}O > Zn_{0.9}Ni_{0.1}O > Zn_{0.98}Ni_{0.02}O > ZnO$, as shown in Figure 3.27.

Co-doped ZnO photocatalysts also performed photocatalytic activity.[236] In the photocatalytic decolourisation of methylene blue (MB) under visible light irradiation, the $Zn_{0.97}Co_{0.03}O$ photocatalyst exhibited the highest photocatalytic decolourisation efficiency, with the MB concentration reduced as much as 100% in 300 min at pH 10.5 (Figure 3.28). When the doping ratio of Co^{2+} ions increased, more oxygen vacancies and defects were generated and became the centres to capture photo-generated electrons so that the recombination can be effectively inhibited. When the ratio of doping ions is higher than the optimal value, deep level defects caused by high concentration doping ions may become the recombination centres for electron–hole pairs, which usually results in the decreased photocatalytic activity.

Due to the high specific surface area, ZnO nanostructures always showed higher photocatalytic efficiency when compared with ZnO thin film. Specifically, the photocatalytic performance of needle-like ZnO nanowire arrays with diameters of around 50 nm and lengths of around 1.1 μm were studied.[237] Each nanowire stood vertically and they were separated from each other, which enabled full contact between the catalyst and reactants. In contrast, for the decolourisation rate of the orange-II dye solution, ZnO

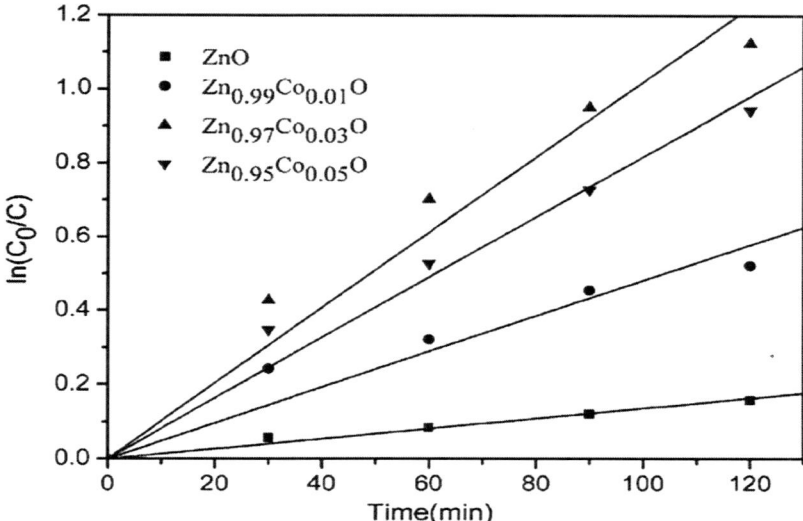

Figure 3.28 Photocatalytic decolourisation kinetics of MB using $Zn_{1-x}Co_xO$ photo-catalysts.
Reprinted from Mater. Sci. Eng.: B, 142(2–3), Q. Xiao, J. Zhang, C. Xiao and X. Tan, Photocatalytic decolorization of methylene blue over $Zn_{1-x}Co_xO$ under visible light irradiation, 121–125. Copyright (2007), with permission from Elsevier.[236]

polycrystalline thin film performance was at least three times lower than that of the fabricated ZnO nanowire arrays.

Besides, the ZnO catalyst concentration and its surface states, the degradable substance concentration and its isothermal adsorption quantity, solution pH, temperature, solvent type, intensity of the illumination may influence the photocatalytic properties as well.[234,238,239]

References

1. A. Mang, K. Reimann and St. Rübenacke, Band gaps, crystal-field splitting, spin-orbit coupling, and exciton binding energies in ZnO under hydrostatic pressure, *Solid State Commun.*, 1995, **94**(4), 251–254.
2. Ü. Özgür, Y. I. Alivov, C. Liu, A. Teke, M. A. Reshchikov, S. Doğan, V. Avrutin, S.-J. Cho and H. Morkoç, A comprehensive review of ZnO materials and devices, *J. Appl. Phys.*, 2005, **98**(4), 041301.
3. C. F. Klingshirn, B. K. Meyer, A. Waag, A. Hoffmann and J. Geurts, *Zinc Oxide: From Fundamental Properties Towards Novel Applications*, Springer, Berlin, 2010.
4. M. Usuda, N. Hamada, T. Kotani and M. van Schilfgaarde, All-electron GW calculation based on the LAPW method: Application to wurtzite ZnO, *Phys. Rev. B*, 2002, **66**(12), 125101.

 5. A. Janotti and C. G. Van de Walle, Fundamentals of zinc oxide as a semiconductor, *Rep. Prog. Phys.*, 2009, **72**(12), 126501.
 6. Q. Yan, P. Rinke, M. Winkelnkemper, A. Qteish, D. Bimberg, M. Scheffler and C. G. Van de Walle, Band parameters and strain effects in ZnO and group-III nitrides, *Semicond. Sci. Technol.*, 2011, **26**(1), 014037.
 7. M. Kobayashi, G. S. Song, T. Kataoka, Y. Sakamoto, A. Fujimori, T. Ohkochi, Y. Takeda, T. Okane, Y. Saitoh, H. Yamagami, H. Yamahara, H. Saeki, T. Kawai and H. Tabata, Experimental observation of bulk band dispersions in the oxide semiconductor ZnO using soft x-ray angle-resolved photoemission spectroscopy, *J. Appl. Phys.*, 2009, **105**(12), 122403.
 8. S.-H. Park, K. J. Kim, S.-N. Yi, D.-Y. Ahn and S.-J. Lee, ZnO/ ZnMgO quantum well lasers for optoelectronic applications in the blue and the UV spectral regions, *J. Korean Phys. Soc.*, 2005, **47**(3), 448–453.
 9. W. Fan, A. Abiyasa, S. T. Tan, S. F. Yu, X. W. Sun, J. B. Xia, Y. C. Yeo, M. F. Li and T. C. Chong, Electronic structures of wurtzite ZnO and ZnO/MgZnO quantum well, *J. Cryst. Growth*, 2006, **287**(1), 28–33.
10. S. L. Chuang and C. S. Chang, k·p method for strained wurtzite semiconductors, *Phys. Rev. B*, 1996, **54**(4), 2491.
11. E. O. Kane, Band structure of indium antimonide, *J. Phys. Chem. Solids*, 1957, **1**(4), 249–261.
12. G. L. Bir and G. E. Pikus, *Symmetry and Strain-Induced Effects in Semiconductors*, Wiley, New York, 1974.
13. M. Suzuki, T. Uenoyama and A. Yanase, First-principles calculations of effective-mass parameters of AlN and GaN, *Phys. Rev. B*, 1995, **52**(11), 8132.
14. D. W. Langer and R. N. Euwema, Spin Exchange in Excitons, the Quasicubic Model and Deformation Potentials in II-VI Compounds, *Phys. Rev. B*, 1970, **2**(10), 4005.
15. S.-H. Wei, Overcoming the doping bottleneck in semiconductors, *Comput. Mater. Sci.*, 2004, **30**(3), 337–348.
16. S. J. Clark, J. Robertson, S. Lany and A. Zunger, Intrinsic defects in ZnO calculated by screened exchange and hybrid density functionals, *Phys. Rev. B*, 2010, **81**(11), 115311.
17. M. R. Wagner, G. Callsen, J. S. Reparaz, R. Kirste, A. Hoffmann, A. V. Rodina, A. Schleife, F. Bechstedt and M. R. Phillips, Effects of strain on the valence band structure and exciton-polariton energies in ZnO, *Phys. Rev. B*, 2013, **88**(23), 235210.
18. M. Leslie and M. J. Gillan, The energy and elastic dipole tensor of defects in ionic crystals calculated by the supercell method, *J. Phys. C: Solid State Phys.*, 1985, **18**(5), 973.
19. A. Janotti and C. G. Van de Walle, Absolute deformation potentials and band alignment of wurtzite ZnO, MgO, and CdO, *Phys. Rev. B*, 2007, **75**(12), 121201.

20. Y.-H. Li, A. Walsh, S. Chen, W.-J. Yin, J.-H. Yang, J. Li, J. L. F. Da Silva, X. G. Gong and S.-H. Wei, Revised ab initio natural band offsets of all group IV, II-VI, and III-V semiconductors, *Appl. Phys. Lett.*, 2009, **94**(21), 212109.

21. Y. Hinuma, A. Gruneis, G. Kresse and F. Oba, Band alignment of semiconductors from density-functional theory and many-body perturbation theory, *Phys. Rev. B*, 2014, **90**(15), 155405.

22. R. Ishii, A. Kaneta, M. Funato and Y. Kawakami, All deformation potentials in GaN determined by reflectance spectroscopy under uniaxial stress: Definite breakdown of the quasi-cubic approximation, *Phys. Rev. B*, 2010, **81**(15), 155202.

23. W. A. Adeagbo, S. Thomas, S. K. Nayak, A. Ernst and W. Hergert, First-principles study of uniaxial strained and bent ZnO wires, *Phys. Rev. B*, 2014, **89**(19), 195135.

24. X. Han, L. Kou, X. Lang, J. Xia, N. Wang, R. Qin, J. Lu, J. Xu, Z. Liao, X. Zhang, X. Shan, X. Song, J. Gao, W. L. Guo and D. P. Yu, Electronic and mechanical coupling in bent ZnO nanowires, *Adv. Mater.*, 2009, **21**(48), 4937–4941.

25. L. Z. Kou, Y. Zhang, C. Li, W. L. Guo and C. Chen, Local-Strain-Induced Charge Carrier Separation and Electronic Structure Modulation in Zigzag ZnO Nanotubes: Role of Built-In Polarization Electric Field, *J. Phys. Chem. C*, 2011, **115**(5), 2381–2385.

26. D. C. Look, J. W. Hemsky and J. R. Sizelove, Residual native shallow donor in ZnO, *Phys. Rev. Lett.*, 1999, **82**(12), 2552.

27. G. W. Tomlins, J. L. Routbort and T. O. Mason, Zinc self-diffusion, electrical properties, and defect structure of undoped, single crystal zinc oxide, *J. Appl. Phys. Rev.*, 2000, **87**(1), 117–123.

28. C. Kittel, *Introduction to Solid State Physics*, New York, Wiley, 8th edn, 2005.

29. C. G. Van de Walle and J. Neugebauer, First-principles calculations for defects and impurities: Applications to III-nitrides, *J. Appl. Phys.*, 2004, **95**(8), 3851–3879.

30. W. J. Lee, J. Kang and K. J. Chang, Defect properties and p-type doping efficiency in phosphorus-doped ZnO, *Phys. Rev. B*, 2006, **73**(2), 024117.

31. S. B. Zhang, The microscopic origin of the doping limits in semiconductors and wide-gap materials and recent developments in overcoming these limits: a review, *J. Phys.: Condens. Matter*, 2002, **14**(34), R881.

32. C. Freysoldt, B. Grabowski, T. Hickel, J. Neugebauer, G. Kresse, A. Janotti and C. G. Van, de Walle. First-principles calculations for point defects in solids, *Rev. Mod. Phys.*, 2014, **86**(1), 253–306.

33. A. Janotti and C. G. Van de Walle, Native point defects in ZnO, *Phys. Rev. B*, 2007, **76**(12), 165202.

34. F. Oba, A. Togo, I. Tanaka, J. Paier and G. Kresse, Defect energetics in ZnO: A hybrid Hartree-Fock density functional study, *Phys. Rev. B*, 2008, **77**(24), 245202.

35. F. Oba, M. Choi, A. Togo, A. Seko and I. Tanaka, Native defects in oxide semiconductors: a density functional approach, *J. Phys.: Condens. Matter*, 2010, **22**(38), 384211.

36. J. Li, S.-H. Wei, S. S. Li and J. B. Xia, Design of shallow acceptors in ZnO: First-principles band-structure calculations, *Phys. Rev. B*, 2006, **74**(8), 081201.

37. P. H. Kasai, Electron Spin Resonance Studies of Donors and Acceptors in ZnO, *Phys. Rev.*, 1963, **130**(3), 989.

38. O. F. Schirmer, The structure of the paramagnetic lithium center in zinc oxide and beryllium oxide, *J. Phys. Chem. Solids*, 1968, **29**(8), 1407–1429.

39. S. Limpijumnong, S. B. Zhang, S. H. Wei and C. H. Park, Doping by Large-Size-Mismatched Impurities: The Microscopic Origin of Arsenic- or Antimony-Doped p-Type Zinc Oxide, *Phys. Rev. Lett.*, 2004, **92**(15), 155504.

40. X. Sun, Y. S. Gu, X. Q. Wang and Y. Zhang, Defects Energetics and Electronic Properties of Li Doped ZnO: A Hybrid Hartree-Fock Density Functional Study, *Chin. J. Chem. Phys.*, 2012, **25**(3), 261–268.

41. M.-H. Du and S. B. Zhang, Impurity-bound small polarons in ZnO: Hybrid density functional calculations, *Phys. Rev. B*, 2009, **80**(11), 115217.

42. M. G. Wardle, J. P. Goss and P. R. Briddon, Theory of Li in ZnO: A limitation for Li-based p-type doping, *Phys. Rev. B*, 2005, **71**(15), 155205.

43. R. Qin, J. Zheng, J. Lu, L. Wang, L. Lai, G. Luo, J. Zhou, H. Li, Z. Gao, G. Li and W. N. Mei, Origin of p-Type Doping in Zinc Oxide Nanowires Induced by Phosphorus Doping: A First Principles Study, *J. Phys. Chem. C*, 2009, **113**(22), 9541–9545.

44. M.-P. Lua, M.-Y. Lu and L. J. Chen, p-Type ZnO nanowires: From synthesis to nanoenergy, *Nano Energy*, 2012, **1**(2), 247–258.

45. D. C. Look, *Electrical Characterization of GaAs Materials and Devices*, Wiley, Singapore, 1989.

46. D. C. Look and R. J. Molnar, Degenerate layer at GaN/sapphire interface: influence on Hall-effect measurements, *Appl. Phys. Lett.*, 1997, **70**(25), 3377–3379.

47. A. R. Hutson, Hall effect studies of doped zinc oxide single crystals, *Phys. Rev.*, 1957, **108**(2), 222.

48. M. A. Seitz and D. H. Whitmore, Electronic drift mobilities and space-charge-limited currents in lithium-doped zinc oxide, *J. Phys. Chem. Sol.*, 1968, **29**(6), 1033–1049.

49. D. C. Look, D. C. Reynold, J. R. Sizelove, R. L. Jones, C. W. Litton, G. Cantwell and W. C. Harsch, Electrical properties of bulk ZnO, *Solid Stat. Commun.*, 1998, **105**(6), 399–401.

50. A. Tsukazaki, A. Ohtomo, T. Onuma, M. Ohtani, T. Makino, M. Sumiya, K. Ohtani, S. F. Chichibu, S. Fuke, Y. Segawa, H. Ohno, H. Koinuma and M. Kawasaki, Repeated temperature modulation epitaxy for p-type doping and light-emitting diode based on ZnO, *Nat. Mater.*, 2005, **4**(1), 42–46.

51. P. Wagner and R. Helbig, Halleffekt und anisotropie der beweglichkeit der elektronen in ZnO, *J. Phys. Chem. Solids*, 1974, **35**(3), 327–335.

52. W. I. Park, J. S. Kim, G. C. Yi, M. H. Bae and H. J. Lee, Fabrication and electrical characteristics of high-performance ZnO nanorod field-effect transistors, *Appl. Phys. Lett.*, 2004, **85**(21), 5052–5054.

53. K. H. Liu, P. Gao, Z. Xu, X. D. Bai and E. G. Wang, In-situ probing electrical response on bending of ZnO nanowires inside transmission electron microscope, *Appl. Phys. Lett.*, 2008, **92**(21), 213105.

54. Z. M. Liao, K. J. Liu, J. M. Zhang, J. Xu and D. P. Yu, Effect of surface states on electron transport in individual ZnO nanowires, *Phys. Lett. A*, 2007, **367**(3), 207–210.

55. X. Li, J. J. Qi, Q. Zhang and Y. Zhang, Temperature-dependent electron transport in ZnO micro/nanowires, *J. Appl. Phys.*, 2012, **112**(8), 084313.

56. R. W. Shao, K. Zheng, B. Wei, Y.-F. Zhang, Y.-J. Li, X.-D. Han, Z. Zhang and J. Zou, Bandgap engineering and manipulating electronic and optical properties of ZnO nanowires by uniaxial strain, *Nanoscale*, 2014, **6**(9), 4936–4941.

57. Y. Yang, J. J. Qi, Q. Liao, W. Guo, Y. Wang and Y. Zhang, Negative differential resistance in PtIr/ZnO ribbon/sexithiophen hybrid double diodes, *Appl. Phys. Lett.*, 2009, **95**(12), 123112.

58. P. X. Gao, Y. Ding and Z. L. Wang, Electronic Transport in Superlattice-Structured ZnO Nanohelix, *Nano Lett.*, 2009, **9**(1), 137–143.

59. C. X. Xu, X. W. Sun and B. J. Chen, Field emission from gallium-doped zinc oxide nanofiber array, *Appl. Phys. Lett.*, 2004, **84**(9), 1540–1542.

60. M. J. Zhou, H. J. Zhu, Y. Jiao, Y. Rao, S. Hark, Y. Liu, L. Peng and Q. Li, Optical and electrical properties of Ga-doped ZnO nanowire arrays on conducting substrates, *J. Phys. Chem. C*, 2009, **113**(20), 8945–8947.

61. V. Bhosle, A. Tiwari and J. Narayan, Electrical properties of transparent and conducting Ga doped ZnO, *J. Appl. Phys.*, 2006, **100**(3), 033713.

62. Q. H. Li, Q. Wan, Y. G. Wang and T. H. Wang, Abnormal temperature dependence of conductance of single Cd-doped ZnO nanowires, *Appl. Phys. Lett.*, 2005, **86**(26), 263101.

63. B. Xiang, P. W. Wang, X. Zhang, S. A. Dayeh, D. P. R. Aplin, C. Soci, D. Yu and D. Wang, Rational synthesis of p-type zinc oxide nanowire arrays using simple chemical vapor deposition, *Nano Lett.*, 2007, 7(2), 323–328.

64. W. Lee, M. C. Jeong and J. M. Myoung, Optical characteristic of arsenic-doped ZnO nanowires, *Appl. Phys. Lett.*, 2004, **85**(25), 6167–6169.

65. M. H. Sun, Q. F. Zhang and J. L. Wu, Electrical and electroluminescence properties of As-doped p-type ZnO nanorod arrays, *J. Phys. D: Appl. Phys.*, 2007, **40**(12), 3798–3802.

66. T. Kamiya, K. Tajima *et al.*, Interface electronic structures of zinc oxide and metals: First-principle study, *Phys. Status Solidi A*, 2008, **205**(8), 1929–1933.

67. Z. Yang, L. Wan, Y. Yu, Y. Wei and J. Wang, Electron transport through Al-ZnO-Al: An *ab initio* calculation, *J. Appl. Phys.*, 2010, **108**(3), 033704.

68. Q. Han, B. Cao, L. Zhou, G. Zhang and Z. Liu, Electrical Transport Study of Single-Walled ZnO Nanotubes: A First-Principles Study of the Length Dependence, *J. Phys. Chem. C*, 2011, **115**(8), 3447–3452.

69. X. Sun, Y. S. Gu, X. Q. Wang and Y. Zhang, First-Principles Studies on Transport Properties and Contact Effects of Cu (111)/ZnO-Nanobelt/Cu (111) systems, *Phys. Chem. Chem. Phys.*, 2013, **15**(31), 13070–13076.

70. X. Sun, Y. S. Gu, X. Q. Wang and Y. Zhang, Strain-modulated transport properties of Cu/ZnO-nanobelt/Cu nanojunctions, *Phys. Status Solidi B*, 2015, **252**(8), 1767–1772.

71. G. Zhang, X. Luo, Y. Zheng and B. Wang, Giant piezoelectric resistance effect of nanoscale zinc oxide tunnel junctions: first principles simulations, *Phys. Chem. Chem. Phys.*, 2012, **14**(19), 7051–7058.

72. W. Liu, A. Zhang, Y. Zhang and Z. L. Wang, First principles simulations of piezotronic transistors, *Nano Energy*, 2015, **14**, 355–363.

73. J. Taylor, H. Guo and J. Wang, Ab initio modeling of quantum transport properties of molecular electronic devices, *Phys. Rev. B*, 2001, **63**(24), 245407.

74. M. Koentopp, C. Chang, K. Burke and R. Car, Density functional calculations of nanoscale conductance, *J. Phys.: Condens. Matter*, 2008, **20**(8), 083203.

75. D. K. Schroder, *Semiconductor Material and Device Characterization*, Wiley, Singapore, 1990.

76. L. J. Brillson and Y. Lu, ZnO Schottky barriers and Ohmic contacts, *J. Appl. Phys.*, 2011, **109**(12), 121301.

77. Y. Dong and L. J. Brillson, First-Principles Studies of Metal (111)/ZnO{0001} Interfaces, *J. Electron. Mater.*, 2008, **37**(5), 743–748.

78. B. Shan and K. Cho, Ab initio study of Schottky barriers at metal-nanotube contacts, *Phys. Rev. B*, 2004, **70**(23), 233405.

79. R. T. Tung, Formation of an electric dipole at metal-semiconductor interfaces, *Phys. Rev. B*, 2001, **64**(20), 205310.

80. M. W. Allen, S. M. Durbin and J. B. Metson, Silver oxide Schottky contacts on n-type ZnO, *Appl. Phys. Lett.*, 2007, **91**(5), 053512.

81. R. T. Tung, Recent advances in Schottky barrier concepts, *Mater. Sci. Eng., R.*, 2001, **35**(1), 1–138.

82. A. Y. Polyakov, N. B. Smirnov, E. A. Kozhukhova, V. I. Vdovin, K. Ip, Y. W. Heo, D. P. Norton and S. J. Pearton, Electrical characteristics of Au and Ag Schottky contacts on n-ZnO, *Appl. Phys. Lett.*, 2003, **83**(8), 1575–1577.

83. M. W. Allen, M. M. Alkaisi and S. Durbin, Metal Schottky diodes on Zn-polar and O-polar bulk ZnO, *Appl. Phys. Lett.*, 2006, **89**(10), 103520.

84. K. Ip and G. T. Thaler, Contacts to ZnO, *J. Cryst. Growth*, 2006, **287**(1), 149–156.

85. N. R. D. Amico, G. Cantele, C. A. Perroni and D. Ninno, Electronic properties and Schottky barriers at ZnO-metal interfaces from first principles, *J. Phys.: Condens. Matter*, 2015, **27**(1), 015006.

86. M. W. Allen, P. Miller, R. J. Reeves and S. Durbin, Influence of spontaneous polarization on the electrical and optical properties of bulk, single crystal ZnO, *Appl. Phys. Lett.*, 2007, **90**(6), 062104.

87. C. A. Mead, Metal-semiconductor surface barriers, *Solid-State Electron.*, 1966, **9**(11), 1023–1033.

88. Y. F. Dong, Z. Q. Fang, D. C. Look, G. Cantwell, J. Zhang, J. J. Song and L. J. Brillson, Zn- and O-face polarity effects at ZnO surfaces and metal interfaces, *Appl. Phys. Lett.*, 2008, **93**(7), 072111.

89. L. J. Brillson, H. L. Mosbacker, M. J. Hetzer, Y. Strzhemechny, G. H. Jessen, D. C. Look, G. Cantwell, J. Zhang and J. J. Song, Dominant effect of near interface native point defects on ZnO Schottky barriers, *Appl. Phys. Lett.*, 2007, **90**(10), 102116.

90. K. Ip, Y. W. Heo, K. H. Baik, D. P. Norton, S. J. Pearton, S. Kim, J. R. LaRoche and F. Ren, Temperature-dependent characteristics of Pt Schottky contacts on n-type ZnO, *Appl. Phys. Lett.*, 2004, **84**, 2835–2837.

91. M. S. Oh, D. K. Hwang, J. H. Lim, Y. S. Choi and S. J. Park, Improvement of Pt Schottky contacts to n-type ZnO by KrF excimer laser irradiation, *Appl. Phys. Lett.*, 2007, **91**(15), 042109.

92. S. H. Kim and T. Y. Seong, Electrical characteristics of Pt Schottky contacts on sulfide-treated n-type ZnO, *Appl. Phys. Lett.*, 2005, **86**(2), 022101.

93. H. L. Mosbacker, S. E. Hage, M. Gonzalez, S. A. Ringel, M. J. Hetzer, D. C. Look, G. Cantwell, J. Zhang, J. J. Song and L. J. Brillson, Role of subsurface defects in metal-ZnO(0001) Schottky barrier formation, *J. Vac. Sci. Technol., B: Microelectron. Nanometer Struct.--Process., Meas., Phenom.*, 2007, **25**(4), 1405–1411.

94. H. V. Wenckstern, G. Biehne, R. A. Rahman, H. Hochmuth, M. Lorenz and M. Grundmann, Mean barrier height of Pd Schottky contacts on ZnO thin films, *Appl. Phys. Lett.*, 2006, **88**(9), 092102.

95. C. A. Mead, Metal-semiconductor surface barriers, *Solid-State Electron.*, 1966, **9**(11), 1023–1033.

96. H. S. Yang, D. P. Norton, S. J. Pearton and F. Ren, Ti/Au n-type Ohmic contacts to bulk ZnO substrates, *Appl. Phys. Lett.*, 2005, **87**, 212106.

97. R. Fiederling, M. Keim, G. Reuscher, W. Ossau, G. Schmidt, A. Waag and L. Molenkamp, Injection and detection of a spin-polarized current in a light-emitting diode, *Nature*, 1999, **402**(21), 787–790.

98. S. Kuroda, N. Nishizawa, K. Takita, M. Mitome, Y. Bando, K. Osuch and T. Dietl, Origin and control of high-temperature ferromagnetism in semiconductors, *Nat. Mater.*, 2007, **6**(6), 440–446.

99. T. Dietl, H. Ohno, F. Matsukura, J. Cibert and D. Ferrand, Zener Model Description of Ferromagnetism in Zinc-Blende Magnetic Semiconductors, *Science*, 2000, **287**(5455), 1019–1022.

100. K. Ueda, H. Tabata and T. Kawai, Magnetic and electric properties of transition-metal-doped ZnO films, *Appl. Phys. Lett.*, 2001, **79**(7), 988–990.

101. M. Venkatesan, C. B. Fitzgerald, J. G. Lunney and J. M. D. Coey, Anisotropic Ferromagnetism in Substituted Zinc Oxide, *Phys. Rev. Lett.*, 2004, **93**(17), 177206.

102. P. Sharma, A. Gupta, K. V. Rao, F. J. Owens, R. Sharma, R. Ahuja, J. M. O. Guillen, B. Johansson and G. A. Gehring, Ferromagnetism above room temperature in bulk and transparent thin films of Mn-doped ZnO, *Nat. Mater.*, 2003, **2**(10), 673–677.

103. D. C. Kundaliya, S. B. Ogale, S. E. Lofland, S. Dhar, C. J. Metting and S. R. Shinde, On the origin of high-temperature ferromagnetism in the low-temperature processed Mn–Zn–O system, *Nat. Mater.*, 2004, **3**(10), 709–714.

104. X. Zhang, Y. Zhang, Y. Gu, J. Qi, Y. Huang and J. Liu, Abnormal magnetic behavior in DMS Zn1 − xMnxO nanowires, *Chin. Sci. Bull.*, 2006, **51**(4), 490–492.

105. Y. Gu, X. Zhang, X. Wang, Y. Huang, J. Qi and Y. Zhang, A quantum explanation of the abnormal magnetic behaviour in Mn-doped ZnO nanowires, *J. Phys.: Condens. Matter*, 2007, **19**(23), 236223.

106. Z. W. Jin, T. Fukumura, M. Kawasaki, K. Ando, H. Saito, T. Sekiguchi, Y. Z. Yoo, M. Murakami, Y. Matsumoto, T. Hasegawa and H. Koinuma, High throughput fabrication of transition-metal-doped epitaxial ZnO thin films: A series of oxide-diluted magnetic semiconductors and their properties, *Appl. Phys. Lett.*, 2001, **78**(24), 3824–3826.

107. J. M. D. Coey, M. Venkatesan and C. B. Fitzgerald, Donor impurity band exchange in dilute ferromagnetic oxides, *Nat. Mater.*, 2005, **4**(2), 173–179.

108. Y. M. Cho, W. K. Choo, H. Kim, D. Kim and Y. E. Ihm, Effects of rapid thermal annealing on the ferromagnetic properties of sputtered $Zn_{1-x}(Co_{0.5}Fe_{0.5})_xO$ thin films, *Appl. Phys. Lett.*, 2002, **80**(18), 3358–3360.

109. X. X. Wei, C. Song, K. W. Geng, F. Zeng, B. He and F. Pan, Local Fe structure and ferromagnetism in Fe-doped ZnO films, *J. Phys.: Condens. Matter*, 2006, **18**(31), 7471–7479.

110. H. J. Lee, S. Y. Jeong, J. Y. Hwang and C. R. Cho, Ferromagnetism in Li co-doped ZnO:Cr, *Europhys. Lett.*, 2003, **64**(6), 797–802.

111. N. H. Hong, J. Sakai, N. T. Huong, N. Poirot and A. Ruyter, Role of defects in tuning ferromagnetism in diluted magnetic oxide thin films, *Phys. Rev. B*, 2005, **72**(4), 045336.

112. T. Wakano, N. Fujimura, Y. Morinaga, N. Abe, A. Ashida and T. Ito, Magnetic and magneto-transport properties of ZnO:Ni films, *Phys. E*, 2001, **10**(1), 260–264.

113. X. Liu, F. Lin, L. Sun, W. Cheng, X. Ma and W. Shi, Doping concentration dependence of room-temperature ferromagnetism for Ni-doped ZnO thin films prepared by pulsed-laser deposition, *Appl. Phys. Lett.*, 2006, **88**(6), 062508.

114. N. H. Hong, J. Sakai and V. Brize, Observation of ferromagnetism at room temperature in ZnO thin films, *J. Phys.: Condens. Matter*, 2007, **19**(3), 036219.

115. S. Banerjee, M. Mandal, N. Gayathri and M. Sardar, Enhancement of ferromagnetism upon thermal annealing in pure ZnO, *Appl. Phys. Lett*, 2007, **91**(18), 182501.

116. H. Pan, J. B. Yi, L. Shen, R. Q. Wu, J. H. Yang, J. Y. Lin, Y. P. Feng, J. Ding, L. H. Van and J. H. Yin, Room-temperature ferromagnetism in carbon-doped ZnO, *Phys. Rev. Lett.*, 2007, **99**(12), 127201.

117. D. B. Buchholz, R. P. H. Chang, J. H. Song and J. B. Ketterson, Room-temperature ferromagnetism in Cu-doped ZnO thin films, *Appl. Phys. Lett.*, 2005, **87**(8), 082504.

118. G. Z. Xing, J. B. Yi, J. G. Tao, T. Liu, L. M. Wong, Z. Zhang, G. P. Li, S. J. Wang, J. Ding, T. C. Sum, C. H. A. Huan and T. Wu, Comparative Study of Room-Temperature Ferromagnetism in Cu-Doped ZnO Nanowires Enhanced by Structural Inhomogeneity, *Adv. Mater.*, 2008, **20**(18), 3521–3527.

119. T. S. Herng, M. F. Wong, D. Qi, J. Yi, A. Kumar, A. Huang, F. C. Kartawidjaja, S. Smadici, P. Abbamonte, C. Sánchez-Hanke, S. Shannigrahi, J. M. Xue, J. Wang, Y. P. Feng, A. Rusydi, K. Zeng and J. Ding, Mutual Ferromagnetic–Ferroelectric Coupling in Multiferroic Copper-Doped ZnO, *Adv. Mater.*, 2011, **23**(14), 1635–1640.

120. J. M. D. Coey, M. Vankatesan, C. B. Fitzgerald, L. S. Dornneles, P. Stamenov and J. Luney, Anisotropy of the magnetization of a dilute magnetic oxide, *J. Magn. Magn. Mater.*, 2005, **290**(2005), 1405–1407.

121. D. Chakraborti, G. Trichy, J. T. Prater and J. Narayan, The effect of oxygen annealing on ZnO:Cu and ZnO:(Cu, Al) diluted magnetic semiconductors, *J. Phys. D: Appl. Phys.*, 2007, **40**(24), 7606–7613.

122. R. P. Borges, R. C. Da Silva, S. Magalhaes, M. M. Cruz and M. Godinho, Magnetism in Ar-implanted ZnO, *J. Phys.: Condens. Matter*, 2007, **19**(47), 476207.

123. Q. Wang, Q. Sun, P. Jena and Y. Kawazoe, Magnetic properties of transition-metal-doped $Zn_{1-x}T_xO$ (T = Cr, Mn, Fe, Co, and Ni) thin films with and without intrinsic defects: A density functional study, *Phys. Rev. B*, 2009, **79**(11), 115407.

124. T. Yao and S.-K. Hong, *Oxide and Nitride Semiconductors: Processing, Properties and Applications*, Springer, Berlin, 2009.

125. T. B. Bateman, Elastic moduli of single-crystal zinc oxide, *J. Appl. Phys.*, 1962, **33**(11), 3309–3312.

126. F. Decremps, J. Zhang, B. Li and R. C. Liebermann, Pressure-induced softening of shear modes in ZnO, *Phys. Rev. B*, 2001, **63**(22), 224105.

127. Y. H. Huang, X. D. Bai and Y. Zhang, In situ mechanical properties of individual ZnO nanowires and the mass measurement of nano-particles, *J. Phys.: Condens. Matter*, 2006, **18**(15), L179–L184.

128. X. D. Bai, P. X. Gao, Z. L. Wang and E. G. Wang, Dual-mode mechanical resonance of individual ZnO nanobelts, *Appl. Phys. Lett.*, 2003, **82**(26), 4806–4808.

129. Y. H. Huang, Y. Zhang, X. Q. Wang, X. D. Bai, Y. S. Gu, X. Q. Yan, Q. L. Liao, J. J. Qi and J. Liu, Size independence and doping of bending modulus in ZnO nanowires, *Cryst. Growth Des.*, 2009, **4**, 1640–1642.

130. C. Q. Chen, Y. Shi, Y. S. Zhang, J. Zhu and Y. J. Yan, Size dependence of Young's modulus in ZnO nanowires, *Phys. Rev. Lett*, 2006, **96**(4), 075505.

131. G. Stan, C. V. Ciobanu, P. M. Parthangal and R. F. Cook, Diameter-dependent radial and tangential elastic moduli of ZnO nanowires, *Nano Lett.*, 2007, 7(12), 3691–3697.

132. S. X. Mao, M. H. Zhao and Z. L. Wang, Nanoscale mechanical behavior of individual semiconducting nanobelts, *Appl. Phys. Lett.*, 2003, **83**(5), 993–995.

133. R. Agrawal, B. Peng, E. E. Gdoutos and H. D. Espinosa, Elasticity size effects in ZnO nanowires A combined experimental-computational approach, *Nano Lett.*, 2008, 8(11), 3668–3674.

134. I. B. Kobiakov, Elastic, piezoelectric and dielectric properties of ZnO and CdS single crystals in a wide range of temperatures, *Solid State Commun.*, 1980, **35**(3), 305–310.

135. A. Zaoui and W. Sekkal, Pressure-induced softening of shear modes in wurtzite ZnO: A theoretical study, *Phys. Rev. B*, 2002, **66**(17), 174106.

136. X. Q. Wang, Y. S. Gu, X. Sun, H. Wang and Y. Zhang, Third order elastic constants of ZnO and size effect in ZnO nanowires, *J. Appl. Phys.*, 2014, **115**(21), 213516.

137. Z. J. Gao, Y. S. Gu and Y. Zhang, Mechanical Properties and Size effects of ZnO Nanowires Studied by First-Principles Calculation, *Mater. Sci. Forum*, 2010, **654**, 1670–1673.

138. J. Song, X. Wang, E. Riedo and Z. L. Wang, Elastic Property of Vertically Aligned Nanowires, *Nano Lett.*, 2005, **5**(10), 1954–1958.

139. B. Wen, J. E. Sader and J. J. Boland, Mechanical Properties of ZnO Nanowires, *Phys. Rev. Lett.*, 2008, **101**(17), 175502.

140. G. Wang and X. Li, Size dependency of the elastic modulus of ZnO nanowires: Surface stress effect, *Appl. Phys. Lett.*, 2007, **91**(23), 231912.

141. J. S. Qi, D. N. Shi and B. Wang, Different mechanical properties of the pristine and hydrogen passivated ZnO nanowires, *Comput. Mater. Sci.*, 2009, **46**(2), 303–306.

142. S. J. Pearton, D. P. Norton, K. Ip, Y. W. Heo and T. Steiner, Recent progress in processing and properties of ZnO, *Prog. Mater. Sci.*, 2005, **50**(3), 293–340.

143. J. Cai and N. Chen, First-principles study of the wurtzite-to-rocksalt phase transition in zinc oxide, *J Phys.*, 2007, **19**(26), 266207.

144. S. Desgreniers, High-density phases of ZnO: structural and compressive parameters, *Phys. Rev. B*, 1998, **58**(21), 14102.

145. H. Liu, Y. Ding, M. Somayazulu, J. Qian, J. Shu, D. Häusermann and H. K. Mao, Rietveld refinement study of the pressure dependence of the internal structural parameter u in the wurtzite phase of ZnO, *Phys. Rev. B*, 2005, **71**(21), 212103.

146. H. Liu, J. S. Tse and H. Mao, Stability of rocksalt phase of zinc oxide under strong compression: synchrotron Xray diffraction experiments and first-principles calculation studies, *J. Appl. Phys.*, 2006, **100**, 093509.

147. F. Decremps, J. Pellicer-Porres, A. M. Saitta, J. C. Chervin and A. Polian, High-pressure Raman spectroscopy study of wurtzite ZnO, *Phys. Rev. B*, 2002, **65**(9), 092101.

148. F. Decremps, F. Datchi, A. M. Saitta, A. Polian, S. Pascarelli, A. Di Cicco, J. P. Itié and F. Baudelet, Local structure of condensed zinc oxide, *Phys. Rev. B*, 2003, **68**(10), 104101.

149. J. Y. Liang, G. Lin, H. B. Xu, L. Jing, L. X. Dong, W. Z. Hua, W. Z. Yu and J. Weber, A novel synthesis route and phase transformation of ZnO nanoparticles modified by DDAB, *J. Cryst. Growth*, 2003, **252**(1), 226–229.

150. S.-Y. Chen, P. Shen and J. Jiang, Polymorphic transformation of dense ZnO nanoparticles: implications for chair/boat-type Peierls distortions of AB semiconductor, *J. Chem. Phys.*, 2004, **121**(22), 1130–11313.

151. X. Wu, Z. Wu, L. Guo, C. Liu, J. Liu, X. Li and H. Xu, Pressure-induced phase transformation in controlled shape ZnO nanorods, *Solid State Commun.*, 2005, **135**(11), 780–784.

152. S. J. Chen, Y. C. Liu, C. L. Shao, C. S. Xu, Y. X. Liu, C. Y. Liu, B. P. Zhang, L. Wang, B. B. Liu and G. T. Zou, Photoluminescence study of ZnO nanotubes under hydrostatic pressure, *Appl. Phys. Lett.*, 2006, **88**(13), 133127.

153. R. S. Kumar, A. L. Cornelius *et al.*, Structure of nanocrystalline ZnO up to 85 GPa, *Curr. Appl. Phys.*, 2007, **7**(2), 135–138.

154. V. L. Solozhenko, O. O. Kurakevych, P. S. Sokolov and A. N. Baranov, Kinetics of the wurtzite-to-rock-salt phase transformation in ZnO at high pressure, *J. Phys. Chem. A*, 2011, **115**(17), 4354–4358.

155. B. J. Wuensch and H. L. Tuller, Lattice diffusion, grain boundary diffusion and defect structure of ZnO, *J. Phys. Chem. Solids*, 1994, **55**(10), 975–984.

156. C. S. Tiwary, D. Vishnu, A. K. Kole, J. Brahmanandam, D. R. Mahapatra, P. Kumbhakar and K. Chattopadhyay, Stabilization of the high-temperature and high-pressure cubic phase of ZnO by temperature-controlled milling, *J. Mater. Sci.*, 2016, **51**(1), 126–137.

157. A. J. Kulkarni, M. Zhou, K. Sarasamak and S. Limpijumnong, Novel phase transformation in ZnO nanowires under tensile loading, *Phys. Rev. Lett.*, 2006, **97**(10), 105502.

158. A. J. Kulkarni, K. Sarasamak, J. Wang, F. J. Ke, S. Limpijumnong and M. Zhou, Effect of load triaxiality on polymorphic transitions in zinc oxide, *Mech. Res. Commun.*, 2008, **35**(1), 73–80.

159. R. S. Koster, C. M. Fang, M. Dijkstra, A. Van Blaaderen and M. A. Van Huis, Stabilization of Rock Salt ZnO Nanocrystals by Low-Energy Surfaces and Mg Additions: A First-Principles Study, *J. Phys. Chem. C*, 2015, **119**(10), 5648–5656.

160. Z. J. Gao, Y. S. Gu and Y. Zhang, First-Principles Studies on the Structural Transition of ZnO Nanowires at High Pressure, *J. Nanomater.*, 2010, **2010**, 13.

161. L. Zhang and H. Huang, Structural transformation of ZnO nanostructures, *Appl. Phys. Lett.*, 2007, **90**(2), 023115.

162. Y. Dai, Y. Zhang, Q. K. Li and C. W. Nan, Synthesis and optical properties of tetrapod-like zinc oxide nanorods, *Chem. Phys. Lett.*, 2002, **358**(1), 83–86.

163. B. K. Meyer, H. Alves, D. M. Hofmann, W. Kriegseis, D. Forster, F. Bertram, J. Christen, A. Hoffmann, M. Straßburg, M. Dworzak, U. Haboeck and A. V. Rodina, Bound exciton and donor-acceptor pair recombinations in ZnO, *Phys. Status Solidi B*, 2004, **241**, 231–260.

164. A. B. Djurisic and Y. H. Leung, Optical properties of ZnO nanostructures, *Small*, 2006, **241**(2), 944–961.

165. S. Xu and Z. L. Wang, One-dimensional ZnO nanostructures: Solution growth and functional properties, *Nano Res.*, 2011, **4**(11), 1013–1098.

166. Z. Bai, X. Yan, X. Chen, H. Liu, Y. Shen and Y. Zhang, ZnO nanowire array ultraviolet photodetectors with self-powered properties, *Curr. Appl. Phys.*, 2013, **13**(1), 165–169.

167. X. Wang, Y. Ding, C. J. Summers and Z. L. Wang, Large-Scale Synthesis of Six-Nanometer-Wide ZnO Nanobelts, *J. Phys. Chem. B*, 2004, **108**(26), 8773–8777.

168. H. Y. Shih, T. T. Chen, Y. C. Chen, T. H. Lin, L. W. Chang and Y. F. Chen, Size-dependent photoelastic effect in ZnO nanorods, *Appl. Phys. Lett.*, 2009, **94**(2), 021908.

169. I. Shalish, H. Temkin and V. Narayanamurti, Size-dependent surface luminescence in ZnO nanowires, *Phys. Rev. B*, 2004, **69**(26), 245401.

170. L. L. Yang, Q. X. Zhao, M. Willander, J. H. Yang and I. Ivanov, Annealing effects on optical properties of low temperature grown ZnO nanorod arrays, *J. Appl. Phys.*, 2009, **105**(5), 053503.

171. C.-C. Lin, S.-Y. Chen and S.-Y. Cheng, RETRACTED: Nucleation and growth behavior of well-aligned ZnO nanorods on organic substrates in aqueous solutions, *J. Cryst. Growth*, 2005, **283**(1–2), 141–146.

172. S.-J. Lee, S. K. Park, C. R. Park, J. Y. Lee, J. Park and Y. R. Do, Spatially Separated ZnO Nanopillar Arrays on Pt/Si Substrates Prepared by Electrochemical Deposition, *J. Phys. Chem. C*, 2007, **111**(32), 11793–11801.

173. J. Cui and U. Gibson, Thermal modification of magnetism in cobalt-doped ZnO nanowires grown at low temperatures, *Phys. Rev. B*, 2006, **74**(4), 045416.

174. K. Samanta, P. Bhattacharya and R. S. Katiyar, Optical properties of $Zn_{1-x}Co_xO$ thin films grown on Al_2O_3 (0001) substrates, *Appl. Phys. Lett.*, 2005, **87**(10), 101903.

175. K. J. Kim and Y. R. Park, Spectroscopic ellipsometry study of optical transitions in $Zn_{1-x}Co_xO$ alloys, *Appl. Phys. Lett.*, 2002, **81**(8), 1420–1422.

176. S. Y. Bae, C. W. Na, J. H. Kang and J. Park, Comparative structure and optical properties of Ga-, In-, and Sn-doped ZnO nanowires synthesized via thermal evaporation, *J. Phys. Chem. B*, 2005, **109**(7), 2526–2531.

177. W. S. Shi, B. Cheng, L. Zhang and E. T. Samulski, Influence of excitation density on photoluminescence of zinc oxide with different morphologies and dimensions, *J. Appl. Phys.*, 2005, **98**(8), 083502.

178. Q. X. Zhao, M. Willander, R. E. Morjan, Q. H. Hu and E. E. B. Campbell, Optical recombination of ZnO nanowires grown on sapphire and Si substrates, *Appl. Phys. Lett.*, 2003, **83**(1), 165–167.

179. W. I. Park, Y. H. Jun, S. W. Jung and G.-C. Yi, Excitonic emissions observed in ZnO single crystal nanorods, *Appl. Phys. Lett.*, 2003, **82**(6), 964–966.

180. B. Cao, X. Teng, S. H. Heo, Y. Li, S. O. Cho, G. Li and W. Cai, Different ZnO Nanostructures Fabricated by a Seed-Layer Assisted Electrochemical Route and Their Photoluminescence and Field Emission Properties, *J. Phys. Chem. C*, 2007, **111**(6), 2470–2476.

181. A. Manoogian and J. C. Woolley, Temperature dependence of the energy gap in semiconductors, *Can. J. Phys.*, 1984, **62**(3), 285–287.

182. C. Klingshirn, Properties of the electron-hole plasma in II–VI semiconductors, *J. Cryst. Growth*, 1992, **117**(1), 753–757.

183. W. Wegscheider, L. Pfeiffer, M. Dignam, A. Pinczuk, K. West, S. McCall and R. Hull, Lasing from excitons in quantum wires, *Phys. Rev. Lett.*, 1993, **71**(24), 4071.

184. J. H. Choy, E. S. Jang, J. H. Won, J. H. Chung, D. J. Jang and Y. W. Kim, Soft Solution Route to Directionally Grown ZnO Nanorod Arrays on Si Wafer: Room-Temperature Ultraviolet Laser, *Adv. Mater.*, 2003, **15**(22), 1911–1914.

185. M. H. Huang, S. Mao, H. Feick, H. Yan, Y. Wu, H. Kind, E. Weber, R. Russo and P. Yang, Room-temperature ultraviolet nanowire nanolasers, *Science*, 2001, **292**(5523), 1897–1899.

186. H. Cao, J. Xu, D. Zhang, S.-H. Chang, S. Ho, E. Seelig, X. Liu and R. P. Chang, Spatial confinement of laser light in active random media, *Phys. Rev. Lett.*, 2000, **84**(24), 5584.

187. C. Liu, J. A. Zapien, Y. Yao, X. Meng, C. S. Lee, S. Fan, Y. Lifshitz and S. T. Lee, High-Density, Ordered Ultraviolet Light-Emitting ZnO Nanowire Arrays, *Adv. Mater.*, 2003, **15**(10), 838–841.

188. Q. Zhang, J. Qi, X. Li, F. Yi, Z. Wang and Y. Zhang, Electrically pumped lasing from single ZnO micro/nanowire and poly(3,4-ethylenedioxythiophene):poly(styrenexulfonate) hybrid heterostructures, *Appl. Phys. Lett.*, 2012, **101**(4), 043119.

189. S. Chu, G. Wang, W. Zhou, Y. Lin, L. Chernyak, J. Zhao, J. Kong, L. Li, J. Ren and J. Liu, Electrically pumped waveguide lasing from ZnO nanowires, *Nat. Nanotechnol.*, 2011, **6**(8), 506–510.

190. J. C. Johnson, H. Yan, R. D. Schaller, P. B. Petersen, P. Yang and R. J. Saykally, Near-Field Imaging of Nonlinear Optical Mixing in Single Zinc Oxide Nanowires, *Nano Lett.*, 2002, **2**(4), 279–283.

191. T. Voss, I. Kudyk, L. Wischmeier and J. Gutowski, Nonlinear optics with ZnO nanowires, *Phys. Status Solidi B*, 2009, **246**(2), 311–314.

192. C. Klingshirn, J. Fallert, H. Zhou, J. Sartor, C. Thiele, F. Maier-Flaig, D. Schneider and H. Kalt, 65 years of ZnO research-old and very recent results, *Phys. Status Solidi B*, 2010, **247**(6), 1424–1447.

193. Z. L. Wang, Nanostructures of zinc oxide, *Mater. Today*, 2004, **7**(6), 26–33.

194. J. L. Gomez and O. Tigli, Zinc oxide nanostructures: from growth to application, *J. Mater. Sci.*, 2013, **48**(2), 612–624.

195. B. Sunandan and D. Joydeep, Hydrothermal growth of ZnO nanostructures, *Sci. Technol. Adv. Mater.*, 2009, **10**(1), 013001.

196. Z. L. Wang, X. Y. Kong, Y. Ding, P. Gao, W. L. Hughes, R. Yang and Y. Zhang, Semiconducting and piezoelectric oxide nanostructures induced by polar surfaces, *Adv. Funct. Mater.*, 2004, **14**(10), 943–956.

197. Y. Hu, Y. Gao, S. Singamaneni, V. V. Tsukruk and Z. L. Wang, Converse piezoelectric effect induced transverse deflection of a free-standing ZnO microbelt, *Nano Lett.*, 2009, **9**(7), 2661–2665.

198. A. Dal Corso, M. Posternak, R. Resta and A. Baldereschi, Ab initio study of piezoelectricity and spontaneous polarization in ZnO, *Phys. Rev. B*, 1994, **50**(15), 10715–10721.

199. Ü. Özgür, Y. I. Alivov, C. Liu, A. Teke, M. A. Reshchikov, S. Doğan, V. Avrutin, S.-J. Cho and H. Morkoç, A comprehensive review of ZnO materials and devices, *J. Appl. Phys.*, 2005, **98**(4), 041301.

200. J. Xin, Y. Zheng and E. Shi, Piezoelectricity of zinc-blende and wurtzite structure binary compounds, *Appl. Phys. Lett.*, 2007, **91**(11), 112902.

201. Z. L. Wang, Progress in piezotronics and piezo-phototronics, *Adv. Mater.*, 2012, **24**(34), 4632–4646.

202. D. A. Bonnell, S. V. Kalinin, A. L. Kholkin and A. Gruverman, Piezoresponse force microscopy: a window into electromechanical behavior at the nanoscale, *MRS Bull.*, 2009, **34**(9), 648–657.

203. C. P. Li and B. H. Yang, Local piezoelectricity and polarity distribution of preferred c-axis-oriented ZnO film investigated by piezoresponse force microscopy, *J. Electron. Mater.*, 2011, **40**(3), 253–258.

204. J. A. Christman, R. R. Woolcott, A. I. Kingon and R. J. Nemanich, Piezoelectric measurements with atomic force microscopy, *Appl. Phys. Lett.*, 1998, **73**(26), 3851–3853.

205. M. Laurenti, S. Stassi, M. Lorenzoni, M. Fontana, G. Canavese, V. Cauda and C. F. Pirri, Evaluation of the piezoelectric properties and voltage generation of flexible zinc oxide thin films, *Nanotechnology*, 2015, **26**(21), 215704.

206. W. Wang, Y. Geng and W. Wu, Background-free piezoresponse force microscopy for quantitative measurements, *Appl. Phys. Lett.*, 2014, **104**(7), 072905.

207. D. A. Scrymgeour, T. L. Sounart, N. C. Simmons and J. W. P. Hsu, Polarity and piezoelectric response of solution grown zinc oxide nanocrystals on silver, *J. Appl. Phys.*, 2007, **101**(1), 014316.

208. M.-H. Hao, Z.-L. Wang and S. X. Mao, Piezoelectric characterization of individual zinc oxide nanobelt probed by piezoresponse force microscope, *Nano Lett.*, 2004, **4**(4), 587–590.

209. D. Tamvakos, S. Lepadatu, V.-A. Antohe, A. Tamvakos, P. M. Weaver, L. Piraux, M. G. Cain and D. Pullini, Piezoelectric properties of template-free electrochemically grown ZnO nanorod arrays, *Appl. Surf. Sci.*, 2015, **356**, 1214–1220.

210. F. Felten, G. A. Schneider, J. M. Saldaña and S. V. Kalinin, Modeling and measurement of surface displacements in BaTiO3 bulk material in piezoresponse force microscopy, *J. Appl. Phys.*, 2004, **96**(1), 563–568.

211. H. J. Fan, W. Lee, R. Hauschild, M. Alexe, G. Le Rhun, R. Scholz, A. Dadgar, K. Nielsch, H. Kalt, A. Krost, M. Zacharias and U. Gösele, Template-assisted large-scale ordered arrays of ZnO pillars for optical and piezoelectric applications, *Small*, 2006, **2**(4), 561–568.

212. B. J. Rodriguez, A. Gruverman, A. I. Kingon, R. J. Nemanich and O. Ambacher, Piezoresponse force microscopy for polarity imaging of GaN, *Appl. Phys. Lett.*, 2002, **80**(22), 4166–4168.

213. E. Broitman, M. Y. Soomro, J. Lu, M. Willander and L. Hultman, Nanoscale piezoelectric response of ZnO nanowires measured using a nanoindentation technique, *Phys. Chem. Chem. Phys.*, 2013, **15**(26), 11113–11118.

214. R. Agrawal and H. D. Espinosa, Giant piezoelectric size effects in zinc oxide and gallium nitride nanowires. A first principles investigation, *Nano Lett.*, 2011, **11**(2), 786–790.

215. C. Qin, Y. Gu, X. Sun, X. Wang and Y. Zhang, Structural dependence of piezoelectric size effects and macroscopic polarization in ZnO nanowires: A first-principles study, *Nano Res.*, 2015, **8**(6), 2073–2081.

216. D. A. Scrymgeour and J. W. P. Hsu, Correlated piezoelectric and electrical properties in individual ZnO nanorods, *Nano Lett.*, 2008, **8**(8), 2204–2209.

217. O. Tomoya, O. Hideo and K. Akira, Decrement of piezoelectric constants caused by screening effect of conduction electrons on the effective charge of CdS crystals. Japan, *J. Appl. Phys.*, 1971, **10**(5), 593–599.

218. J. I. Sohn, S. N. Cha, B. G. Song, S. Lee, S. M. Kim, J. Ku, H. J. Kim, Y. J. Park, B. L. Choi, Z. L. Wang, J. M. Kim and K. Kim, Engineering of efficiency limiting free carriers and an interfacial energy barrier for an enhancing piezoelectric generation, *Energy Environ. Sci.*, 2013, **6**(1), 97–104.

219. R. Hinchet, S. Lee, G. Ardila, L. Montès, M. Mouis and Z. L. Wang, Performance optimization of vertical nanowire-based piezoelectric nanogenerators, *Adv. Funct. Mater.*, 2014, **24**(7), 971–977.

220. A. Rodolfo and F. Christian, Lateral bending of tapered piezo-semiconductive nanostructures for ultra-sensitive mechanical force to voltage conversion, *Nanotechnology*, 2013, **24**(26), 265707.

221. K. Y. Lee, B. Kumar, J.-S. Seo, K.-H. Kim, J. I. Sohn, S. N. Cha, D. Choi, Z. L. Wang and S.-W. Kim, P-type polymer-hybridized high-

performance piezoelectric nanogenerators, *Nano Lett.*, 2012, **12**(4), 1959–1964.

222. M. Y. Soomro, O. Nur and M. Willander, Enhancing the piezopotential from zinc oxide (ZnO) nanowire using p-type polymers, *Mater. Lett.*, 2014, **124**, 123–125.

223. X. Xue, Y. Nie, B. He, L. Xing, Y. Zhang and Z. L. Wang, Surface free-carrier screening effect on the output of a ZnO nanowire nanogenerator and its potential as a self-powered active gas sensor, *Nanotechnology*, 2013, **24**(22), 225501.

224. Y. C. Yang, C. Song, X. H. Wang, F. Zeng and F. Pan, Giant piezoelectric d33 coefficient in ferroelectric vanadium doped ZnO films, *Appl. Phys. Lett.*, 2008, **92**(1), 012907.

225. D. Karanth and H. Fu, Large electromechanical response in ZnO and its microscopic origin, *Phys. Rev. B*, 2005, **72**(6), 064116.

226. Y. Q. Chen, X. J. Zheng and X. Feng, The fabrication of vanadium-doped ZnO piezoelectric nanofiber by electrospinning, *Nanotechnology*, 2010, **21**(5), 055708.

227. W. L. Ong, H. Huang, J. Xiao, K. Zeng and G. W. Ho, Tuning of multifunctional Cu-doped ZnO films and nanowires for enhanced piezo/ferroelectric-like and gas/photoresponse properties, *Nanoscale*, 2014, **6**(3), 1680–1690.

228. Y. Yang, W. Guo, X. Wang, Z. Wang, J. Qi and Y. Zhang, Size Dependence of Dielectric Constant in a Single Pencil-Like ZnO Nanowires, *Nano Lett.*, 2012, **12**(4), 1919–1922.

229. C. Staii, A. T. Johnson and N. J. Pinto, Quantitative analysis of scanning conductance microscopy, *Nano Lett.*, 2004, **4**(5), 859–862.

230. R. Tsu and D. Babic, Simple model for the dielectric constant of nanoscale silicon particle, *J. Appl. Phys.*, 1997, **82**(3), 1327–1329.

231. Z. Fan and J. G. Lu, Zinc oxide nanostructures: synthesis and properties, *J. Nanosci. Nanotechnol.*, 2005, **5**(10), 1561–1573.

232. A. Fujishima and K. Honda, Electrochemical Photolysis of Water at a Semiconductor Electrode, *Nature*, 1972, **238**, 37–38.

233. Q. Wan, T. H. Wang and J. C. Zhao, Enhanced photocatalytic activity of ZnO nanotetrapods, *Appl. Phys. Lett.*, 2005, **87**(8), 083105.

234. N. Daneshvar, D. Salari and A. R. Khataee, Photocatalytic degradation of azo dye acid red 14 in water on ZnO as an alternative catalyst to TiO_2, *J. Photochem. Photobiol., A*, 2004, **162**(2–3), 317–322.

235. J. Zhao, L. Wang, X. Yan, Y. Yang, Y. Lei, J. Zhou, Y. Huang, Y. Gu and Y. Zhang, Structure and photocatalytic activity of Ni-doped ZnO nanorods, *Mater. Res. Bull.*, 2011, **46**(8), 1207–1210.

236. Q. Xiao, J. Zhang, C. Xiao and X. Tan, Photocatalytic decolorization of methylene blue over $Zn_1 − xCo_xO$ under visible light irradiation, *Mater. Sci. Eng., B*, 2007, **142**(2–3), 121–125.

237. J. L. Yang, S. J. An, W. I. Park, G. C. Yi and W. Choi, Photocatalysis using ZnO thin films and nanoneedles grown by metal-organic chemical vapor deposition, *Adv. Mater.*, 2004, **16**(18), 1661–1664.

238. J. Nishio, M. Tokumura, H. T. Znad and Y. Kawase, Photocatalytic decolorization of azo-dye with zinc oxide powder in an external UV light irradiation slurry photoreactor, *J. Hazard. Mater.*, 2006, **138**(1), 106–115.

239. V. Kandavelu, H. Kastien and K. R. Thampi, Photocatalytic degradation of isothiazolin-3-ones in water and emulsion paints containing nanocrystalline TiO_2 and ZnO catalysts, *Appl. Catal., B*, 2004, **48**(2), 101–111.

CHAPTER 4

Electromechanical Devices

ZHENG ZHANG AND YUE ZHANG*

University of Science and Technology Beijing, Beijing, China
*Email: yuezhang@ustb.edu.cn

4.1 Individual Nanostructure-based Electromechanical Devices

To measure nano- and micro-scale strain/stress, various sensors have been fabricated based on one-dimensional nanomaterials, including carbon nanotubes (CNT),[1] Si[2] and ZnO nanostructures.[3] Amongst the variety of ZnO nanostructures, one-dimensional structures, such as nanowires, nanorods, nanobelts and nanowire arrays, are attractive for electromechanical device construction. To maximise the sensing ability of nanostructures, it was worth designing strain/stress sensors utilising a single nanostructure and these have been widely studied in the last decade.

4.1.1 Individual Nanowire-based Mess Sensor

The mechanical properties of individual ZnO nanowires, synthesised by a solid–vapour phase thermal evaporation process, were studied *in situ* by transmission electron microscopy (TEM).[4] The mechanical resonance was electrically induced by applying an oscillating voltage, and *in situ* imaging has been achieved simultaneously. The results indicated that the elastic bending modulus of individual ZnO nanowires was measured to be ~58 GPa. The damping time constant of the resonance in a vacuum of 10^{-8} Torr was ~14 ms. The ZnO nanowires were promising in potential applications as nanocantilevers and nanoresonators. A ZnO nanowire as a

Nanoscience & Nanotechnology Series No. 43
ZnO Nanostructures: Fabrication and Applications
By Yue Zhang
© Yue Zhang 2017
Published by the Royal Society of Chemistry, www.rsc.org

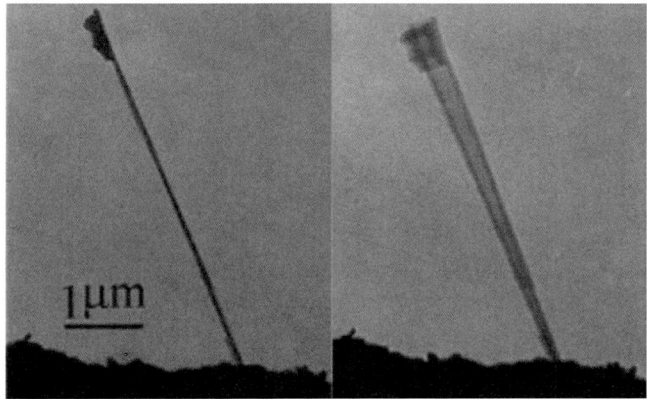

Figure 4.1   Nanobalance using a ZnO nanowire cantilever.
Reproduced with permission from Y. Huang, X. Bai and Y. Zhang, *In situ* mechanical properties of individual ZnO nanowires and the mass measurement of nanoparticles, *J. Phys.: Condens. Matter*, 2006, **18**(15), L179, IOP Publishing, Ltd.[4]

cantilever can be used as a force sensor or a nanobalance to measure the mass of the nanoparticle attached at the tip of a nanowire. Figure 4.1 shows a ZnO cantilever with a nanoparticle attached at the tip. The mass of the nanoparticle was calculated through measuring the resonance frequency (ν), the diameter (D) and length (L). Like a pendulum, the mass of particle is given by

$$\nu = D^2/16\pi\ L(3E/Lm_{particle})^{1/2} \qquad (4.1)$$

By taking the nanobalance in Figure 4.1 as an example, where ν was 149 kHz, D and L of the nanowire were 48 nm and 5.0 μm, respectively, and E was 58 GPa, the mass of the nanoparticle could be calculated as 3.3×10^{-15} g. Per the results, a nanobalance built using ZnO nanowires was feasible for applications to measure the mass of attached nanoparticles. The single crystalline, structurally-controlled nanowires could be used as a new type of nanoresonator and nanocantilever, which was useful in nanoelectro-mechanical systems (NEMS) and highly functional nanodevices.

4.1.2   Individual Nanostructure-based Strain Sensor

The considerable potential of an individual ZnO piezoelectric fine-wire (PFW) as a flexible strain sensor has been demonstrated.[5] The strain sensor was fabricated by transferring a ZnO PFW to the polystyrene (PS) substrate and bonding two ends of ZnO utilising a silver paste that served as source and drain electrodes. Then, a thin layer of polydimethylsiloxane (PDMS) was introduced to package the device for protection. The *I–V* characteristic of the device was highly sensitive to strain mainly due to the change in the Schottky barrier height, which was due to the strain induced band structure change and piezoelectric effect. The current response of the sensor over many cycles of repeated compressing and stretching were also tested. The highly regular response curves indicated high reproducibility and good stability for this type of flexible sensor device. The gauge factor, which is an important performance value for the strain sensor, was 1250, which was higher than the best gauge factor achieved for a piezoresistive CNT strain sensor.

Furthermore, to detect the frequency of the vibration, a piezotronic strain sensor based on a single ZnO wire was constructed.[6] An ultra-long ZnO nanowire with a diameter of about 600 nm was fixed with the silver paste as source and drain electrodes. The *I–V* characteristic of the sensor was measured in the atmosphere when the ZnO wire was stretched by the tip of an atomic force microscope (AFM), as shown in Figure 4.2(a). The *I–V* curves shifted upward with increasing tensile strain and fully recovered when the strain was relieved, as shown in Figure 4.2(b), illustrating that the resistance

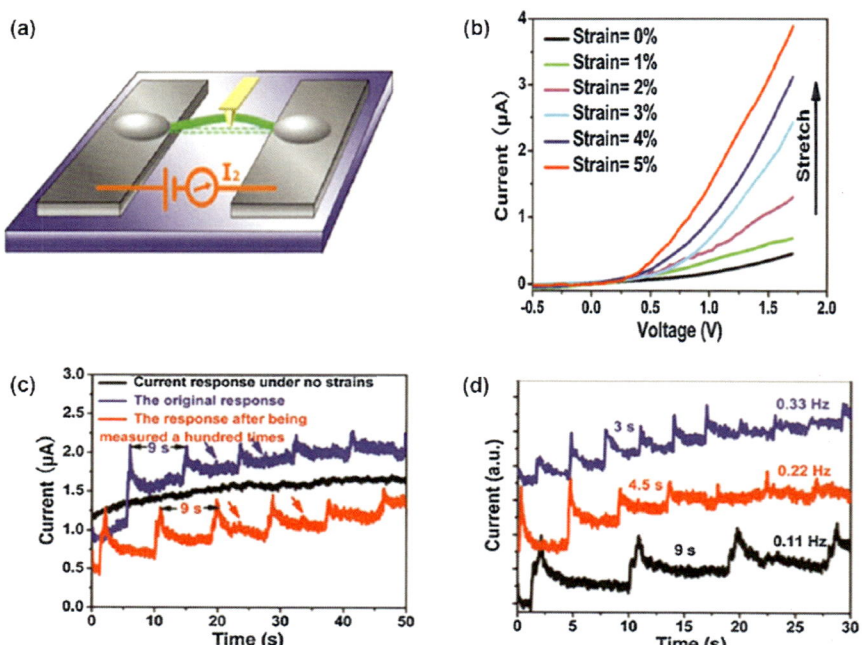

Figure 4.2 (a) Schematic diagram of the device under the tensile strains induced by an AFM tip; (b) typical *I–V* characteristics of the detector at different tensile strains; (c) current response of a sensor device that was subjected to periodic tensile strains under a fixed bias of 3 V. The dark line corresponds to the original value with no tensile strains and the maximum tensile strain in the red and blue lines is about 0.5%; (d) current response of the fabricated device when the ZnO nanowire was subjected to tensile strains with different frequencies.
Reproduced with permission from John Wiley and Sons Copyright © 2009 WILEY-VCH Verlag GmbH & Co. KGaA, Weinheim.[6]

of the wire decreased as the tensile strain increased. The results, calculated by first principles density functional theory, indicated that the bandgap decreased with increasing tensile stress in the ZnO wire. The bandgap change under tensile strains did not result in a change of Schottky barrier heights, but did change the intrinsic resistance of the single ZnO wire due to a piezoresistance effect. The device was also used to detect the movement frequencies of an AFM tip. The obtained frequencies of 0.11, 0.22 and 0.33 Hz from the current response curves were completely consistent with the real movement frequencies of the AFM tip as shown in Figure 4.2(c) and (d). The devices were found to have a high sensitivity (about 200%) and a fast response time to the tensile strains induced by the AFM tip. These results supported the applications of ZnO nanowires in nanoscale piezoelectric and piezoresistance strain sensors.

Recently, researchers have constructed a new type of performance-tunable force-sensor based on a cantilevered ZnO wire.[7] A specially chosen ZnO wire was transferred to a piece of clean glass with most of the wire extending out from the edge of the glass. A pad of silver paste was used to fix the end of the ZnO wire onto the glass tightly as well as the electrode. A schematic diagram of the device is given in Figure 4.3(a). The applied force was tuned by adjusting the deformation of the AFM cantilever through setting different deflection voltages of the cantilever as per Hooke's law. Figure 4.3(b) shows that the device had rectifying *I–V* behaviour, which resulted from the large dissimilar value between the work function of ZnO and the work function of Pt/Ir (5.5 eV). By introducing a force at different positions along the *c* axis of the ZnO wire, a device with tunable performance was realised.

Figure 4.3(c) illustrates the absolute and relative currents as the force was introduced at different places along the *c* axis. It was interesting to find that at a fixed force, the further the distance between the position of the applied force and the Ag electrode, the higher the detecting current. That was to say, if the current was too small to be detected, then by increasing the distance between the Ag electrode and the position of the detecting force, the current was enhanced enough to be detected. It was reasonable to believe that the detection limit or the sensitivity was improved by increasing the distance between the electrodes. This was further confirmed by the slope of the ln *I–F* relationship (shown in Figure 4.3(d)) which showed that the sensitivity increased step-by-step from 50.6 lnA/pN to 60.1 lnA/pN to 70.2 lnA/pN as the force was applied from point '1' to '2' to '3'. When compared with previous sensors, the benefit of this kind of force sensor lay in the simplicity of the fabrication process and the uniqueness of the tunable performance, which gave the possibility to monitor the spatial force in different ranges or when applied at any position.

The synthesis and transverse electromechanical properties of ZnO nanoleaves have also been reported.[8] As shown in Figure 4.4(a), the as-synthesised products consist of a large quantity of leaf-like nanostructures with typical lengths in the range of several hundreds of micrometres and the X-ray diffraction (XRD) pattern of the grown nanoleaves is presented in Figure 4.4(b).

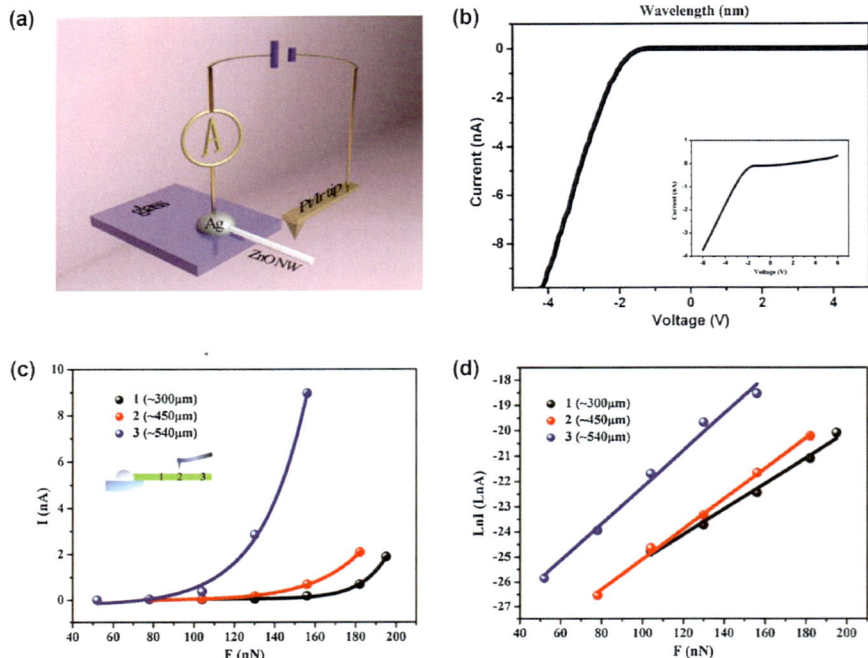

Figure 4.3 (a) A schematic diagram of the device; (b) the *I–V* characteristic of the M–S–M structure. The inset shows another *I–V* characteristic of the fabricated device; (c) the relative current responses; (d) the natural logarithm of the current and the change of the SBH.
Reproduced from ref. 7 with permission from The Royal Society of Chemistry.

The diffraction peaks have been indexed to the hexagonal wurtzite structure of ZnO ($a = 0.3246$ nm and $c = 0.52$ nm). The structure of the nanoleaves was further characterised by high-resolution transmission electron microscopy and energy dispersive X-ray spectroscopy. Therefore, we investigated the transverse electrical transport properties of single ZnO nanoleaf under external forces. When a positive bias was applied, the current flowed along the AFM tip, the nanoleaf, and the graphite substrate as shown in Figure 4.4(c). At the bottom surface of the nanoleaf, since the electron affinity of ZnO was 4.5 eV and the work function of graphite was 4.4 eV, the ZnO–graphite contact should be Ohmic. At the top surface of the nanoleaf, since the work function of the Pt/Ir tip was 5.5 eV, the Pt/Ir–ZnO contact was Schottky with a barrier height of about 1 eV. Under different loading forces, the resulting current density–voltage (*J–V*) data was collected (shown in Figure 4.4(d)). At a low loading force of about 20 nN, a clear rectifying behaviour was observed, which was consistent with the Schottky contact between the Pt/Ir tip and the ZnO nanobelt. However, at the same negative bias, the negative current density was found to increase significantly as the loading force increased. In this device, the increase of loading forces resulted in the tension strains

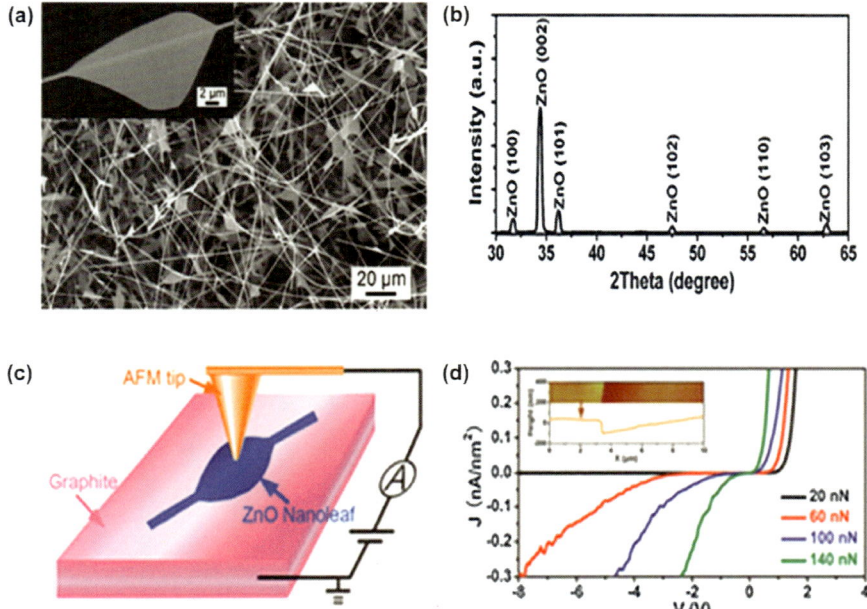

Figure 4.4 (a) SEM image of the synthesised sample. The inset image shows a high-resolution image of single nanoleaves; (b) XRD pattern of the synthesised nanoleaves; (c) schematic diagram of the transverse electromechanical characterisation measurements; (d) *J–V* curves of single ZnO nanoleaves lying on a graphite substrate under different loading forces. The inset is an AFM image of a part of the nanoleaf with its line profile.
Reproduced from ref. 8 with permission from The Royal Society of Chemistry.

along the [0001] direction, contributing to the decrease of the bandgap and the decrease of the Schottky barrier height. The slight decrease of the Schottky Barrier Height induced a large increase of the negative current density due to the exponential function relation between them. The synthesis of ZnO nanoleaves opened a door to the applications of these nanostructures as torsional nanowire/nanotube-based NEMS devices.

4.1.3 Individual Doped Nanostructure-based Strain Sensors

Some related studies reported the fabrication of the high-quality Sb-doped ZnO nanobelts by using a simple chemical vapour deposition method.[9] The synthesised products were thoroughly characterised by FESEM, HRTEM, XRD and EDS, as shown in Figure 4.5(a) and (b). The transverse electromechanical characterisation of a single Sb-doped ZnO nanobelt was examined using a standard C-AFM Pt/Ir tips at room temperature. The AFM tip was fixed on the top of the nanobelt on the graphite substrate. As shown in Figure 4.5(c), the *I–V* curve indicated good Ohmic contacts at both sides of

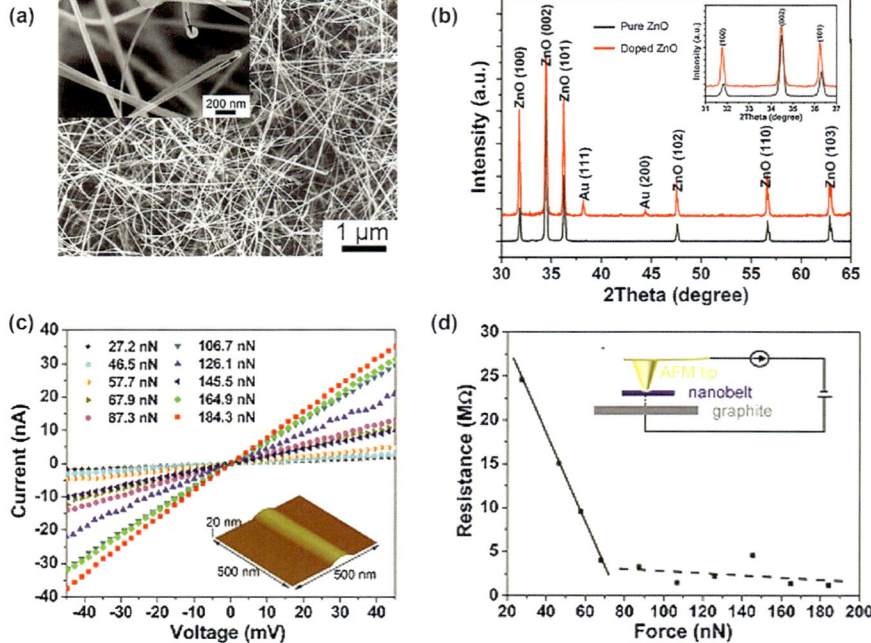

Figure 4.5   (a) SEM image of the Sb-doped ZnO nanobelts grown on the silicon substrate. The inset is an enlarged image of the nanobelts; (b) XRD pattern taken from the pure and Sb-doped ZnO nanobelts. The inset is the enlarged (100), (002), and (101) diffraction peaks; (c) I–V curve of the nanobelt under different stresses; (d) the relationship between the loading stresses and the resistance. The inset is the schematic diagram of the measurement structure.
Reproduced from ref. 9 with permission of AIP Publishing.

the nanobelt. The Ohmic contact between the ZnO and the AFM tip attributed to the high doping in the nanobelt. The schematic diagram of the electromechanical characterisation measurements is shown in Figure 4.5(d), and an almost linear relationship between the applied loading force and the resistance was found at small deformation regions, which demonstrated that the nanobelts have potential applications as force/pressure sensors for measuring the nano-Newton forces.

In addition, the longitudinal electromechanical property of a single Sb-doped ZnO nanobelt under different compressed strains was investigated.[10] A single Sb-doped ZnO nanobelt was placed on the PS substrate *via* a probe station under optical microscopy, and the two ends of the nanobelt were tightly fixed by the silver paste, as shown in Figure 4.6(a). Figure 4.6(c) shows the corresponding *I–V* curves of the Sb-doped ZnO nanobelt under different compressed strains. It was clearly seen that the current through the nanobelt increased with increasing compressed strains. In the experiments, the direction of the compressed strain was along the growth direction of the Sb-doped ZnO nanobelt, which has no piezoelectric effect as it was not

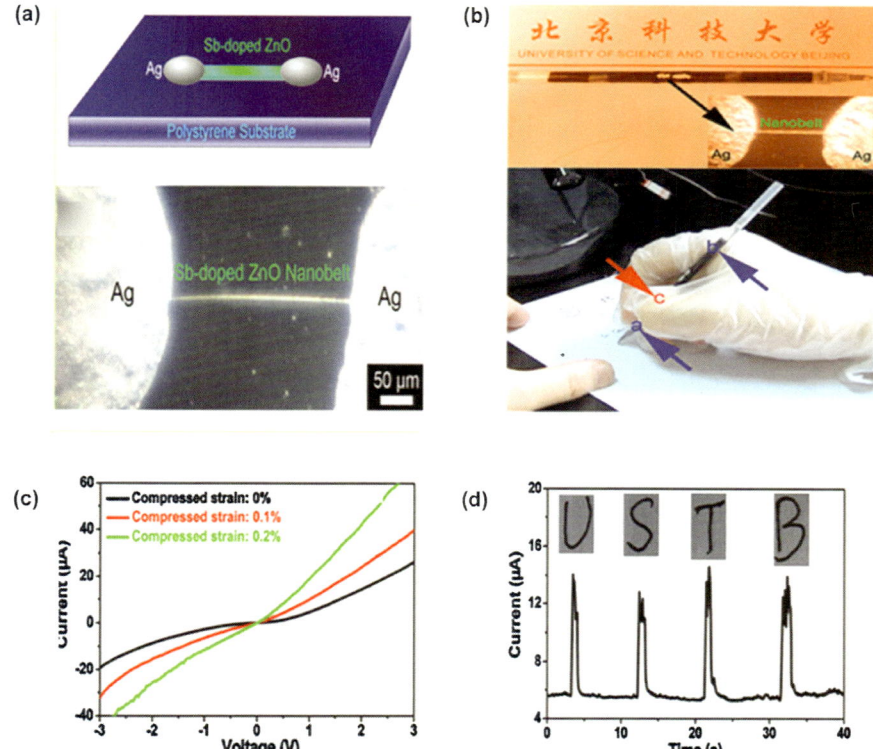

Figure 4.6 (a) Schematic of a single Sb-doped ZnO nanobelt fixed by silver paste on a PS substrate and optical image of a single Sb doped ZnO nanobelt fixed by silver paste on a PS substrate; (b) optical image of the fabricated flexible piezoresistive strain sensor in a signature pen. The inset shows a single Sb-doped ZnO nanobelt fixed by silver paste on the signature pen (up) and an optical image of the work process of the fabricated device (bottom); (c) typical *I–V* curves of the Sb-doped ZnO nanobelt under different compressed stains; (d) current response of the fabricated device when the four letters (U, S, T, and B) were written one by one. Reproduced from ref. 10 with permission of AIP Publishing.

the polar direction of ⟨0001⟩. The electrical transport of the nanobelt was modulated by the compressed strains due to the piezoresistance effect.

Furthermore, a flexible piezoresistive strain sensor in a signature pen based on a single Sb-doped ZnO nanobelt was successfully fabricated (Figure 4.6(b)). When different words were written, different compressed stains appeared in the nanobelt, which induced a change of the current through the Sb-doped ZnO nanobelt. Figure 4.6(d) shows the time resolved measurement of current response to the different compressed strains. It could be clearly seen that the current through the Sb-doped ZnO nanobelt increased with increasing compressed strains, which was consistent with the experimental results shown in Figure 4.6(c). Based on this finding, this flexible piezoresistive strain sensor in a signature pen could be used to detect the corresponding compressed strains when the characters were re-corded. The fabricated flexible piezoresistive strain sensor was used to detect the strains in a signature pen. The results supported the applications of Sb-doped ZnO nanobelts in electromechanical devices.

A flexible piezotronic strain sensor based on an indium-doped ZnO nanobelt has also been developed, of which the top surface was the mono-polar surface.[11] Unlike comparable nanowire devices, the source and drain electrodes for this sensor were connected to the same monopolar surface of a ZnO nanobelts as shown in Figure 4.7(a). Doped with indium for improved mechanical performance, the nanobelt was fixed on a flexible PS substrate and packaged with PDMS. The nanobelt dimensions were approximately 20 µm long by 180 nm wide. The identification of the polar direction was not required thanks to the monopolar surface configuration, simplifying device fabrication. The device was placed under strain by fixing one end and ap-plying bending. A bias voltage was swept from −3 V to +3 V and the current output was then measured to map the *I–V* curves for strains from −0.4% to 0.3% as seen in Figure 4.7(b). The change in Schottky barrier height for a

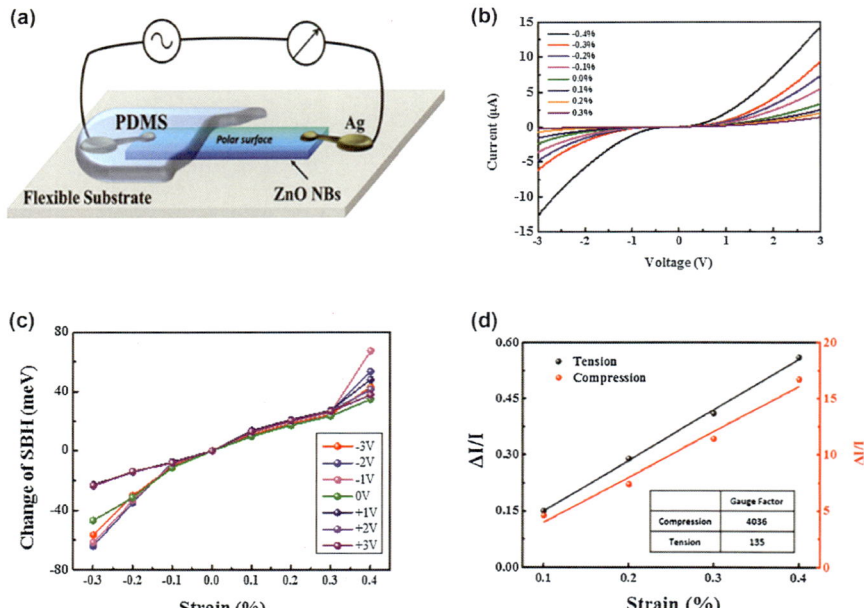

Figure 4.7   (a) The schematic diagram of the strain sensor; (b) the *I–V* performance of the strain sensor under different strains; (c) the relationship between the Schottky barrier height change and strain under different biases; (d) the relationship between the normalised current and strain. The red line is the linear fitting of the compressive strain corresponding to the right axis, and the black line is the linear fitting of the tensile strain corres-ponding to the left axis.
Reproduced from ref. 11 with permission from The Royal Society of Chemistry.

given strain is shown in Figure 4.7(c) as a function of bias. Normalised current *vs.* strain and gauge factor under various strain conditions are presented in Figure 4.7(d) for both tension and compression. A gauge factor of 4036 was reported for compressive strain, which was the highest reported gauge factor in the strain sensors that we reviewed. For tensile strain, a gauge factor of 135 was reported. The difference in gauge factors for tensile and compressive strain was hypothesised to be caused by screening from increased electron mobility due to the piezoresistance effect. A response time of 120 ms was measured for the device under periodic compressive and tensile strains of varying intensities. This type of piezotronic strain sensor with a polar surface facing upward presented a high performance and easier fabrication, showing promise for applications in electrical mechanical sensors and MEMS.

4.2 Nanowire Array-based Electromechanical Devices

When compared with single nanowires, nanowire arrays provide robust and stable devices due to the redundancy of multiple wires. A highly sensitive strain sensor with vertically aligned ZnO nanowire arrays on polyethylene terephthalate (PET) film was fabricated.[12] The setup was used to test the performance of the strain sensor. The typical *I–V* characteristics of the strain sensor and its response to the strain were measured. Due to the Poisson effect, the normal strain along the surface of the PET induced the strain along the axial direction of the nanowire that grew perpendicular to the surface, and caused significantly increased current when the nanowire was compressed, especially when a forward bias was applied. With a forward bias of 1.5 V, the gauge factor of the strain sensor was as high as 1813 at a normal strain $\varepsilon = 0.6\%$. The high sensitivity was ascribed to the high quality ZnO nanowire arrays and the piezotronic effect. In addition, stability and fast response of the device had been observed. The 'stress' state and 'release' state correspond respectively to the state when the device was mechanically bent and the state when the device returned to its straight and strain-free condition, which demonstrated the stability of the strain sensor. The instant response to 2.5 Hz excitation indicated a quick switch of the strain sensor. The sensitive and robust strain sensor is expected to find applications in civil, medical and other fields.

4.2.1 Nanowire Network-based Strain Sensors

A new type of flexible strain sensor based on a ZnO nanowires network structure with a polyimide (PI) substrate was constructed.[13] Figure 4.8(a) shows the characterisation of the morphology and structure of ZnO microwires and the testing system. In the network structure devices, every ZnO wire was considered as an electric resistance, which constituted a complex circuit. As shown in Figure 4.8(b), the current characteristics of the sensor were investigated by applying tensile strains. When a tensile strain was

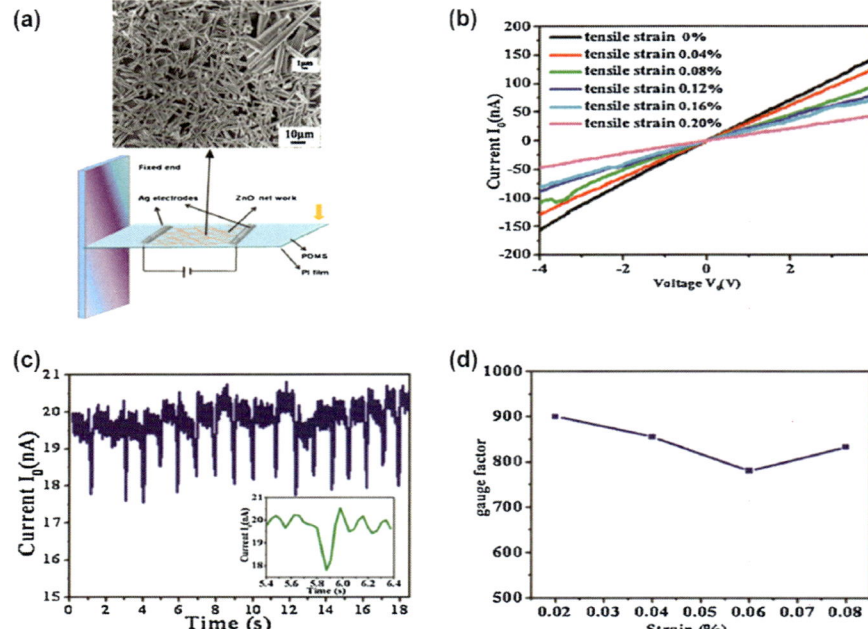

Figure 4.8   (a) The SEM image of ZnO microwire networks (up) and the testing system (bottom); (b) the sensor's electric properties under tensile strain; (c) current signal under tensile strain at frequencies of 1.0 Hz; inset shows the details of a typical current response of one vibration; (d) gauge factors of the strain sensor under compressed strain.
Copyright (2012) The Japan Society of Applied Physics.[13]

applied to the sensor, the current clearly decreased, which was ascribed to both piezoelectric potential and piezoresistance effect. In Figure 4.8(c), the sensor exhibited excellent response properties to the periodic strain with a frequency of 1.0 Hz. Figure 4.8(d) illustrates that the highest gauge factor of the sensor was up to 900 under compressive stress. The results provided a new design method for high-performance strain sensors based on piezoelectric materials.

4.2.2   Nanowire Array-based Vibration Sensors

By utilising the piezoelectric and piezoresistance properties of ZnO nanostructures, a ZnO nanowire array based vibration sensor has been fabricated by a hydrothermal process, which was employed to detect vibration in both self-powered (SP) and external-powered (EP) modes.[14] The cross-section view under a FESEM shows that the Pt electrode was fixed on top of the ZnO nanowire arrays, as shown in Figure 4.9(a). A schematic diagram of the test system including the FNG and a vibration output system is shown in Figure 4.9(b), where the frequency and magnitude of vibration can be controlled by a frequency generator. The FNG was affixed on an immobilised

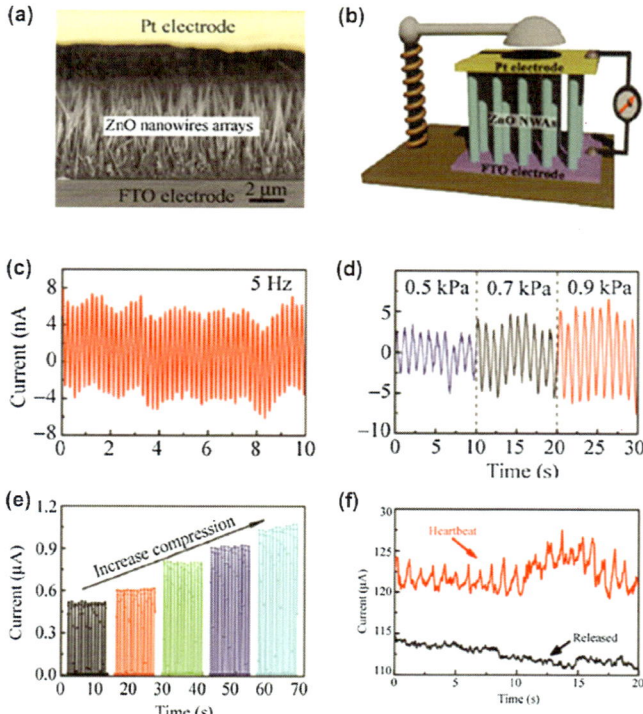

Figure 4.9 (a) The cross-section view of the device under a field-emission scanning electron microscope (FESEM); (b) a schematic diagram of the test system; (c) the current responses at 5 Hz respectively in 10 s under compressive strain of 0.9 kPa; (d) the current responses under increased strain, where the relative changes of current responses are about 5.35, 8.14, and 10.98 nA in average at each compressive strain; (e) the current response under increasing amplitude of the vibration cantilever, when the sensor was applied with a fixed voltage of 3 V and a frequency of 1 Hz; (f) the constructed sensor used as a sphygmic sensor in EP mode.

Reproduced from Nano Res., Functional nanogenerators as vibration sensors enhanced by piezotronic effects, 7, 2014, 190–198, Z. Zhang, Q. Liao, X. Yan, Z. L. Wang, W. Wang, X. Sun, P. Lin, Y. Huang and Y. Zhang, (© Tsinghua University Press and Springer-Verlag Berlin Heidelberg 2014) With permission from Springer.[14]

table. As shown in Figure 4.9(c), the FNG was used as a vibration sensor in SP mode and provided with regular vibration frequencies 5 Hz during 10 s. Then the sensor was subjected to different compressive strains of 0.5, 0.7 and 0.9 kPa. In Figure 4.9(d), the corresponding average relative changes of current response (ARCCs) were 5.35, 8.14 and 10.98 nA. In EP mode, the current responses of FNG were significantly enhanced *via* the piezotronic effect. When the sensor was subjected to different strains at 3 V and 1 Hz, the ARCCs increased almost linearly with the strain (Figure 4.9(e)). Figure 4.9(f) shows that the FNG was used as a pulse sensor working in EP mode. From the information in the curves, the dynamics of the heartbeat and the person's health condition were analysed. The results showed that this type of functional strain sensor could detect vibration in both SP and EP modes as per the demands of the applications.

4.2.3 Nanowire Array-based Pressure Sensor

A piezotronic pressure sensor based on ZnO nanowire arrays was designed with an on-off ratio of 10^5 by introducing an ultra-thin magnesium oxide (MgO) insulating/barrier layer.[15] Figure 4.10(a) illustrates a model of the device with i-MgO. In Figure 4.10(b), the current values of the sensors fluctuated from 10^{-11} to 10^{-8} A scales between 0 to +5 V under 0 gf (1 gf = 0.0098 N). The pressure sensor with 10 nm i-MgO demonstrated excellent performance and featured a small 'off' current, and an ultrahigh on-off ratio of 7.2×10^5. It is worth noting that the MgO nanolayer played a crucial role in modulating carrier transport and improving sensitivity of the sensors. The electromechanical properties of the sensor were measured by monitoring the current under different force level. The current value was

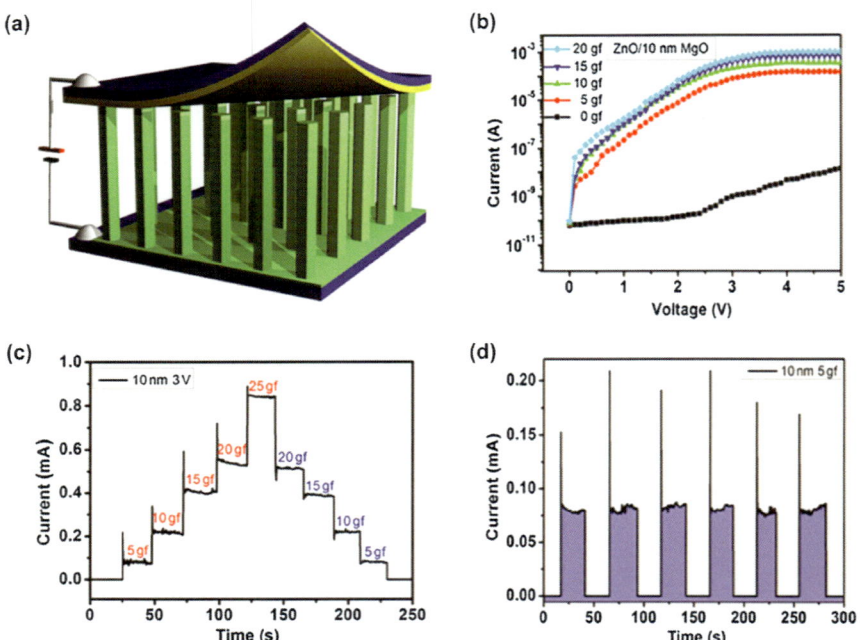

Figure 4.10   (a) The model of a pressure sensor; (b) *I–V* characteristics of the sensors with 10 nm MgO layer; (c) two characteristic relationships between current and applied force. One is the current increasing with applied force, which adds from 0 to 25 gf at 5 gf each step, and the other is the current with applied force decreased; (d) current response under circular loading at 5 gf.
Reproduced with permission from ref. 15. Copyright (2015) American Chemistry Society.

changed by controlling the force to modulate the potential barrier height. As it was applied with a force, the current was fixed at a certain value (Figure 4.10(c)). Furthermore, the sensor responded rapidly at every turning point within 128 ms, and the current remained identical under the same force cycles, as shown in Figure 4.10(d). The innovation of piezoelectric effect combined with electron tunnelling modulation provided an alternative route to improve the performance of other nanosensors.

4.3 Hybrid Structure-based Electromechanical Devices

The hybrid structure has a combination of different materials to meet the demand of flexibility and adaptability in next generation electronics. The presence of the paper matrix makes the composite material very flexible. A nanocomposite material consisting of ZnO nanostructures embedded in a paper matrix was synthesised by a low-temperature solvothermal method.[16] Strain sensing was demonstrated under both static and dynamic loading using this nanocomposite material and good strain sensitivity has been found for both static and dynamic loading with very low power input.

A cellulose paper after coating with ZnO based strain sensor was demonstrated. The initial paper was highly porous and composed of micrometer-sized fibres that were randomly oriented. The fibres were quite smooth before coating with ZnO but became very rough after deposition of ZnO. The high-resolution TEM image of a single ZnO nanorods grown on the cellulose paper indicated the uniform crystalline nature of the rods, with the fast Fourier transform (FFT) depicting the (002) plane of the ZnO crystal. These rods grew perpendicular to the cellulose fibres. The current response from the nanocomposite sensor and the strain reading from the commercial strain gauge for a stepped tensile loading were detected. The loading was carried out at a very low rate, and hence no dynamic effects of the brass beam were observed in the response. The behaviour of the nanocomposite sensor suffering from constant load was tested by applying a tensile load to the specimen in increments. It was evident from the electrical response that the nanocomposite sensor demonstrates a stable response both when the load was increasing and when it remained constant at a given level. The electrical response from the nanocomposite sensor was found to be in keeping with the strain measured by a foil strain gauge for static loading. The response of the sensor was also tested for dynamic loading without the application of any input voltage. Under the frequency of 1 Hz, the excitation was applied in repeated sets of cycles of sinusoidal excitation. And the frequency of the response fitted well with the frequency of excitation; also, no drift was observed in the response at the end of each cycle of excitation.

Meanwhile, a novel strain sensor based on a ZnO nanowire/polystyrene nanofiber hybrid structure has been demonstrated.[17] The device withstood strain up to 50%, with high durability, fast response, and high gauge factors.

With flexibility and stretchability, the device showed potential applications in nanosensing systems for precision measurements and e-skins.

Inspired by the hierarchical and interlocked structures of biological systems, highly-sensitive and rapidly-responsive e-skins based on the interlocked geometry of hierarchical micro- and nanostructures of polydimethylsiloxane (PDMS) micropillars decorated with ZnO nanowires array were developed.[18] In particular, this demonstrated that these e-skins can be utilised in the detection of both static and dynamic tactile signals due to the piezoresistive and piezoelectric properties, respectively, of ZnO nanowires.

An interlocked geometry of hierarchical micro- and nanostructures for highly sensitive piezoresistive e-skins has been fabricated. The relative resistance of e-skins decreased rapidly with the increase in normal pressure at low-pressure regimes (below 2 kPa), then decreased slowly at a high-pressure regime (over 2 kPa). For different pitch sizes of micropillar arrays, the largest decrease in resistance was observed in the sensor with the smallest pitch size (20 µm), which was about 3.7 times more pressure-sensitive than a planar nanowire array at pressures below 0.3 kPa. The temperature stability of hierarchical and planar structured ZnO NWs has been investigated in a temperature range of 30–90 °C. For planar structures, the thermal expansion of PDMS with temperature variation resulted in crack formation and a subsequent fluctuation in pressure sensitivity, which depended on the temperature. However, for the same temperature range, the increased thermal stability of hierarchical ZnO nanowires array with minimal crack formations resulted in a stable pressure sensitivity (error rate <0.05%) compared with the planar one (error rate <8.5%). A pushing tester was utilised to provide a mechanical vibration with variable frequency on the e-skins. While the piezoresistive e-skins could not distinguish the vibration frequency at 3.6 Hz, the output current signals of the piezoelectric e-skins increased with the increase in frequency from 0.5 to 3.6 Hz and clearly detected the frequency at 3.6.

Furthermore, a highly sensitive pressure sensor was constructed using a PVDF thin film, with vertically grown ZnO nanorods.[19] A schematic diagram of the flexible multilayer device is shown in Figure 4.11(a). Large area ZnO nanorods were embedded in the PVDF on a reduced graphene oxide (rGO)-treated flexible polyethylene terephthalate (PET) thin film. The performance was measured from the change in the electrical resistance; *i.e.* $\Delta R = R_{Loading} - R_{Unloading}$. Figure 4.11(b) shows how the composite device formed of PVDF, ZnO nanorods and ZnO nanodisks responded to the application of constant pressure. Because of the 1D structure of the vertically grown ZnO nanorods, they generated an enhanced piezoelectric response to mechanical displacement compared to the ZnO nanodisks. The pressure-sensing capability of the system was evaluated from the response ΔR, as shown in Figure 4.11(c). Typically, the smallest detectable pressure using the ZnO nanorods/PVDF film was 10 Pa, which corresponded to 1 mg mm^{-2} and was more sensitive than the minimum requirements for artificial skin. In addition, the response times of the output signals from ZnO nanorods/PVDF

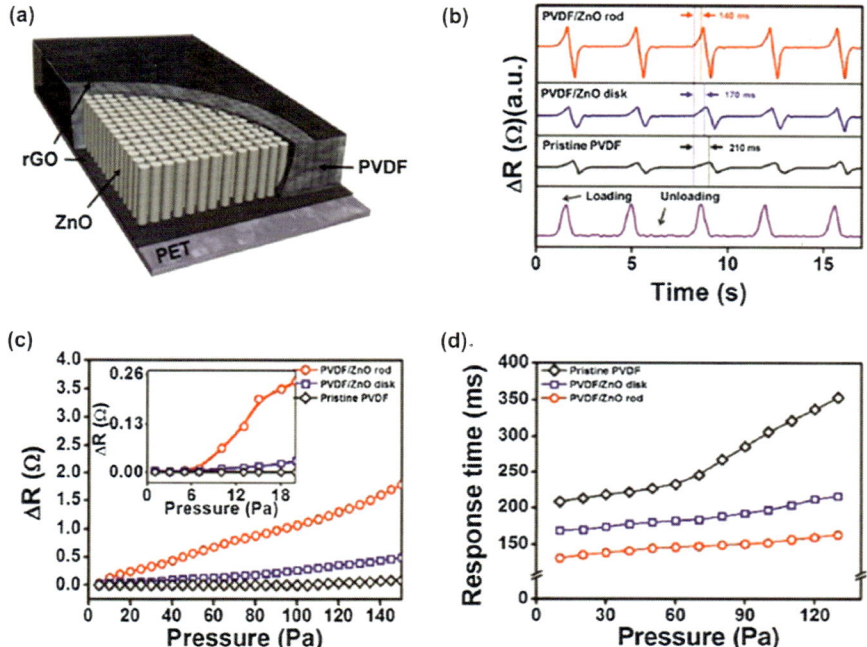

Figure 4.11 (a) Schematic diagram of the device consisting of the ZnO/PVDF composite film and rGO electrodes; (b) in response to an applied pressure of 30 Pa, showing the response time of the PVDF film, the ZnO nanodisks/PVDF film, and the ZnO nanorods/PVDF; (c) the sensitivity of the films at 20 °C, using various Pt weights; (d) the response times with various pressure.
Reproduced by permission from Macmillan Publishers Ltd: *Sci. Rep.*, (ref. 19) copyright (2015).

device were obtained under various pressures at 20 °C, as shown in Figure 4.11(d). The results demonstrated that the device had the potential for artificial skin.

Based on previous research, a highly sensitive, wearable and wireless pressure sensor was successfully fabricated based on a ZnO nanoneedle/poly(vinylidene difluoride) (PVDF) hybrid film.[20] Due to its high permittivity, low polarisation response time and outstanding durability, the hybrid film was applied for a real-time pressure sensor to monitor heart rate. Notably, the lowest detectable pressure of the hybrid film was as small as 4 Pa.

In other recent examples, a highly robust textile-based wireless flexible strain sensor was developed by integrating functional hybrid carbon nanomaterials and ZnO nanowires with piezoresistive effect into the textile substrate.[21] A rapid response to the diverse motions of the human body with accurate detections of the quantity of the applied bending strain was observed. The performance of the sensor was extended when connected to a wireless transmitter, which demonstrated the viability for wireless communication and remote monitoring. In addition, a piezotronic pressure sensor based on ZnO/NiO core/shell nanorods array was constructed by a simple method.[22] The performance of the pressure sensors was significantly enhanced by coupling of the piezoelectric effect and photo-excitation of ZnO nanorods. The enhancement in switch ratio and sensitivity of the pressure sensor was about 353% and 445% under UV illumination. These results indicated the potential for the piezophototronic effect to effectively optimise the performances of pressure sensors. Our developments were carried out based on the following designs.

4.3.1 Carbon Fibre/ZnO Nanowire Array-based Flexible Strain Sensors

As shown in Figure 4.12(a), the flexible piezotronic strain sensors was fabricated by carbon fibre-ZnO NW hybrid structures by arranging the carbon fibre-ZnO NW hybrid structures in an orderly manner on a flexible

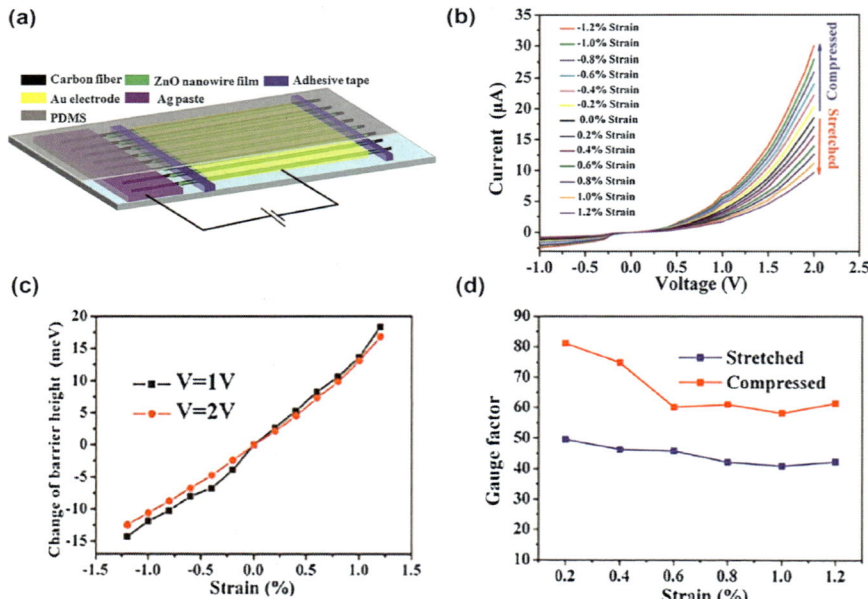

Figure 4.12 (a) The schematic diagram depicting the structure of the hybrid strain sensor; (b) typical *I–V* characteristics of the fibre-based strain sensor as a function of static loading and the decreased current indicates an increase in SBH; (c) the calculated change in SBH based on the thermionic emission-diffusion model, as a function of stain at bias of *V* 1/4 1.0 and 2.0 V, respectively; (d) gauge factors derived from the *I–V* curves as a function of strain.
Reproduced from ref. 23 with permission from The Royal Society of Chemistry.

substrate.[23] The sensor was based on a textured ZnO NW film grown on the surface of a carbon fibre. The top and bottom of the ZnO NW film were connected electrically. Figure 4.12(b) presents the *I–V* curves of the fibre-based strain sensor investigated at different static strains. The *I–V* characteristics of the device were sensitive to strain because of the change in Schottky barrier height. Apparently, the non-linear *I–V* curves were caused by the Schottky barriers formed at the semiconductor-metal contact interface. The quite asymmetric shape of the *I–V* curve demonstrated that the ZnO NWs had different Schottky barrier height contact at their top and bottom ends. In addition, the *I–V* curves shift upward with compressive strain and downward with tensile strain. The *I–V* curve fully recovered without strain. The change of Schottky barrier height $\Delta\Phi_s$ with strain for bias of 1.0 V and 2.0 V is shown in Figure 4.12(c); the relationship between $\Delta\Phi_s$ and strain approached linear. As to the practical application, gauge factor was vital to the performance of strain sensors. From the measured *I–V* curves, the gauge factors of the strain sensor at different strains are shown in Figure 4.12(d). The highest gauge factor of the sensor device was 81, which was higher than the values of the reported ZnO-paper nanocomposite strain sensor and commercial metal strain sensors. Moreover, the sensor had a higher gauge factor under compressive strain than that under tensile strain, because the compressive strain can produce more effective strain than the same tensile strain.

Figure 4.13 shows the current response of the sensor over several cycles in compression at a frequency of 1.0 Hz under a fixed bias of 1.0 V and dynamic cyclic loading conditions. It demonstrated that the strain sensor had high reproducibility and good stability. The strain sensor also showed a quick current response to the dynamic strain just like its properties for the static strain.

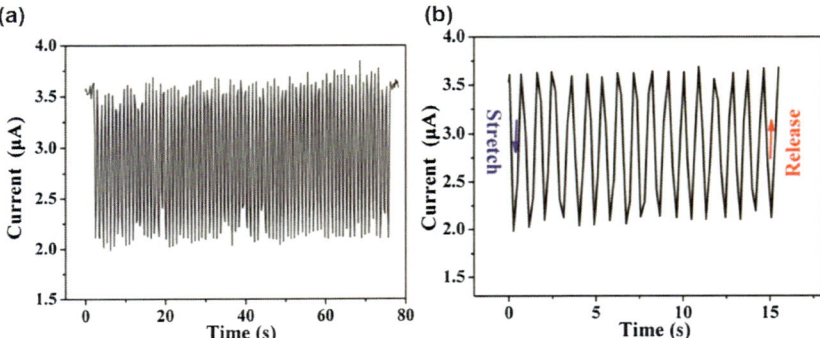

Figure 4.13 (a) Dynamic current response of the hybrid strain sensor that was cyclically tensile at a frequency of 1.0 Hz under a fixed bias of 1.0 V; (b) enlarged figure of (a).

Reproduced from ref. 23 with permission from The Royal Society of Chemistry.

4.3.2   PU Fibre/ZnO Nanowire Array-based Multifunctional Strain Sensor

A relatively low-cost, stretchable, and multifunctional sensor based on ZnO nanowires with PU fibres was fabricated.[24] Typically, the stretchable and multifunctional sensor could be stretched up to 150% and maintained the ability to detect strain. Figure 4.14(a) schematically illustrates the fabrication processes of the multifunctional sensor. As a stretchable strain sensor in Figure 4.14(b), the electrical signals responded fast and reset rapidly (38 ms). Figure 4.14(c) showed that the stability can be maintained over more than 10 000 cyclic loading tests. Regarding the frequency response in Figure 4.14(d), the input load signals were rapidly detected by the stretchable strain sensor without evident electrical response amplitude distortion. The performances of the stretchable strain sensor were favourably comparable to the properties of the recent strain sensing devices.

The multifunctional sensor was also used as a stretchable temperature sensor. The current of the stretchable temperature sensor increased with temperature under 0%, 25%, 50%, and 100% strain, as shown in Figure 4.15(a). A stretchable UV sensor is another application of the multifunctional sensor. Typically, the sensing mechanism was due to the processes of oxygen adsorption and desorption, as shown in Figure 4.15(b).

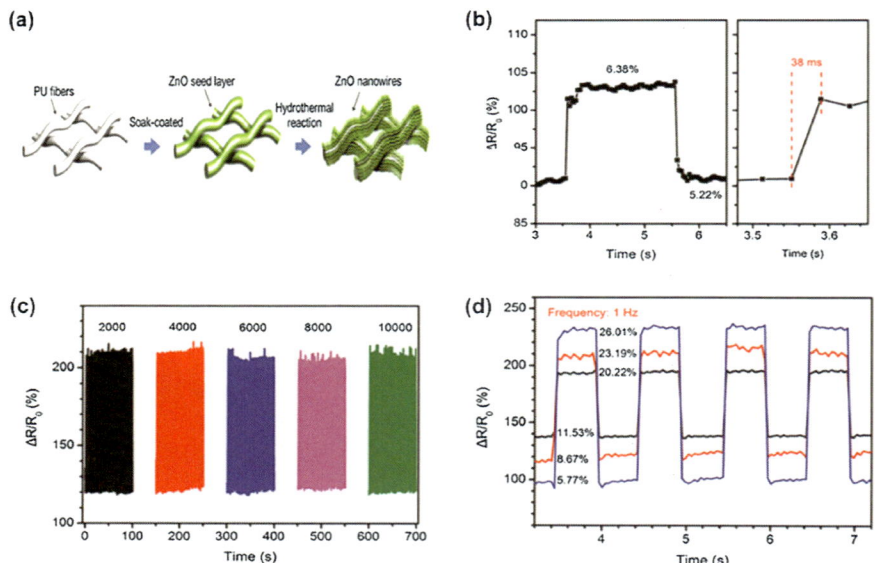

Figure 4.14   (a) The fabrication steps of the multifunctional sensor; (b) response time measurement of the stretchable strain sensor applied strain from 5.22% to 6.38% (left) and a magnification illustrating the response time (right); (c) multiple-cycle strain tests at a frequency of 3 Hz; (d) various strain response at 1 Hz under different static strain loading.
Reproduced with permission from John Wiley and Sons Copyright © 2016 WILEY-VCH Verlag GmbH & Co. KGaA, Weinheim.[24]

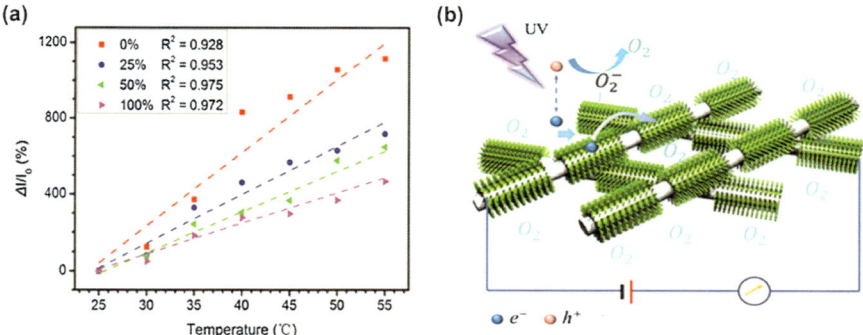

Figure 4.15 (a) Normalised current changes of the stretchable temperature sensor *vs.* temperature under 0%, 25%, 50%, and 100% strain; (b) schematic diagram of sensing mechanism of the stretchable UV sensor. Reproduced with permission from John Wiley and Sons Copyright © 2016 WILEY-VCH Verlag GmbH & Co. KGaA, Weinheim.[24]

Notably, due to the soft characteristic, portability, and fibre layout, the stretchable and multifunctional sensor was applied in diverse smart sensing applications and multi-parametric sensing platforms in the future.

We reviewed the fast-developing field of the electromechanical devices based on ZnO. Various sensors have been demonstrated with very enhanced performances. The development of ZnO nanomaterials has a promising future for creating multifunctional sensors. However, practical applications of this new technology ask for more research efforts to address the challenges that remain. First, the cost of nanomaterial growth and device fabrication needs to be reduced. Second, the stability and the performance of the devices should be improved. Finally, the integration of multifunctional sensors with traditional electronics will be critical for the realisation of commercial applications.

Furthermore, some new phenomena about ZnO under strain had been discovered. For example, the effect of uniaxial tensile strain on individual ZnO nanowires was systematically investigated by Raman spectroscopy.[25] It was found for the first time that the tensile and compressive strains result in a linear downshift and upshift of the phonon frequencies. In addition, point defect-induced giant anelasticity in single-crystalline ZnO nanowire was discovered under bending.[26] The ZnO nanowire can exhibit an elastic behaviour that was up to four orders of magnitude larger than the largest anelasticity observed in bulk materials, with a recovery time scale in the order of minutes. These results may offer unprecedented opportunities for advanced stress sensors based on ZnO.

References

1. T. W. Tombler, C. Zhou, L. Alexseyev, J. Kong, H. Dai, L. Liu, C. Jayanthi, M. Tang and S. Y. Wu, Reversible electromechanical characteristics of

carbon nanotubes under local-probe manipulation, *Nature*, 2000, **405**(6788), 769–772.

2. R. He and P. Yang, Giant piezoresistance effect in silicon nanowires, *Nat. Nanotechnol.*, 2006, **1**(1), 42–46.

3. X. Wang, J. Zhou, J. Song, J. Liu, N. Xu and Z. L. Wang, Piezoelectric field effect transistor and nanoforce sensor based on a single ZnO nanowire, *Nano Lett.*, 2006, **6**(12), 2768–2772.

4. Y. Huang, X. Bai and Y. Zhang, In situ mechanical properties of individual ZnO nanowires and the mass measurement of nanoparticles, *J. Phys.: Condens. Matter*, 2006, **18**(15), L179.

5. J. Zhou, Y. Gu, P. Fei, W. Mai, Y. Gao, R. Yang, G. Bao and Z. L. Wang, Flexible piezotronic strain sensor, *Nano Lett.*, 2008, **8**(9), 3035–3040.

6. Y. Yang, J. Qi, Y. Gu, X. Wang and Y. Zhang, Piezotronic strain sensor based on single bridged ZnO wires, *Phys. Status Solidi RRL*, 2009, **3**(7–8), 269–271.

7. S. Lu, J. Qi, Z. Wang, P. Lin, S. Liu and Y. Zhang, Size effect in a cantilevered ZnO micro/nanowire and its potential as a performance tunable force sensor, *RSC Adv.*, 2013, **3**(42), 19375–19379.

8. Y. Yang, Q. Liao, J. Qi, W. Guo and Y. Zhang, Synthesis and transverse electromechanical characterization of single crystalline ZnO nanoleaves, *Phys. Chem. Chem. Phys.*, 2010, **12**(3), 552–555.

9. Y. Yang, J. Qi, Y. Zhang, Q. Liao, L. Tang and Z. Qin, Controllable fabrication and electromechanical characterization of single crystalline sb-doped ZnO nanobelts, *Appl. Phys. Lett.*, 2008, **92**(18), 183117.

10. Y. Yang, W. Guo, J. Qi and Y. Zhang, Flexible piezoresistive strain sensor based on single sb-doped ZnO nanobelts, *Appl. Phys. Lett.*, 2010, **97**(22), 223107.

11. Z. Zhang, Q. Liao, X. Zhang, G. Zhang, P. Li, S. Lu, S. Liu and Y. Zhang, Highly efficient piezotronic strain sensors with symmetrical schottky contacts on the monopolar surface of ZnO nanobelts, *Nanoscale*, 2015, **7**(5), 1796–1801.

12. W. Zhang, R. Zhu, V. Nguyen and R. Yang, Highly sensitive and flexible strain sensors based on vertical zinc oxide nanowire arrays, *Sens. Actuators, A*, 2014, **205**, 164–169.

13. P. Li, Q. Liao, Z. Zhang, Y. Zhang, Y. Huang and S. Ma, Flexible micro-strain sensors based on piezoelectric ZnO microwire network structure, *Appl. Phys. Express*, 2012, **5**(6), 061101.

14. Z. Zhang, Q. Liao, X. Yan, Z. L. Wang, W. Wang, X. Sun, P. Lin, Y. Huang and Y. Zhang, Functional nanogenerators as vibration sensors enhanced by piezotronic effects, *Nano Res.*, 2014, **7**(2), 190–198.

15. X. Liao, X. Yan, P. Lin, S. Lu, Y. Tian and Y. Zhang, Enhanced performance of ZnO piezotronic pressure sensor through electron-tunneling modulation of mgo nanolayer, *ACS Appl. Mater. Interfaces*, 2015, **7**(3), 1602–1607.

16. H. Gullapalli, V. S. Vemuru, A. Kumar, A. Botello-Mendez, R. Vajtai, M. Terrones, S. Nagarajaiah and P. M. Ajayan, Flexible piezoelectric ZnO-paper nanocomposite strain sensor, *Small*, 2010, **6**(15), 1641–1646.

17. X. Xiao, L. Yuan, J. Zhong, T. Ding, Y. Liu, Z. Cai, Y. Rong, H. Han, J. Zhou and Z. L. Wang, High-strain sensors based on ZnO nanowire/polystyrene hybridized flexible films, *Adv. Mater.*, 2011, **23**(45), 5440–5444.

18. M. Ha, S. Lim, J. Park, D. S. Um, Y. Lee and H. Ko, Bioinspired interlocked and hierarchical design of ZnO nanowire arrays for static and dynamic pressure-sensitive electronic skins, *Adv. Funct. Mater.*, 2015, **25**(19), 2841–2849.

19. J. S. Lee, K. Y. Shin, O. J. Cheong, J. H. Kim and J. Jang, Highly sensitive and multifunctional tactile sensor using free-standing ZnO/PVDF thin film with graphene electrodes for pressure and temperature monitoring, *Sci. Rep.*, 2015, **5**, 7887.

20. K. Y. Shin, J. S. Lee and J. Jang, Highly sensitive, wearable and wireless pressure sensor using freestanding ZnO nanoneedle/PVDF hybrid thin film for heart rate monitoring, *Nano Energy*, 2016, **22**, 95–104.

21. T. Lee, W. Lee, S. W. Kim, J. J. Kim and B. S. Kim, Flexible textile strain wireless sensor functionalized with hybrid carbon nanomaterials supported ZnO nanowires with controlled aspect ratio, *Adv. Funct. Mater.*, 2016, **26**(34), 6206–6214.

22. B. Yin, H. Zhang, Y. Qiu, Y. Chang, J. Lei, D. Yang, Y. Luo, Y. Zhao and L. Hu, Piezo-phototronic effect enhanced pressure sensor based on ZnO/NiO core/shell nanorods array, *Nano Energy*, 2016, **21**, 106–114.

23. Q. Liao, M. Mohr, X. Zhang, Z. Zhang, Y. Zhang and H. J. Fecht, Carbon fiber-ZnO nanowire hybrid structures for flexible and adaptable strain sensors, *Nanoscale*, 2013, **5**(24), 12350–12355.

24. X. Liao, Q. Liao, Z. Zhang, X. Yan, Q. Liang, Q. Wang, M. Li and Y. Zhang, A highly stretchable ZnO@ fiber-based multifunctional nanosensor for strain/temperature/UV detection, *Adv. Funct. Mater.*, 2016, **26**(18), 3074–3081.

25. X. W. Fu, Z. M. Liao, R. Liu, J. Xu and D. Yu, Size-dependent correlations between strain and phonon frequency in individual ZnO nanowires, *ACS Nano*, 2013, **7**(10), 8891–8898.

26. G. Cheng, C. Miao, Q. Qin, J. Li, F. Xu, H. Haftbaradaran, E. C. Dickey, H. Gao and Y. Zhu, Large anelasticity and associated energy dissipation in single-crystalline nanowires, *Nat. Nanotechnol.*, 2015, **10**(8), 687–691.

CHAPTER 5

Photoelectrical Devices

ZHIMING BAI, YANWEI SHEN, PEI LIN, GUANGJIE ZHANG
AND YUE ZHANG*

University of Science and Technology Beijing, Beijing, China
*Email: yuezhang@ustb.edu.cn

5.1 Light-emitting Diodes

Electroluminescence is a phenomenon in which light is generated by a current flowing through a material under a bias. In 1907, H. J. Round[1] reported the yellow electroluminescence phenomenon in SiC crystals, which is the earliest report of solid-state light-emitting diodes. As an important component of semiconductor optoelectronic devices, light-emitting diodes play an important role in society. The semiconducting light-emitting diode is the third-generation light source after the incandescent lamp and fluorescent tube light source, with advantages of high luminous efficiency, good monochromaticity, less heat generation, and small size, which have greatly improved and changed living standards. In recent years, the blue-violet light-emitting diode has been a hot research topic internationally, due to its urgent application requirements in scientific research, military, optical storage and optical communication. Zinc oxide is a promising candidate for short-wavelength ultraviolet light-emitting diodes (LEDs) and lasing diodes (LDs), since it has both a direct wide bandgap of 3.37 eV and a very large exciton binding energy of 60 meV. In this chapter, we review recent research progress in ZnO-based light-emitting devices in view of different device structures.

Nanoscience & Nanotechnology Series No. 43
ZnO Nanostructures: Fabrication and Applications
By Yue Zhang
© Yue Zhang 2017
Published by the Royal Society of Chemistry, www.rsc.org

5.1.1 Homojunction LED

p–n Homojunction LEDs are the most efficient light emitters. However, ZnO materials usually exhibit *n*-type conductivity due to the presence of intrinsic defects and the self-compensation effect of impurities, which makes it a big challenge to fabricate *p*-type ZnO thin films with high repeatability, high efficiency and low resistance. Until now, research on the ZnO electrically pumped light-emitting devices has mostly been devoted to the preparation of *p*-ZnO materials for realising ZnO-based *p–n* homojunction electroluminescent devices. In 2000, an excimer-laser doping technique was first utilised to fabricate a *p*-type phosphorous-doped ZnO, which was then used to form ZnO *p–n* homojunction LEDs.[2] The experimental results showed that the device can emit white-violet light at 110 K, has a low luminous efficiency and a strong noise signal in the spectrum. Nevertheless, the near-band ultraviolet radiation from ZnO is still visible. In 2004, a repeated temperature modulation epitaxy method was proposed as a reproducible way to grow a *p*-type nitrogen-doped ZnO using laser MBE.[3] A typical ZnO *p–i–n* homojunction dioded was fabricated on lattice-matched ScAlMgO$_4$ substrates through a layer-by-layer growth mode. Figure 5.1(a) and (b) present the schematic of the device and its *I–V* curve, respectively. The electroluminescence spectrum (shown in Figure 5.1(c)) covers the wavelength range from violet to green and shows an obvious redshift compared with the exciton emission at 3.2 eV from the undoped ZnO, which may be due to the low hole concentration in *p*-type ZnO. In 2006, a high-temperature radiofrequency sputtering technique was used to grow ZnO *p–n* homojunction LEDs composed of 0.4 μm phosphorus-doped *p*-type ZnO and 1.5 μm gallium-doped *n*-type ZnO.[4] The rapid thermal annealling process in a nitrogen atmosphere was considered to be important to activate the *p*-type ZnO layers. Excellent diode-rectifying behaviour with a threshold of 3.2 V and a violet band-edge emission at 380 nm was achieved at room temperature. Later, researchers from Hong Kong found that by simply changing the seed layer preparation, a ZnO nanorod array with a different conductivity type could be grown using a hydrothermal method, which may result from the different concentration of zinc vacancies and different incorporation of compensation donor defects. Furthermore, *p–n* homojunction LEDs were fabricated using as-grown *p*-ZnO nanorod arrays and a *n*-ZnO single crystal substrate. A blue and broad red-near infrared emission appears in the electroluminescence spectrum when the device is under forward bias. Lithium–nitrogen as a dual-acceptor dopant was also demonstrated as an effective and reproducible doping source to obtain *p*-ZnO, which could be used to fabricate ZnO *p–i–n* homojunction LEDs.[5] The 12 K eltroluminescence spectrum shows a dominant blue emission at around 425 nm when the diode is under the injection current of 23 mA at a forward bias of 23 V. Two years later, researchers from the same group demonstrated that Ag nanoparticle surface plasmon can significantly enhance the emission of the obtained ZnO homojunction LEDs.[6]

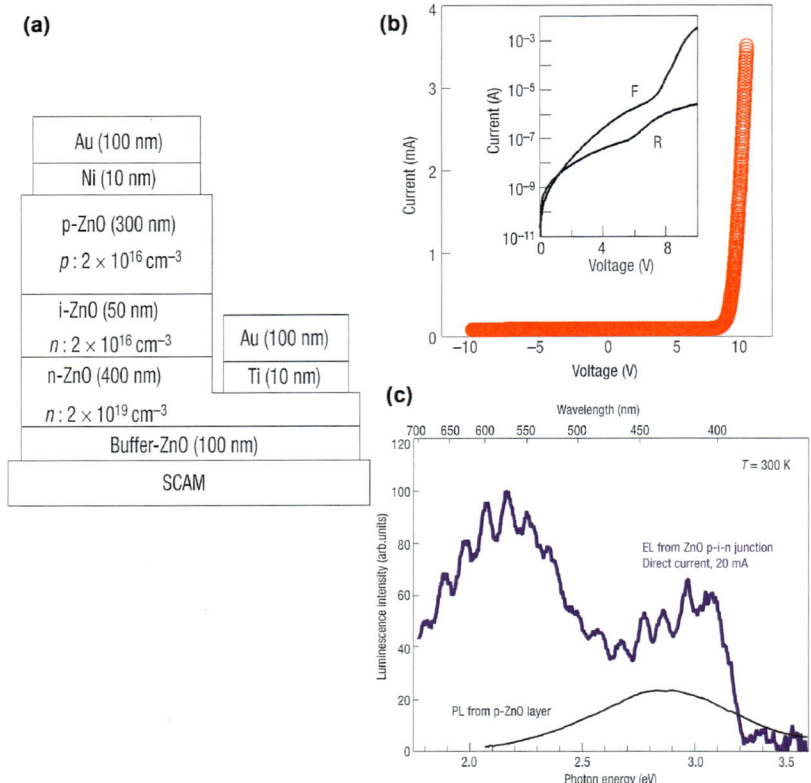

Figure 5.1 The structure (a) of the ZnO *p–i–n* homojunction LED and its current–voltage curve (b) and electroluminescence spectrum (c) under forward bias at room-temperature.
Reproduced by permission from Macmillan Publishers Ltd: *Nat. Mater.*[3] Copyright (2004).

5.1.2 *p–n* Heterojunction LED

Although much progress has been made in the research of *p*-type doping of ZnO and its homojunction preparation, there are still some problems in *p*-ZnO, such as low carrier concentration, poor repeatability, poor stability and so on. However, ZnO homojunction devices usually have poor luminescent properties featuring in poor luminescent monochromaticity, co-existence of the ZnO band-edge and defect-related emission. Therefore, to improve the performance of ZnO based ultraviolet light-emitting diodes, researchers must look for other effective *p*-type materials as a hole injection layer to substitute *p*-ZnO. So far, inorganic semiconducting materials such as *p*-GaN,[7–12] *p*-NiO,[13,14] *p*-Si,[15,16] *p*-SiC,[17] *p*-CuAlO$_2$[18] and *p*-SrCu$_2$O[19] have been used to fabricate heterojunction LED devices with *n*-type ZnO.

n-ZnO/*p*-Si film heterojunction LEDs could emit weak visible light emission.[15] Researchers studied the origin of the weak electroluminescence in view of the energy band alignment and interfacial microstructure of heterojunction. The results showed that there is high ZnO/Si valence band offset and unavoidable SiO_x interfacial layer formed at the ZnO/Si interface, so the electroluminescence spectrum exhibits only weak visible light emission related to defects. Double side light emission had been achieved in *p*-NiO film/*n*-ZnO nanowire heterojunction LEDs, which began to emit light at a forward bias of 7 V.[13] When the forward bias voltage is 15 V, the electroluminescence spectrum is mainly composed of two narrow emission peaks, one near band-edge emission at 385 nm and another wide visible light emission at 570 nm. When compared with the yellow-green band, the UV-emission peak can be enhanced with the increase of forward bias voltage, which proves that ZnO is a high-efficiency ultraviolet light-emitting material.

In early 2003, *p*-type GaN was used to fabricate a *n*-ZnO/*p*-GaN heterojunction using chemical vapor deposition method.[7] This *n*-ZnO/*p*-GaN film heterojunction LEDs exhibit good *I–V* rectification characteristics and emit blue-violet light with a peak wavelength of 430 nm. This work demonstrated the feasibility of *p*-type GaN instead of *p*-type ZnO to make blue-violet light-emitting diodes. However, the luminescence properties of thin film/thin film LEDs are usually poor due to the total internal reflection, so to increase the extraction efficiency of the LEDs, *n*-ZnO nanorods/*p*-GaN thin film heterostructures have been studied further. Many research groups have used different methods to prepare *n*-ZnO nanorods/*p*-GaN heterojunction LEDs, and have studied their photoelectric properties. Researchers synthesised ZnO nanorods on a *p*-GaN substrate using catalyst-free metal–organic vapour phase epitaxy and fabricated *n*-ZnO nanorod/*p*-GaN heterojunction LEDs, which gave out light only when under a reverse bias voltage.[8] At a reverse-bias voltage of 3 V, the electroluminescence spectrum presents a wide yellow emission band centered at 2.2 eV, which is mainly attributed to the high concentrations of defects formed in ZnO nanorod at the interface, and the emission intensity increases as the reverse-bias increases from 3 to 7 V. A blue emission peak at 2.8 eV relating to the Mg acceptors in GaN and a UV emission peak at 3.35 eV from GaN band edge emission for the reverse-bias voltages of 4 V and 5 V can also be observed, respectively. It was demonstrated that a ZnO-nanowire-inserted GaN/ZnO heterojunction LED can effectively enhance the electroluminescence emission and injection current, due to the existence of nanosized junctions with a low density of interfacial defects.[20] Moreover, the Al-doped ZnO film deposited on the *n*-ZnO nanowires can provide more electrons to the nano-heterojunctions. We had grown a high-quality *n*-type ZnO nanowire array on a commercially available *p*-type GaN substrate by chemical vapour deposition, and constructed a high-brightness ZnO/GaN heterojunction LED.[9] Its structure is shown in the inset of Figure 5.2. It can be observed from Figure 5.2 that the ZnO/GaN junction exhibits clear *p–n* junction rectification characteristics and has a current turn-on voltage of about 3 V, which almost approaches the bandgap of ZnO. When the

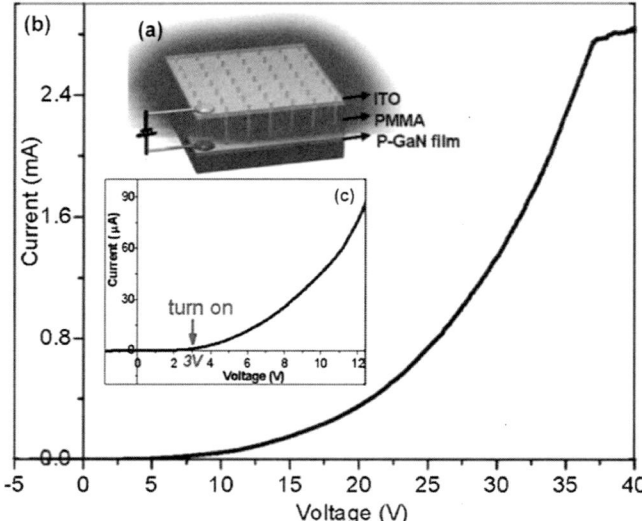

Figure 5.2 (a) A schematic of the *n*-ZnO nanowire/*p*-GaN film heterojunction LED and its current–voltage characteristics; (b) *I–V* curve of the ZnO/GaN heterojunction diodes from −5 to 40 V; the current reached saturation beyond 37 V; (c) *I–V* curve of the ZnO/GaN heterojunction diodes from −5 to 12 V, where the turn on voltage is around 3 V.
Reproduced with permission from John Wiley and Sons Copyright © 2009 WILEY-VCH Verlag GmbH & Co. KGaA, Weinheim.[9]

forward bias voltage reaches a certain value, the device's electroluminescence spectrum shows a dominant emission peak from UV to blue, as shown in Figure 5.3(a); meanwhile, a blue shift can be observed in the spectra when the applied voltage increases. Furthermore, UV illumination not only increases the carrier density, but also reduces the height of the Schottky barrier at the interface. The overall electroluminescence intensity increases within 30 min after UV illumination is switched off, which is probably due to the high concentration of residual charge carriers created by UV illumination. Then, after the UV illumination is off for a period, the residual carriers are largely recombined, resulting in the drop of the EL intensity. Thus, the electroluminescence intensity can be tuned by UV illumination in these hybrid heterojunctions, as shown in Figure 5.3(b). In order to further investigate the physical origin and light output process in ZnO-based LEDs for performance improvement, we also constructed ZnO/GaN LEDs by simply transferring a single ZnO micro/nanowire onto *p*-type GaN substrate, as shown in Figure 5.4.[10] As the injection current increases, a saturated electroluminescence phenomenon can be observed.

Compared with the ZnO nanowire arrays with random spatial distribution, the controlled spatial distribution of ZnO nanowire arrays can be improved by the combination of electron beam lithography and wet chemical method for the construction of LEDs, so researchers synthesised position-controlled arrays of *n*-ZnO nanowires on a *p*-GaN substrate, and fabricated ordered ZnO

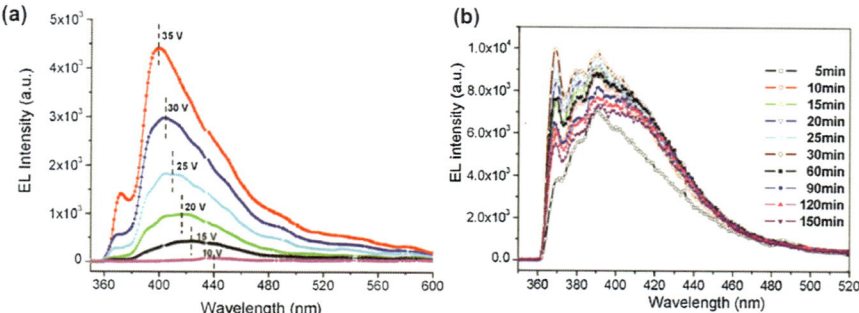

Figure 5.3 (a) The electroluminescence spectra of the device at different forward bias; (b) the time-dependent electroluminescence spectra after turning off the UV illumination.
Reproduced with permission from John Wiley and Sons Copyright © 2009 WILEY-VCH Verlag GmbH & Co. KGaA, Weinheim.[9]

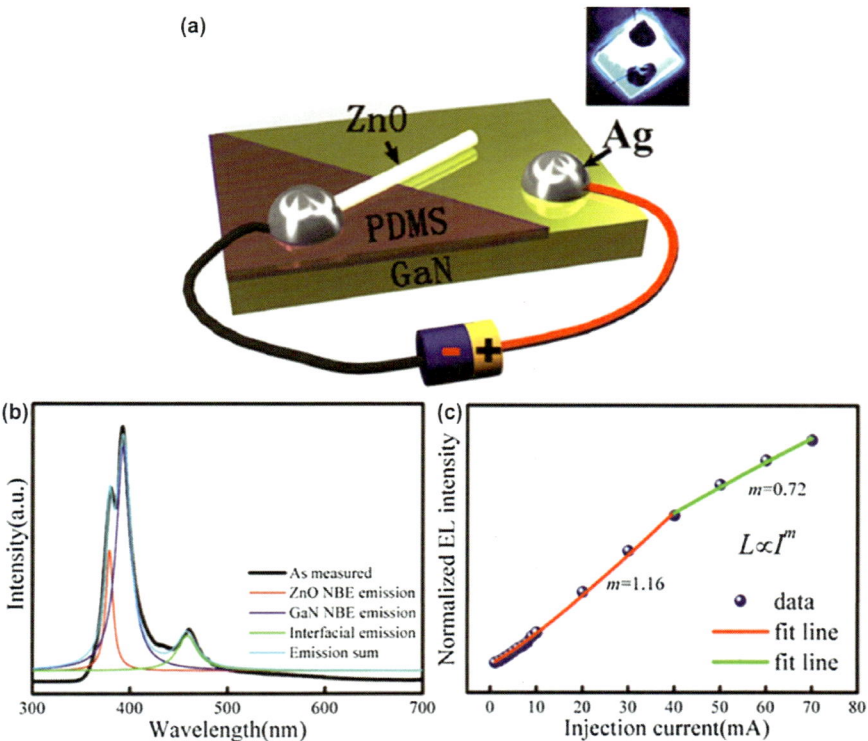

Figure 5.4 (a) Schematic diagram of the device structure; (b) the electroluminescence spectra under different injection current; (c) relationship between the light output intensity and injection current of the device. Reproduced from ref. 10 with permission from AIP Publishing.

nanowire array blue/near-UV LEDs.[11] Each single nanowire is a light emitter under forward bias voltage, and therefore a resolution of as high as 6350 dpi can be achieved because the pitch between each light spot is 4 µm. The fabricated LEDs have an external quantum efficiency of 2.5%.

In addition, some researchers have found that some *p*-type organic materials can also substitute *p*-type ZnO. These inorganic/organic hybrid LEDs can combine the high flexibility of polymers with the chemical stability of inorganic nanostructures. A hybrid LED composed of electrodeposited ZnO nanorods and PEDOT: PSS was first reported.[21] Devices obtained from the as-grown ZnO nanorods present a broad-band EL emission across the spectral region from 350 to 850 nm, while devices from the ZnO nanorods annealed at 300 °C show a narrow UV EL emission centered at 393 nm. Afterwards, a highly flexible LED using electrodeposited *n*-type ZnO nanowires and a *p*-type polymer was also realised.[22] The *p*-type conducting polymer poly(fluorine) (PF) was used to construct a hybrid white-light LED with ZnO nanorods.[23] The devices exhibit a broad emission band covering the entire visible range from 400 to 800 nm, which corresponds to the significantly enhanced ZnO surface defect emission and PF-related emission, respectively. The study provided us a convenient and low-cost way to fabricate ZnO-based white-light-emitting devices.

In recent years, it has been demonstrated that the piezophototronic effect can be used to largely enhance the performance of ZnO-based *p–n* heterojunction LEDs.[24–27] The first ZnO microwire/GaN LED performance enhancement with the presence of piezopotential was reported in 2011.[24] Positive piezoelectric ions generated at the interface of ZnO–GaN under strain is equivalent to applying an extra forward bias at the device, leading to reduced depletion width and internal field. Meanwhile, a dip in the band and a trapping channel for holes near the ZnO–GaN interface will be created, which enhances the electron-hole radiative recombination process. Thus, it was demonstrated that the emission intensity and the injection current at a fixed applied forward voltage have been enhanced by a factor of 17 and 4, respectively, by applying a 0.093% compressive strain, and the conversion efficiency was improved by a factor of 4.25. In addition, it was also demonstrated that this enhancement in electroluminescence intensity is attributed to the polar piezoelectric effect rather than other non-polar factors, such as piezoresistive effect or photoelastic effect. The positive effect of piezopotential on the performance improvement of ZnO microwire/GaN LED was further extended to ZnO nanowire-array-based LEDs. Researchers demonstrated that the piezoelectric ZnO nanowire array/GaN LED could be used for high-resolution EL imaging of pressure distribution.[26] The device was based on a lithography-patterned array of *n*-ZnO nanowire arrays grown on *p*-type GaN film. The NAs-LED can be uniformly lit up at a bias of 5 V and has a pixel resolution of 6350 dpi. Because of the preferred +*c* axis growth direction, when the device is under compressive strain, permanent non-mobile, positive piezoelectric charges will be generated near the interface of ZnO–GaN. The positive piezopotential may result in a local dip in the band, which temporarily traps holes at the vicinity interface, improving the carrier injection rates toward the heterojunction. Thus, an increasing recombination of electrons and holes could be obtained, leading to higher light emitting density. Based on the EL intensity change of the emitting nano-LEDs, the pressure/strain across the entire device could be mapped, showing a mapping rate of about 90 ms. Furthermore, the spatial resolution can be further improved by optimizing the size of the ZnO nanorod.

5.1.3   MIS Heterojunction LED

For the *p–n* heterojunction devices, their electroluminescence inevitably contains emission from the *p*-type layer due to the presence of band mismatch and the difference in electron and hole mobility. Therefore, it is often difficult to obtain pure ZnO UV electroluminescence. Thus, another heterostructure device, a metal-insulator-semiconductor (MIS) heterojunction device, was developed. The barrier layer of the insulator in the device can effectively confine the carriers in ZnO, and so the intrinsic ultraviolet emission from ZnO can be achieved. Until now, two types of ZnO MIS heterojunction LEDs have been developed upon different insulating layers. One is the metal/*i*-ZnO/*n*-ZnO heterojunction LEDs, in which high-resistance intrinsic ZnO is selected as the insulating layer. However, there are usually dominant visible light emission in the electroluminescence spectra. Another type use large bandgap insulating materials (AlN, MgO, SiO_2) to serve as the insulating layer. Pure ZnO intrinsic ultraviolet emission can be realised in these devices. The working mechanism can be explained with impact-ionisation process in the insulation layer.[28] When the device is under forward bias, the local electric field strength in the insulating layer would be as high as $\sim10^7$ V m^{-1} due to the dielectric nature of insulating materials. Therefore, many electrons and holes could be generated under such high electric field. The generated holes could transport to ZnO under forward bias, recombine with electrons accumulated at the interface and produce ZnO near-band-edge emission. In early 1973, a ZnO MIS heterojunction LEDs had been constructed using a Ag/SiO$_x$/ZnO structure, in which ZnO was a ZnO: Li bulk single crystal.[29] A dominant blue emission at $\sim$420 nm appears in the electroluminescence spectrum. Au/SiO$_x$/ZnO MIS heterojunction LEDs was also fabricated on a silicon substrate using reactive direct current sputtering and electron beam evaporation.[30] A pure ultraviolet emission originated from the near-band-edge emission of ZnO can be observed in the forward-biased ZnO MIS device with $\sim$100 nm thick SiO$_x$ layer. Afterwards, electrically pumped ZnO lasing was successfully achieved in the same ZnO MIS heterojunction devices.

5.2   UV Detectors

Photodetectors, which convert optical signals into electrical ones, have wide-ranging applications in light communication, binary switches used in imaging, optoelectronic circuits, and memory storage. It is known that the UV radiation emitted by the sun falls in the range 200–400 nm. Most of UV-C (200–290 nm) light and UV-B (290–320 nm) light can be absorbed by the molecules in sunscreen lotions and the Earth's atmosphere, and UV-A (320–400 nm) light can reach the Earth's surface, leading to skin cancer. Nowadays, commercial silicon-based photodetectors and photo-multipliers have high sensitivity and fast response speed. However, they always have low quantum efficiency and huge volume, and need high vacuum environment and larger voltage, which limits their practical application. To overcome these disadvantages, a new generation of wide band-gap semi-conductors with excellent electrical and optical properties, strong radiation resistance and high chemical stability, have drawn growing attention in the UV detection field. As a typical wide bandgap (3.37 eV) semiconductor, ZnO has a large exciton binding energy (60 meV), and is one of the most important materials for UV-A light photodetectors. Photodetectors based on 1D semi-conductor nanomaterials usually have high internal gain and large optical absorption, owing to their large surface-to-volume ratio. Besides, mono-crystalline 1D nanostructures provide direct pathways for charge carrier transportation, which benefits the improvement of response speed. Currently, ZnO nanocrystals with a diversity of morphologies have been extensively investigated and widely used to fabricate nanoscale UV photodetectors.

Due to the large specific surface area and similar size to Debye length, one-dimensional ZnO nanomaterials exhibit different photoresponsive prop-erties to bulk ZnO. It is well known that oxygen vacancy plays an important role in the photoconductivity.[31] In the dark, the absorbed oxygen molecules on the surface of ZnO nanomaterials trap free electrons ($O_{2(g)} + e^- = O_{2(ad)}^-$), and a depletion layer forms, leading to low conductivity. Upon UV light, the photogenerated holes combine with the negative oxygen ions ($h^+ + O_{2(ad)}^- = O_{2(g)}$). The surplus photogenerated electrons drift towards anodes or combine with oxygen molecules. Owing to the existence of dan-gling bonds, ZnO nanomaterials have abundant surface states, which greatly enhances their photoconductivity through the photogenerated hole-trapping mechanism.[32]

According to the working principle and the device structure, photo-detectors based on ZnO nanomaterials can be divided into three categories: photoconductivity type, Schottky type and *p*–*n* junction type. The photo-conductivity type photodetector is based on the photoconductive effect, and is commonly composed of a photosensitive semiconductor and two Ohmic electrodes. When the energy of the incident photon is larger than the bandgap of the semiconductor, a lot of electron–hole pairs are gener-ated. Under applied bias, these photogenerated electron-hole pairs are separated and form a photocurrent, converting the light signal into electric signal. The Schottky type photodetectors are constructed based on the photovoltaic effect, and their space charge region is made up of the Schottky junction. These type photodetectors are usually composed of a semi-conductor, a Schottky electrode, and an Ohmic electrode. For the *n*-ZnO, its

Fermi level is higher than that of some metal electrodes with large work function. So the free electrons in ZnO can flow into the metal electrode, resulting in band bending at the interface and the formation of built-in electric field at ZnO zone. Upon UV light illumination, the photogenerated electron–hole pairs can be separated by this internal electric field and form a photocurrent. The ZnO based UV detectors have many advantages, such as low dark current, high photosensitivity, and fast response time. Like the Schottky type photodetectors, the *p–n* type photodetectors are also fabricated based on the photovoltaic effect. They use internal *p–n* junction separate photogenerated charge carriers. The *p–n* type photodetectors can be derived by low bias, and have low saturated current and large working frequency.

In this section, we present the recent development in UV photodetectors based on single ZnO nanotetrapod, single ZnO nanowire, ZnO nanowire arrays, and graphene deposit/ZnO composites with emphasis on the efforts for improvement in the performance of the devices.

5.2.1 UV Detectors Based on a Single ZnO Nanowire

In 2002, single ZnO nanowires were first found to have extreme photosensitivity to UV light illumination.[33] The light-induced conductivity allows one to reversibly switch the nanowires between OFF and ON states, indicating that they are good candidates for optoelectronic switches. Under 365 nm UV light irradiation, the conduction of the single ZnO nanowire was remarkably increased by 4–6 orders of magnitude, and it also showed strong power dependence and excellent wavelength selectivity. The rise and recovery times of the ZnO nanowire switches are less than 1 s. After that, different fabrication strategies were used to optimise the ZnO nanowire photodetectors, such as decreasing the space between the electrodes or increasing the mobility of the nanomaterials.[34] Focused ion beam technology was used to deposit Pt electrodes on the ZnO nanowire surface, and the photoconductivity gain was increased to about 10^8 through reducing contact resistance.[35] The optical-electrical characteristics of ZnO nanowires can be optimised by element doping. Due to the avalanche light multiplication effect, Cu-doping can greatly enhance the UV-visible light sensitivity of the device.[36] It was also reported that Co-doped ZnO nanobelt can be used to detect tiny amounts of water vapor under UV light.[37]

ZnO nanotetrapods, consisting of a pyramid-like zinc blend-structured core with four needle-shaped wurtzite-structured legs protruding out at tetrahedral angles, are of particular interest due to their unique nanoscale 3D architectures, and can deliver the functions of delicate multi-terminal nanodevices.[38,39] Due to their novel structure with an occupied 3D tetrahedral space and ease of synthesis with controllable size and shape, ZnO nanotetrapods have been utilised to construct many types of prototypical devices such as photoelectric sensors. In contrast to the crooked Si nanowire, ZnO nanotetrapods can ensure strong practicability for building sophisticated sensors for microenvironment detection in biological cells with one free terminal inserted, due to their simple preparation process and novel structure.

To monitor the UV detecting performance of ZnO nanotetrapod based photodetectors, one leg of a tetrapod was locally irradiated using UV light and measured by recording the photocurrent *vs.* time curves. The results indicate that for the detectors with Ohmic contacts or Schottky contacts, localised UV irradiation can be probed in real time by recording the photocurrent of the detectors and the sensitivity of the detectors changes as UV light density increases. Due to the interface effect and the carrier transfer effect, the detectors with Schottky contacts have superior performance compared with the ones with Ohmic contacts. In addition, ZnO nanotetrapod-based photodetectors also show excellent operation repeatability and reversibility, as well as advantages of fast response and quick recovery. It is further proven by the experimental outcome that detectors with Schottky contact characteristic are a better candidate for applications in monitoring localised irradiation.[40]

The piezoelectric properties of single ZnO nanowires have been extensively investigated; the coupling between piezoelectric properties and optoelectronic properties is currently attracting a great deal of interest as it has potential applications in optoelectronic and sensor technologies. We have investigated the localised ultraviolet photoresponse in a single bent ZnO nanowire bridging two Ohmic contacts.[41] At the straight (marked with '4', bending strain: 0%) region, the sensitivity of the ZnO wire is about 50%. When the bending stain is about 4% at the localised region (marked with '6'), the sensitivity of the ZnO nanowire is up to 190%, as shown in Figure 5.5. The results demonstrate that the localised photoresponse sensitivity increases by increasing the bending strains of single ZnO nanowires. By using a double-exponential rise and decay functions to fit the experimental data, the rise and decay time of the wire at both the straight and bent regions can be calculated to be about 18 s and 30 s, respectively. The rise and decay time constants are almost the same in the straight and bent regions of the ZnO nanowire.

A photoconduction mechanism in the bent ZnO wire can be used to illustrate the above phenomenon. When the UV laser with photon energy larger than the bandgap of ZnO focuses on the nanowires, the electron–hole pairs will be generated. The photogenerated holes were trapped at the outer layer of the ZnO nanowires, and the unpaired photogenerated electrons enhanced the conductivity under an applied bias. Owing to the stretching and compression on the surfaces, the bending of a single ZnO wire can produce a positively charged and negatively charged surface at outer and inner bending surfaces, respectively. The piezoelectric electric field would be generated across the width of the bent ZnO wire. The carrier trapping effect and the creation of a space charge region will reduce the conductance of the wire. In the bent region of the ZnO wire, it could be considered as a wire field effect transistor with the *n*-channel depletion-mode. The piezoelectric electric field acts as a gate electrode. It has been reported that the UV

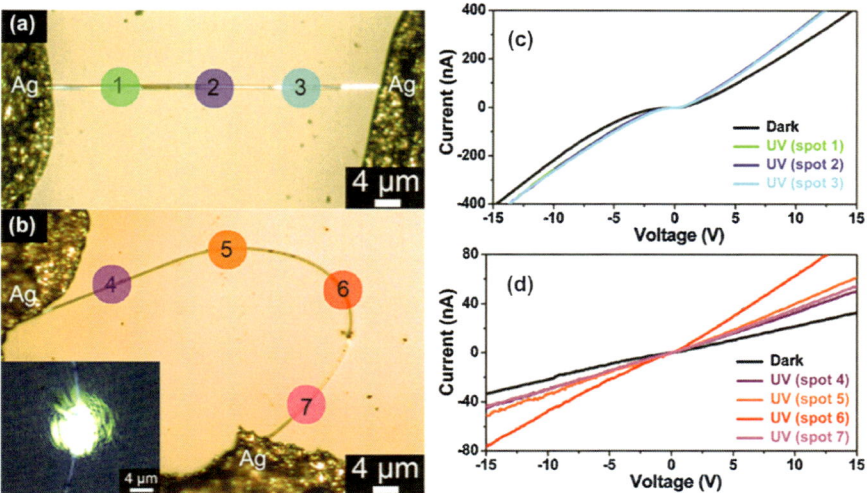

Figure 5.5 (a) Optical image of a single ZnO wire between two Ag electrodes; (b) optical image of a single bent ZnO wire between two Ag electrodes. The inset shows that the UV laser is focused on the wire; (c)–(d) *I–V* characteristics of the straight ZnO wire under dark and UV illumination at the different regions in (a); *I–V* characteristics of the bent ZnO wire under dark and UV illumination at the different regions in (b). Reproduced from ref. 41 with permission from AIP Publishing.

photoresponse can be largely enhanced under a gate-bias condition for a depletion mode, which is consistent with our observed experimental results. The piezoelectric electric field can be neutralised by the photogenerated electron–hole pairs, which could result in the reduction of the trapping effect and enhancement in photoresponse. The results indicate that the bent ZnO micro/nanowires could be potentially useful for fabricating the coupled piezoelectric and optoelectronic nanodevices.

For photoconductive type UV photodetectors, these always have ultrahigh photoconductive gain, due to the presence of oxygen related hole trapping states at the surface. The carrier trapping effect prevents charge carrier recombination and prolongs their lifetime. Nevertheless, the adsorption and desorption of oxygen molecules are both very slow processes, resulting in a long response time.

Surface functionalisation of nanomaterials is the efficient method to promote the separation of the charge and shorten the response time in photoconductive UV detectors.[42,43] A FET-detector with PEDOT: PSS/ZnO NW junction as the gate have been fabricated and investigated to overcome this disadvantage. The depletion layer around the ZnO NWs increases the resistance and greatly reduces the dark current, thereby improving the sensitivity with two orders of magnitude (from 0.2 to 27.5). Upon 325 nm illumination, the FET photodetector shows a fast rise time of about 0.8 s and a decay time of 3.8 s. The bare ZnO nanowire based sensors have a long decay time of 645 s. A possible physical mechanism based on band energy theory is proposed to explain the origin of the enhancement in the sensing performance. The PEDOT on the surface of ZnO can lead to the formation of a localised *p–n* junction, which generates a charge space region in the ZnO nanowire. The *p*-PEDOT works as a gate in a field-effect transistor. When UV light illuminates the PEDOT: PSS/ZnO NR interface, the charge carriers are generated. The electron–hole pairs are efficiently separated by the internal potential in the depletion region at the interface. The electrons grafted to the ZnO side, and the holes move to the PEDOT side, which changes the interface charge distribution and dramatically reduces the capacitance of the interface region. Thus, the width of depletion layer decreases, resulting in a great increase in the photoconductance of the UV detector. To further confirm the UV gate effect on the depletion channel, the photocurrent *vs.* incident light power curves were measured. As per the charge carrier diffusion mode, the thickness of the depletion layer was calculated from the typical I_{ds}–V_{ds} characteristics of the detectors, and the results indicated that the FET channel was linearly related to the light power.[44]

5.2.2   UV Detectors Based on ZnO Nanowire Arrays

ZnO nanowire arrays integrate abundant single nanowires and hence have huge surface area and low reflectance. So, they are regarded as one of the most promising materials for nanoscale UV photodetectors. It was reported that photodetectors based on ZnO nanowire arrays with Au top contact electrodes have a dark current density of only 1.37×10^{-7} A cm^{-2} at 1 V applied bias. The extremely small dark current should be attributed to the highly resistive nature of the undoped ZnO nanowires. It was found that the UV-to-visible rejection ratio of the device is about 1000. The time constant of the fabricated photodetectors is about 0.44 ms.[45] Metal-semiconductor-metal structured UV detector with ZnO nanowire arrays also show high UV photoresponse.[46] The ZnO nanowire arrays were selectively grown between the space of interdigitated electrodes by a chemical solution method through a photolithography process. The devices have a responsivity of 41.22 A/W and a UV-to-visible rejection ratio of 336.65. The nanowire array based detectors show superior photoresponse compared to the one without nanowires. This can be attributed to the very large surface-to-volume ratios of ZnO nanowire arrays, easily promoting oxygen adsorption and desorption at the nanowire surfaces.

Lithography, spin-coating technique, and the pick-and-place method have been used to fabricate UV sensors based on 1D ZnO nanostructures. Those technologies have promoted the development of nanodevices. However, these preparation process are very complicated and uneconomical, which blocks their way to large-scale applications. In addition, many UV detectors based on ZnO nanowire or nanowire arrays have been reported. So, a novel and simple method was created to fabricate high-performance UV photodetectors based on ZnO nanowire arrays. Figure 5.6 illustrates the schematic

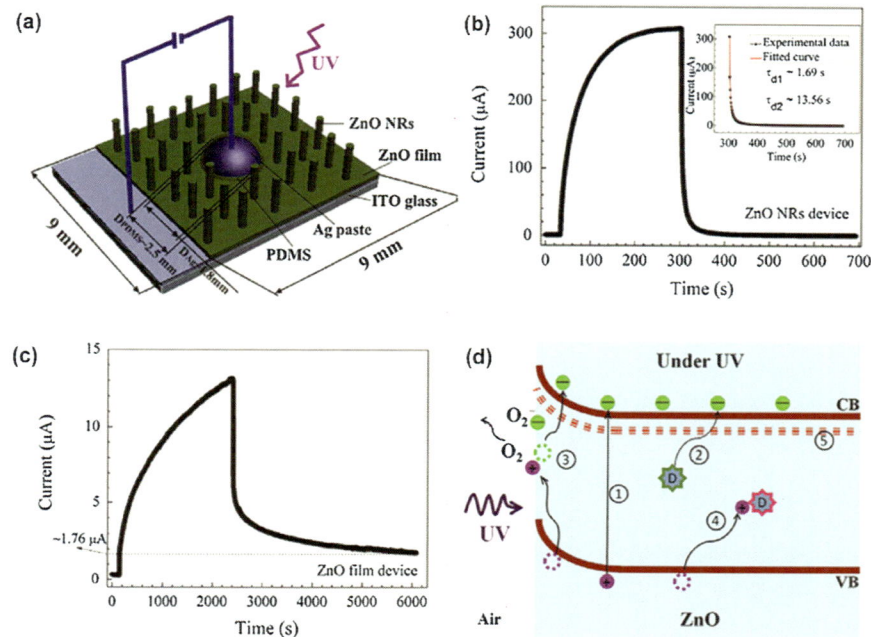

Figure 5.6 (a) Schematic representation of the ZnO nanowire array device; (b) time-resolved photocurrent response of the ZnO nanowire array device illuminated by UV light at 1 V bias. The inset shows the experimental curve and fitted curve of the photocurrent decay process; (c) time-resolved photoresponse of the ZnO film device; (d) schematic diagram for the carrier transport processes in ZnO nanostructures. (1) Electrons are excited from VB to CB; (2) electrons from photosensitive defects are excited to CB; (3) photogenerated holes discharge ionised oxygen on the surface; (4) holes are trapped by defects; and (5) percolation states created by metastable defects.
Reprinted from Phys. E, 61, F. Yi, Q. Liao, X. Yan, Z. Bai, Z. Wang, X. Chen, Q. Zhang, Y. Huang and Y. Zhang, Simple fabrication of a ZnO nanorod array UV detector with a high performance, 180–184. Copyright (2014), with permission from Elsevier.[47]

diagram of the fabricated metal-semiconductor-metal UV photodetector. Upon 365 nm UV illumination ($\sim$2 mW cm^{-2}), the ZnO nanowire based detector showed a photocurrent of about 308 µA with an on/off ratio of 5.13×10^2 and a fast decay time constant of 1.69 s at 1 V applied bias. The performance of the ZnO nanowire device was much better than that of the conventional ZnO film device.

The superior photoresponse of the ZnO NRs device compared to that of the ZnO film device could be attributed to the larger surface-to-volume ratio and the better crystal quality of ZnO nanowire arrays. First, the bigger specific surface area of the ZnO nanowire arrays enables the device to absorb more UV light, thereby increasing the photocurrent. However, the larger surface-to-volume ratio also improves the recombination of charge carriers, contributing to a shorter decay time. Second, the much better crystal quality of the ZnO nanowire arrays reduces the density of charge trapping sites caused by defects, and therefore dramatically improves the photoresponse. The ZnO NRs device has great potential for application in UV detection and is promising for large-scale production.[47]

Although the 1D ZnO nanostructures have large surface area and excellent charge collection ability, the visible emissions associated with native defects in photoluminescence (PL) have been often observed, and have negative effects on optoelectronic devices based on ZnO nanomaterials. Therefore, it is necessary to have a full understanding of the origin of defect-related emissions from ZnO nanostructures under various treatments. Besides, although many reports have studied the photoresponse of 1D ZnO nanostructures to UV light, the wavelength selectivity has not attracted significant attention yet.

Here, hydrothermal synthesis was used to grow ZnO nanowire arrays on the Si substrate. The PL spectrum of the nanowires measured at room temperature show that after annealing in air, the green emission enhances whereas the UV emission quenches along with the near infrared emission. XPS spectra were used to illustrate the mechanisms for the changes of emissions. Because all the intrinsic defects in ZnO, oxygen vacancies have the lowest formation energy among donors which can supply electrons to *n*-type ZnO, the reason why the green emission increases after annealing in air could be that the O atoms evaporate out of the ZnO lattice during the annealing process, causing O–Zn bonds to be broken and leaving O vacancies in the crystal structure. The escaping rate of the O atoms becomes larger as temperature goes up, inducing more O vacancies to appear in the structure. On the other hand, the growing temperature could lead to a change in the charge state of the defect, creating more ionised O vacancies. These O vacancies should be the main luminescent centres of the amplifying green emission as the annealing temperature rises.

Based on the prepared ZnO nanowire arrays, a metal-semiconductor-metal structured UV photodetector was fabricated. The current–voltage curve of the device shows double Schottky diode characteristics in the dark. Upon UV light illumination, the *I–V* curve transforms to Ohmic characteristics. The photogenerated current under 365 nm UV illumination is almost 25 times higher than that under 254 nm UV illumination. The 365 nm UV light has a wavelength near the absorption edge, while the 254 nm UV possesses a much shorter wavelength above the absorption edge. Moreover, the photons of 254 nm UV have a shorter penetration depth than the photons of 365 nm UV, which means that the photogenerated electron–hole pairs under 254 nm UV have a longer distance to travel across before they get to the electrodes, and therefore more electron–hole pairs recombine during the longer travel. The larger percentage of recombined electron–hole pairs leads to the much lower photocurrent when the device is exposed to 254 nm UV. The device exhibited highly different photoresponses between short-wavelength UV and near-band-edge UV, which indicates that ZnO nanowires have the potential for UV detection with wavelength selectivity.[48]

In contrast to the photoconductive type and *p–n* junction type photo-detectors, the Schottky type photodetectors have high performance and simple fabrication. Therefore, studies of ZnO UV detectors have been focused on the Schottky type. Here, Schottky type UV photodetectors based on ZnO nanowire arrays were fabricated with Pt as electrodes. The ZnO nano-wires synthesised by the hydrothermal method were characterised by field-emission scanning electron microscopy, Raman and PL spectroscopy. Photoelectric properties under 254 and 365 nm UV light were investigated.

The on/off ratio of the UV photodetectors under 254 nm and 365 nm UV light was 1.4 and 1.9, respectively. Besides, the rise time and recovery time of the device under 254 nm illumination are 14 s and 126 s, while upon 365 nm illumination the corresponding values were calculated to be 8 s and 125 s, respectively. These differences are illustrated by light transmission theory and energy-band diagram. When a monochromatic light irradiates the detector, a photocurrent may be generated. The schematic setup for detection and the energy band diagram for photoexcitation processes are demonstrated in Figure 5.7. In Schottky type photodetectors, there are two kinds of excitation process: excitation over the barrier and band-to-band excitation.

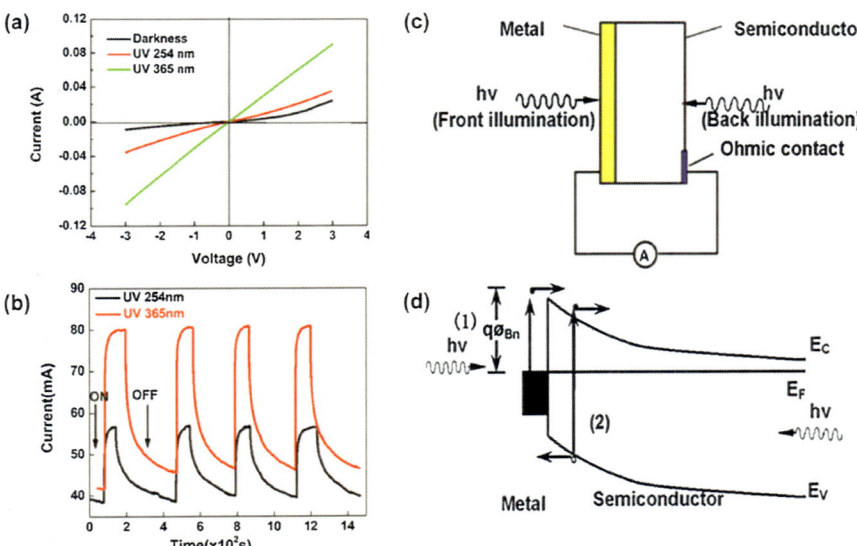

Figure 5.7 UV response behaviors of the detectors. (a) *I–V* characteristics of the ZnO nanowire array detectors in the dark and under illumination at UV (254, 365 nm), respectively; (b) spectral responsivity measurements of the detectors upon UV (254, 365 nm) illumination being turned on and off at −3 V bias; (c) schematic setup for detection; (d) energy-band diagram for photoexcitation processes.

Reprinted from Solid State Commun., 151(24), W. Lin, X. Yan, X. Zhang, Z. Qin, Z. Zhang, Z. Bai, Y. Lei and Y. Zhang, The comparison of ZnO nanowire detectors working under two wavelengths of ultraviolet, 1860–1863. Copyright (2011), with permission from Elsevier.[49]

For the two excitation processes, the most useful wavelength should be in the range of $q\emptyset B_n < h\nu < E_g$ and $h\nu > E_g$. In our case, both processes can occur for UV 254 and 365 nm light because of $h\nu > E_g$ (3.37 eV) ZnO > $q\emptyset B_n$. The long wavelength of UV light easily penetrates the metal film. So, when compared with 254 nm UV light, 365 nm UV light more easily penetrates the Pt film and can be absorbed at the Pt/ZnO interface, inducing higher excitation over the barrier. The results demonstrate that ZnO NWs detectors with selectivity to near-UV light are promising candidates for photoelectric devices.[49]

5.2.3   UV Detectors Based on a Graphene/ZnO Hybrid

With ultrahigh mobility, optical transparency and elongation, graphene is a promising building block for future flexible and wearable electronic and optoelectronics. ZnO coupling with graphene is applicable in the manufacture of high performance flexible UV photodetectors.[50] The transmittance of graphene on PET is about 97.6% over a wide spectrum range. This value is consistent with the reported ~2.3% light absorption by single layer graphene, further verifying the clean surface of our sample. After graphene modification with ZnO nanoparticles, an obvious absorption peak appears at 368 nm, attributed to the 3.37 eV energy band gap of ZnO. The photosensing properties were investigated under a 2.8 $\mu w\,cm^{-2}$ UV illumination system. The device shows a fast and stable response to UV light, while no photocurrent was observed for bare graphene, due to the poor absorption of bare graphene monolayers. Furthermore, the durability of the device was analysed through consecutive bending and releasing cycles. After 1000 bending and releasing cycles, the photocurrent decreased less than 10%, indicating the outstanding flexible properties of the graphene/ZnO detectors.

Tuning the properties of semiconductors to optimise their application performance has always been the focus of many research areas. Due to the higher toughness of materials in the nanoscale, the strain modulation effect could be much more prominent in nanostructures. Thus, the performance of nanodevices could be efficiently tuned by the elastic strain. Here, a graphene/ZnO nanowire array Schottky type photodetector has been fabricated. It exhibits dramatic response and fast on-off switch to the UV illumination. Through utilizing the piezopotential induced by the atom displacement in ZnO under the compressive strain, 17% enhanced photosensing property is achieved in this hybrid structure when applying −0.349% strain.

The mechanism of the outstanding photosensing property of the graphene/ZnO nanowire array Schottky type photodetector and performance enhancement by the strain modulation can be understood by a schematic diagram of the electronic band structure in the interface, as shown in Figure 5.8. While graphene and ZnO are in contact with each other, electron transfer occurs for the different energy level between the work function of graphene and the electron affinity of ZnO. Thus, electrons in ZnO flow into graphene until the Fermi level is aligned, forming a depletion region in ZnO

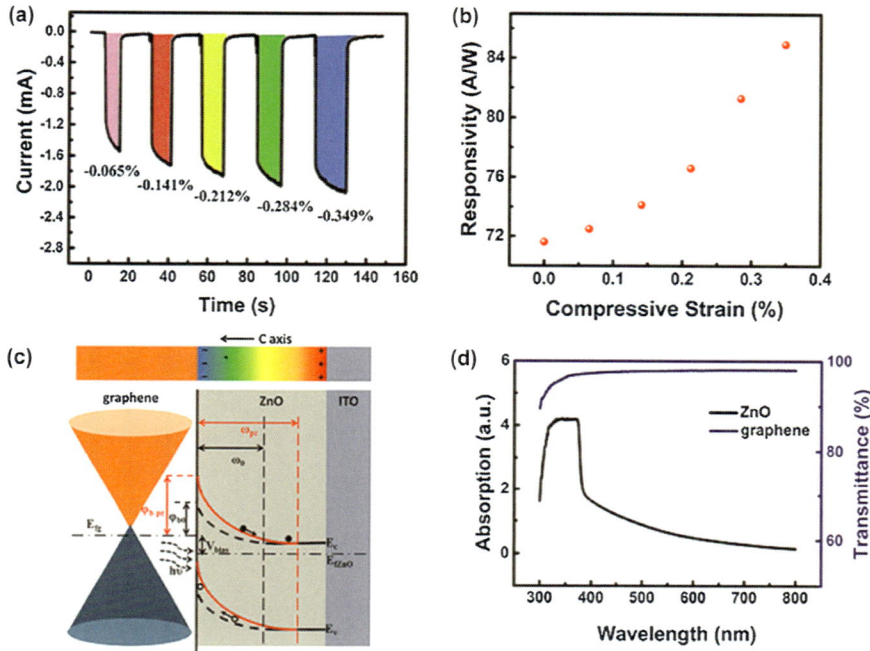

Figure 5.8 (a) The performance of the graphene/ZnO photodetector under different strain with a defined illumination density of 25.12 μW cm^{-2}; (b) the relationship between the calculated responsivity and strain; (c) schematic band diagrams of graphene/ZnO Schottky junction with compressive strain; (d) the absorption spectra of ZnO (black) and transmittance spectra of graphene (blue).
Reproduced with permission from John Wiley and Sons Copyright © 2016 WILEY-VCH Verlag GmbH & Co. KGaA, Weinheim.[51]

and a built-in electric field. Under reversed bias and UV illumination, photocarriers are generated by absorbing photons and are separated by the electric field, resulting in the photocurrent. It should be noticed that the optoelectronic process mostly happens in ZnO, which is essential for the piezopotential modulation. The absorption spectra of ZnO and transmittance spectra of graphene are measured to prove this. It is obvious that the graphene is almost transparent ($T \approx 97\%$) at all wavelengths, while ZnO shows an obvious absorption peak in the UV region, indicating that the photocarriers are mostly generated in the ZnO side at the local interface. Under a compressive strain, negative polarisation charges are generated in the graphene/ZnO interface for the relative displacement of Zn and O atoms in ZnO. The resulted piezoelectric field can lower the conduction band energy. Therefore, the Schottky barrier height raises and the depletion region in ZnO expands, which is favourable for the photogenerated electron–hole pairs separation and thus, improves the photoresponse of the device. In summary, the performance improvement is due to the barrier height modification by the strain-induced piezopotential, resulting in the improvement in electron–hole separation in the graphene/ZnO interface. The results here provide a facile approach to boost the optoelectronic performance of graphene/ZnO heterostructure, which may also be applied to other Schottky junction based hybrid devices.[51]

Over the past decade, UV photodetectors based on one-dimensional ZnO nanomaterials have made fascinating progress. However, they are still far from practical application. More research efforts should be devoted to making these changes. For the synthesis of ZnO nanomaterials, the thermodynamic and kinetic mechanism in the growth process is still unclear. It is necessary to realise the precise control of the morphology of ZnO nanomaterials, including height, radius, density, crystallisation, and so on. For the fabrication of the photodetectors, it is critical to achieve a high quality of Schottky or Ohmic electrode contacts to ensure high photosensing performance of the devices. The responsivity, UV-to-visible rejection ratio and stability require substantial improvement. Surficial decoration should be done to maintain the stability of ZnO nanomaterials in harsh environments. In addition, much effort should be made to promote the compatibility of the nanodevices with modern electronic fabrication technologies.

5.3   Solar Cells

5.3.1   Dye Sensitised Solar Cells

In the early 1990s, dye sensitised solar cells (DSSCs) with high conversion efficiency were first demonstrated by Grätzel and co-workers.[52] Since then DSSCs have begun to attract a significant amount of interest over the last two decades or more as a promising high-efficiency and low-cost alternative to conventional silicon-based photovoltaic devices. TiO_2 nanoporous films were the most commonly used dye-loading and electron-transporting material for DSSCs. However, other metal oxide semiconductors such as ZnO, SnO_2, In_2O_3 and Nb_2O_5 can also be applied for the electrodes of DSSCs.[53–56] In particular, ZnO with a similar conduction band level to TiO_2 and higher electron mobility in conjunction with diverse nanostructures has been expected to be an alternative to TiO_2. It was found that the electron lifetime was much longer in ZnO than that in TiO_2, while ZnO based DSSCs suffered inferior performance due to chemical instability which can result in a lower electron injection efficiency and/or a lower dye regeneration efficiency.[57] Acidic sensitiser dye is destructive to ZnO due to the presence of carboxylic acid binding groups on the molecules.[58] Thus, the dye concentration and immersion time should be synergistically controlled to reach optimal sensitisation and maximum efficiencies.[59] Severe Zn^{2+}/dye glomeration can occur at the surface of ZnO electrodes after a prolonged immersion in dye solutions, which hinders electron injection and light absorption efficiency. Consequently, it is of vital importance to avoid destruction of ZnO during sensitisation and prevent dye aggregation to enhance the performance of ZnO based DSSCs.

To take advantage of the superb electron transport properties of ZnO as well as reducing recombination and avoiding dye aggregation on the surface, one promising solution is to form a blocking layer on the surface of ZnO electrodes. Peidong Yang's group has made extensive efforts in revealing the effect of passivation layers on ZnO nanowires for DSSCs.[60] They fabricated conformal Al_2O_3 or TiO_2 layers on ZnO by atomic layer deposition to form a core–shell structure. It was found that thin Al_2O_3 shells with a thickness of up to 22 nm can increase the open-circuit voltage (V_{OC}) of DSSCs but decrease the short-circuit current (I_{SC}) and thus the overall efficiency (η) of the DSSCs. This is because electron injection efficiency is lowered upon inserting an insulating layer of Al_2O_3. However, the addition of TiO_2 coating with a thickness of 10–35 nm can have a positive effect on both the V_{OC} and I_{SC} due to suppressed recombination rate.

We observed Zn^{2+}/dye aggregates on the surface of ZnO electrodes which were immersed in dye solution for a prolonged time.[61,62] As shown in Figure 5.9, an amorphous compound layer can be identified on the surface of ZnO nanoparticles and nanowires. Drastic degradation of the cell performance was observed, accompanied by the occurrence of the amorphous compounds due to the formation of Zn^{2+}/dye aggregates that can restrain electron injection. Modifying the ZnO electrode with a passivation layer can effectively improve its acid stability. For ZnO nanoparticulate film electrodes, a protective layer of Al_2O_3 or SiO_2 can be modified on the surface by spin-coating,[61] as shown in Figure 5.10. As the sensitisation time was

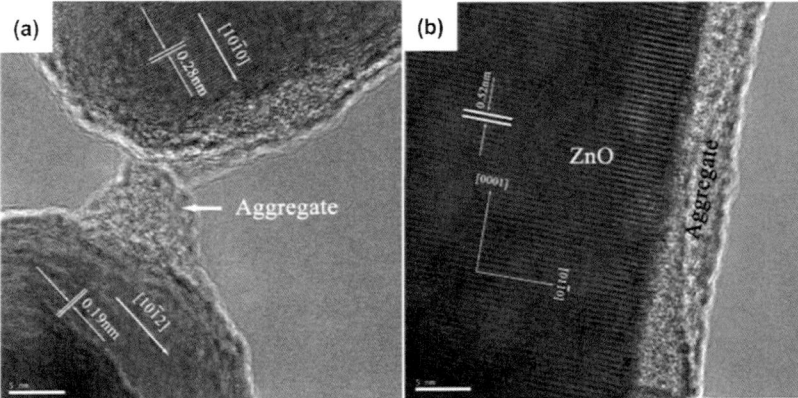

Figure 5.9 Zn^{2+}/dye aggregates on the surface of (a) ZnO nanoparticles and (b) nanowires.
(a) Reprinted from Colloids Surf. A, 386(1–3), Z. Qin, Y. Huang, J. Qi, L. Qu and Y. Zhang, Improvement of the performance and stability of the ZnO nanoparticulate film electrode by surface modification for dye-sensitized solar cells, 179–184. Copyright (2011), with permission from Elsevier. (b) Reprinted from Mater. Lett., 65(23–24), Z. Qin, Y. Huang, J. Qi, Q. Liao, W. Wang and Y. Zhang, Surface destruction and performance reduction of the ZnO nanowire array electrode in dye sensitisation process, 3506–3508. Copyright (2011), with permission from Elsevier.[61,62]

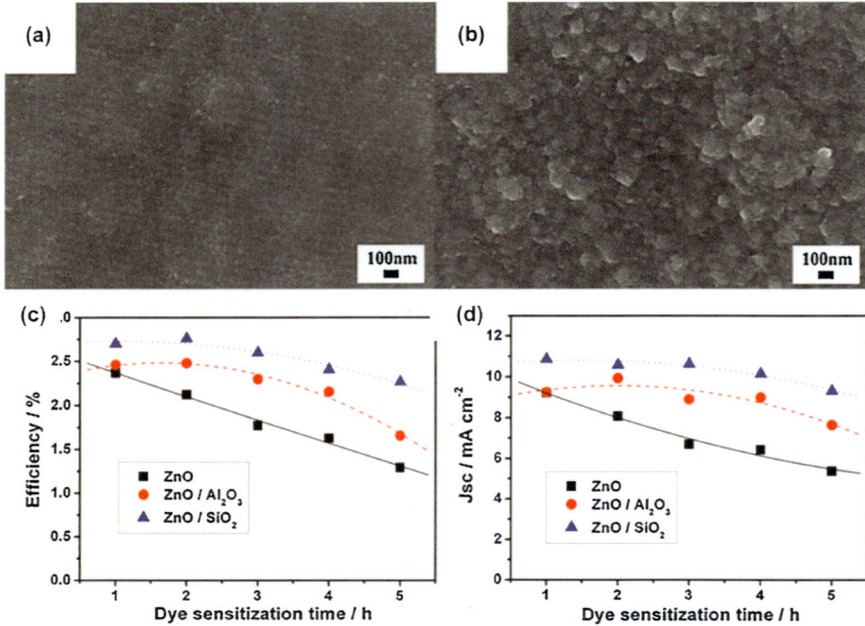

Figure 5.10 (a) Top view of the Al$_2$O$_3$ coated ZnO nanoparticulate film; (b) top view of the SiO$_2$ coated ZnO nanoparticulate film; (c) comparison of efficiency of ZnO nanoparticulate film based DSSCs modified with Al$_2$O$_3$ and SiO$_2$ with different sensitisation time; (d) comparison of short-circuit current densities of ZnO nanoparticulate film based DSSCs modified with Al$_2$O$_3$ and SiO$_2$ with different sensitisation time.

Reprinted from Colloids Surf. A, 386(1–3), Z. Qin, Y. Huang, J. Qi, L. Qu and Y. Zhang, Improvement of the performance and stability of the ZnO nanoparticulate film electrode by surface modification for dye-sensitized solar cells, 179–184. Copyright (2011), with permission from Elsevier.[61]

prolonged from 1 h to 5 h, the pure ZnO nanoparticulate based DSSCs demonstrated a fast decrease in performance, while the DSSCs modified with Al$_2$O$_3$ or SiO$_2$ layers showed much improved stability, as shown in Figure 5.11. Meanwhile, both the I_{SC} and η of the DSSCs are enhanced upon the modification of Al$_2$O$_3$ or SiO$_2$ due to reduced Zn^{2+}/dye aggregates. For ZnO nanowire array electrodes, the acid stability of the electrodes can also be improved upon modification of Al$_2$O$_3$ or SiO$_2$ layer on the surface.[63] As shown in Figure 5.11(a) and (b), the Al$_2$O$_3$ and SiO$_2$ layer can be observed clearly on the surface of ZnO nanowire arrays. As the sensitisation time was prolonged from 20 min to 120 min, the I_{SC} and η were much more stable for the modified DSSCs compared with those for pure ZnO nanowire array based DSSCs. Another effective way to suppress Zn^{2+}/dye aggregates is to reduce the residence time that ZnO electrodes stay in dye solution. By using an electrophoresis based dye adsorption technique, the sensitisation time can be shortened to several minutes without any compromise in DSSC performance.[64]

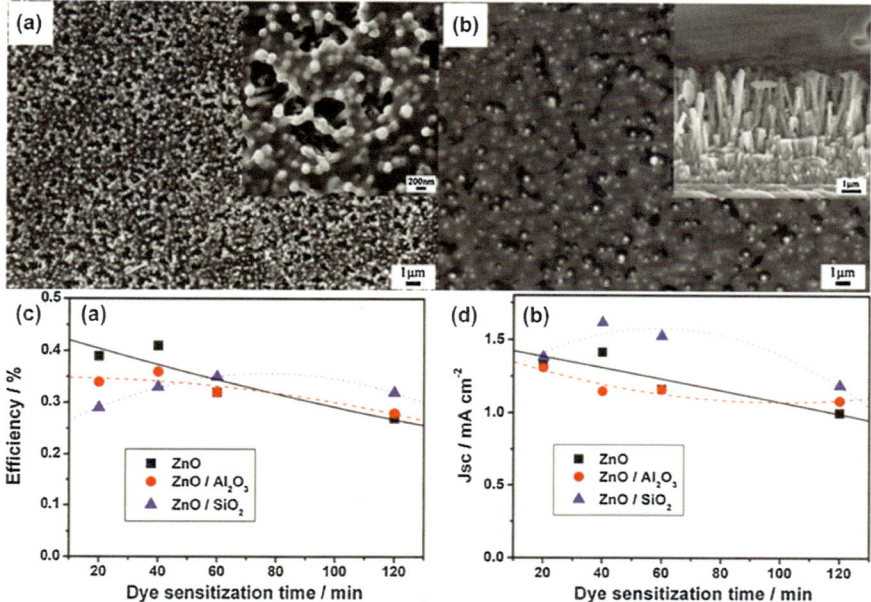

Figure 5.11 (a) Top view of the Al$_2$O$_3$ coated ZnO nanowire array film; (b) top view of the SiO$_2$ coated ZnO nanowire array film; (c) comparison of efficiency of ZnO nanowire array film based DSSCs modified with Al$_2$O$_3$ and SiO$_2$ with different sensitisation time; (d) comparison of short-circuit current densities of ZnO nanowire array film based DSSCs modified with Al$_2$O$_3$ and SiO$_2$ with different sensitisation time.
Reprinted from Mater. Lett., 70, Z. Qin, Y. Huang, Q. Liao, Z. Zhang, X. Zhang and Y. Zhang, Stability improvement of the ZnO nanowire array electrode modified with Al$_2$O$_3$ and SiO$_2$ for dye-sensitized solar cells, 177–180. Copyright (2012), with permission from Elsevier.[63]

5.3.2 Perovskite Solar Cells

Recently, perovskite solar cells (PSCs) have drawn a lot of attention due to their relatively high efficiencies.[65–69] The ZnO material has excellent properties including a simpler deposition process, milder sintering temperature and higher electron mobility compared to other materials. Therefore, ZnO is used as an effective electron transport layer (ETL) in high efficiency perovskite solar cells (PSCs).[70–73] A low-temperature solution-phase hydrothermal growth of ZnO nanorod (NR) array method was demonstrated.[74] The ZnO morphologies were effectively tuned by adjusting growth conditions such as growth time, temperature and precursor concentration. To improve the electron transport properties, nitrogen-doped ZnO was used to enhance the charge-carrier concentration. In addition, polyethyleneimine (PEI) capping limited the lateral growth and increased the aspect ratio which effectively enhanced the infiltration of the perovskite into the ZnO NR array. What's more, PEI coating reduced the work function which enhanced the electron extraction. Such a ZnO NR array with large NR aspect ratio,

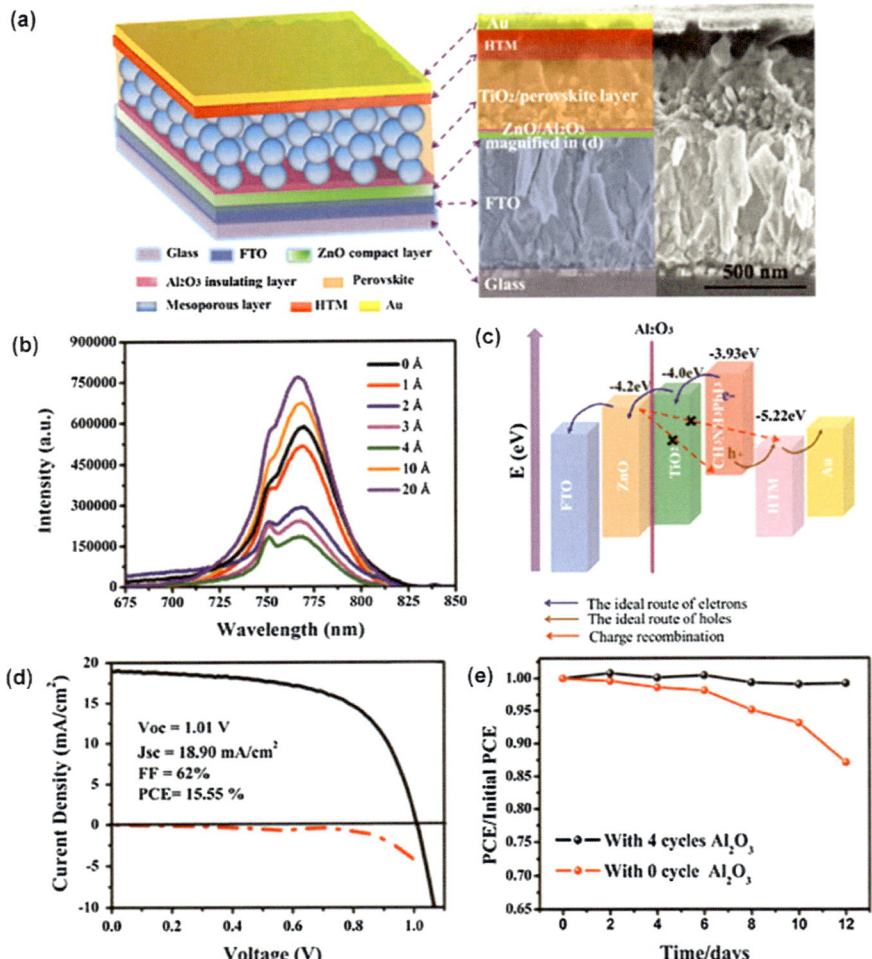

Figure 5.12 (a) Schematic diagram and cross-sectional FESEM image of the designed PSC; (b) steady-state PL emission spectra of FTO/ZnO/mp-TiO$_2$/perovskite with Al$_2$O$_3$ insulating layer of different thickness; (c) schematic of band energy level and carrier transportation route; (d) $J–V$ characteristics for champion device (solid line) and corresponding dark current (dashed line); (e) the stability test of PSC with insulating layer at different deposition thickness in the same stored conditions.

Reprinted from Nano Energy, 22, H. Si, Q. Liao, Z. Zhang, Y. Li, X. Yang, G. Zhang, Z. Kang and Y. Zhang, An innovative design of perovskite solar cells with Al$_2$O$_3$ inserting at ZnO/perovskite interface for improving the performance and stability, 223–231. Copyright (2016), with permission from Elsevier.[75]

increased electron density, and reduced work function altogether led to an improved efficiency, from 10% up to over 16.1% with minimal hysteresis. By doping and surface modification the efficiency of ZnO-based PSCs is significantly improved.

We have demonstrated that the efficiency of optimised ZnO-based PSCs was up to 15.55% and the stability issue was improved by interface modification.[75] In this work, the ZnO compact layer was fabricated by atom layer deposition (ALD). The ALD-fabricated ZnO ETL is homogeneous and flat with a low root mean squared surface roughness. After the deposition of ZnO, an ultra-thin Al_2O_3 insulating film was precisely inserted to separate the ZnO and perovskite layer based on the degradation mechanism. Subsequently, the perovskite film was fabricated by two-step sequential deposition method. Then the hole transport layer and Au were deposited. The designed structure and morphology of the ZnO-based PSCs are shown in Figure 5.12(a). In addition, the thickness of insulating layer was optimised by a quantum tunnelling model based on effective-mass approximation theory. The application of an Al_2O_3 layer not only suppressed the carrier recombination process in the cells but also effectively decreased the degradation rate of perovskite, which simultaneously enhanced both the stability and efficiency of PSCs as shown in Figure 5.12(b) and (c). Through tuning the thickness of Al_2O_3 layer, the optimised PSC performance was significantly increased by 44% and the stability was improved as shown in Figure 5.12(d) and (e).

References

1. H. J. Round, A note on carborundum, *Electr. world*, 1907, **49**(6), 309.
2. T. Aoki, Y. Hatanaka and D. C. Look, ZnO diode fabricated by excimer-laser doping, *Appl. Phys. Lett.*, 2000, **76**(22), 3257–3258.
3. A. Tsukazaki, A. Ohtomo, T. Onuma, M. Ohtani, T. Makino, M. Sumiya, K. Ohtani, S. F. Chichibu, S. Fuke, Y. Segawa, H. Ohno, H. Koinuma and M. Kawasaki, Repeated temperature modulation epitaxy for p-type doping and light-emitting diode based on ZnO, *Nat. Mater.*, 2004, **4**(1), 42–46.
4. J. H. Lim, C. K. Kang, K. K. Kim, I. K. Park, D. K. Hwang and S. J. Park, UV Electroluminescence Emission from ZnO Light-Emitting Diodes Grown by High-Temperature Radiofrequency Sputtering, *Adv. Mater.*, 2006, **18**(20), 2720–2724.
5. F. Sun, C. X. Shan, B. H. Li, Z. Z. Zhang, D. Z. Shen, Z. Y. Zhang and D. Fan, A reproducible route to p-ZnO films and their application in light-emitting devices, *Opt. Lett.*, 2011, **36**(4), 499–501.
6. H. Shen, C. X. Shan, Q. Qiao, J. S. Liu, B. H. Li and D. Z. Shen, Stable surface plasmon enhanced ZnO homojunction light-emitting devices, *J. Mater. Chem. C*, 2013, **1**(2), 234–237.
7. Y. I. Alivov, J. E. Van Nostrand, D. C. Look, M. V. Chukichev and B. M. Ataev, Observation of 430 nm electroluminescence from ZnO/GaN heterojunction light-emitting diodes, *Appl. Phys. Lett.*, 2003, **83**(14), 2943.
8. W. I. Park and G. C. Yi, Electroluminescence in n-ZnO Nanorod Arrays Vertically Grown on p-GaN, *Adv. Mater.*, 2004, **16**(1), 87–90.

9. X. M. Zhang, M. Y. Lu, Y. Zhang, L.-J. Chen and Z. L. Wang, Fabrication of a High-Brightness Blue-Light-Emitting Diode Using a ZnO-Nanowire Array Grown on p-GaN Thin Film, *Adv. Mater.*, 2009, **21**(27), 2767–2770.

10. X. Li, J. Qi, Q. Zhang, Q. Wang, F. Yi, Z. Wang and Y. Zhang, Saturated blue-violet electroluminescence from single ZnO micro/nanowire and p-GaN film hybrid light-emitting diodes, *Appl. Phys. Lett.*, 2013, **102**(22), 221103.

11. S. Xu, C. Xu, Y. Liu, Y. Hu, R. Yang, Q. Yang, J. H. Ryou, H. J. Kim, Z. Lochner, S. Choi, R. Dupuis and Z. L. Wang, Ordered nanowire array blue/near-UV light emitting diodes, *Adv. Mater.*, 2010, **22**(42), 4749–4753.

12. S. Jha, O. Kutsay, I. Bello and S. Lee, ZnO nanorod based low turn-on voltage LEDs with wide electroluminescence spectra, *J. Lumin.*, 2013, **133**, 222–225.

13. J.-Y. Wang, C.-Y. Lee, Y.-T. Chen, C.-T. Chen, Y.-L. Chen, C.-F. Lin and Y.-F. Chen, Double side electroluminescence from p-NiO/n-ZnO nanowire heterojunctions, *Appl. Phys. Lett.*, 2009, **95**(13), 131117.

14. R. Deng, B. Yao, Y. Li, Y. Xu, J. Li, B. Li, Z. Zhang, L. Zhang, H. Zhao and D. Shen, Ultraviolet electroluminescence from n-ZnO/p-NiO heterojunction light-emitting diode, *J. Lumin.*, 2013, **134**, 240–243.

15. J. B. You, X. W. Zhang, S. G. Zhang, H. R. Tan, J. Ying, Z. G. Yin, Q. S. Zhu and P. K. Chu, Electroluminescence behavior of ZnO/Si heterojunctions: Energy band alignment and interfacial microstructure, *J. Appl. Phys.*, 2010, **107**(8), 083701.

16. J. Ye, S. Gu, S. Zhu, W. Liu, S. Liu, R. Zhang, Y. Shi and Y. Zheng, Electroluminescent and transport mechanisms of n-ZnO/p-Si heterojunctions, *Appl. Phys. Lett.*, 2006, **88**(18), 2112.

17. Y.-T. Shih, M. Wu, M. Chen, Y. Cheng, J. Yang and M. Shiojiri, ZnO-based heterojunction light-emitting diodes on p-SiC (4H) grown by atomic layer deposition, *Appl. Phys. B*, 2010, **98**(4), 767–772.

18. B. Ling, X. W. Sun, J. L. Zhao, S. T. Tan, Z. L. Dong, Y. Yang, H. Yu and K. Qi, Electroluminescence from a n-ZnO nanorod/p-CuAlO$_2$ heterojunction light-emitting diode, *Phys. E*, 2009, **41**(4), 635–639.

19. H. Ohta, K.-I. Kawamura, M. Orita, M. Hirano, N. Sarukura and H. Hosono, Current injection emission from a transparent p–n junction composed of p-SrCu$_2$O2/n-ZnO, *Appl. Phys. Lett.*, 2000, **77**(4), 475–477.

20. M. C. Jeong, B. Y. Oh, M. H. Ham, S. W. Lee and J. M. Myoung, ZnO-Nanowire-Inserted GaN/ZnO Heterojunction Light-Emitting Diodes, *Small*, 2007, **3**(4), 568–572.

21. R. Könenkamp, R. Word and M. Godinez, Ultraviolet electroluminescence from ZnO/polymer heterojunction light-emitting diodes, *Nano Lett.*, 2005, **5**(10), 2005–2008.

22. A. Nadarajah, R. C. Word, J. Meiss and R. Könenkamp, Flexible inorganic nanowire light-emitting diode, *Nano lett.*, 2008, **8**(2), 534–537.

23. C. Lee, J. Wang, Y. Chou, C. Cheng, C. Chao, S. Shiu, S. Hung, J. Chao, M. Liu and W. Su, White-light electroluminescence from

ZnO nanorods/polyfluorene by solution-based growth, *Nanotechnology*, 2009, **20**(42), 425202.

24. Q. Yang, W. Wang, S. Xu and Z. L. Wang, Enhancing light emission of ZnO microwire-based diodes by piezo-phototronic effect, *Nano Lett.*, 2011, **11**(9), 4012–4017.
25. Q. Yang, Y. Liu, C. Pan, J. Chen, X. Wen and Z. L. Wang, Largely enhanced efficiency in ZnO nanowire/polymer hybridized inorganic/organic ultraviolet light-emitting diode by piezo-phototronic effect, *Nano Lett.*, 2013, **13**(2), 607–613.
26. C. Pan, L. Dong, G. Zhu, S. Niu, R. Yu, Q. Yang, Y. Liu and Z. L. Wang, High-resolution electroluminescent imaging of pressure distribution using a piezoelectric nanowire LED array, *Nat. Photonics*, 2013, 7(9), 752–758.
27. Y. Zhang and Z. L. Wang, Theory of piezo-phototronics for light-emitting diodes, *Adv. Mater.*, 2012, **24**(34), 4712–4718.
28. Y. Liu, H. Xu, C. Liu and W. Liu, Recent progress in ZnO-based heterojunction ultraviolet light-emitting devices, *Chin. Sci. bull.*, 2014, **59**(12), 1219–1227.
29. B. Thomas and D. Walsh, Metal-insulator-semiconductor electroluminescent diodes in single-crystal zinc oxide, *Electron. Lett.*, 1973, **16**(9), 362–363.
30. P. Chen, X. Ma and D. Yang, Fairly pure ultraviolet electroluminescence from ZnO-based light-emitting devices, *Appl. Phys. Lett.*, 2006, **89**(11), 111112.
31. T. Zhai, X. Fang, M. Liao, X. Xu, H. Zeng and B. Yoshio, A comprehensive review of one-dimensional metal-oxide nanostructure photodetectors, *Sensors*, 2009, **9**(8), 6504–6529.
32. C. Soci, A. Zhang, B. Xiang, S. A. Dayeh, D. P. R. Aplin, J. Park *et al.*, ZnO Nanowire UV Photodetectors with High Internal Gain, *Nano Lett.*, 2007, 7(4), 1003–1009.
33. H. Kind, H. Q. Yan, B. Messer, M. Law and P. D. Yang, Nanowire Ultraviolet Photodetectors and Optical Switches, *Adv. Mater.*, 2002, **14**(2), 158–160.
34. J. D. Prades, R. Jimenez-Diaz, F. Hernandez-Ramirez, L. Fernandez-Romero, T. Andreu, A. Cirera *et al.*, Toward a Systematic Understanding of Photodetectors Based on Individual Metal Oxide Nanowires, *J. Phys. Chem. C*, 2008, **112**(37), 14639–14644.
35. J. H. He, P. H. Chang, C. Y. Chen and K. T. Tsai, Electrical and opto-electronic characterization of a ZnO nanowire contacted by focused-ion-beam-deposited Pt, *Nanotechnology*, 2009, **20**(13), 135701.
36. N. Kouklin, Cu-Doped ZnO Nanowires for Efficient and Multispectral Photodetection Applications, *Adv. Mater.*, 2008, **20**(11), 2190–2194.
37. L. Peng, J.-L. Zhai, D.-J. Wang, P. Wang, Y. Zhang, S. Pang *et al.*, Anomalous photoconductivity of cobalt-doped zinc oxide nanobelts in air, *Chem. Phys. Lett.*, 2008, **456**(4–6), 231–235.
38. Z. X. Zhang, L. F. Sun, Y. C. Zhao, Z. Liu, D. F. Liu, L. Cao, B. S. Zou *et al.*, ZnO tetrapods designed as multiterminal sensors to distinguish false responses and increase sensitivity, *Nano Lett.*, 2008, **8**(2), 652–655.

39. Y. Dai, Y. Zhang and Z. L. Wang, The octa-twin tetraleg ZnO nano-structures, *Solid State Commun.*, 2003, **126**(11), 629–633.
40. W. Wang, J. Qi, Q. Wang, Y. Huang, Q. Liao and Y. Zhang, Single ZnO nanotetrapod-based sensors for monitoring localized UV irradiation, *Nanoscale*, 2013, **5**(13), 5981–5985.
41. W. Guo, Y. Yang, J. Qi, J. Zhao and Y. Zhang, Localized ultraviolet photoresponse in single bent ZnO micro/nanowires, *Appl. Phys. Lett.*, 2010, **97**(13), 133112.
42. C. S. Lao, M. C. Park, Q. Kuang, Y. Deng, A. K. Sood, D. L. Polla *et al.*, Giant enhancement in UV response of ZnO nanobelts by polymer surface-functionalization, *J. Am. Chem. Soc.*, 2007, **129**(40), 12096–12097.
43. M. L. Lu, C. W. Lai, H. J. Pan, C. T. Chen, P. T. Chou and Y. F. Chen, A facile integration of zero- (I-III-VI quantum dots) and one- (single SnO$_2$ nanowire) dimensional nanomaterials: fabrication of a nanocomposite photodetector with ultrahigh gain and wide spectral response, *Nano Lett.*, 2013, **13**(5), 1920–1927.
44. X. Zheng, Y. Sun, X. Yan, X. Chen, Z. Bai, P. Lin *et al.*, Tunable channel width of a UV-gate field effect transistor based on ZnO micro-nano wire, *RSC Adv.*, 2014, **4**(35), 18378–18381.
45. C.-Y. Lu, S.-J. Chang, S.-P. Chang, C.-T. Lee, C.-F. Kuo, H.-M. Chang *et al.*, Ultraviolet photodetectors with ZnO nanowires prepared on ZnO:Ga/glass templates, *Appl. Phys. Lett.*, 2006, **89**(15), 153101.
46. L. W. Ji, S. M. Peng, Y. K. Su, S. J. Young, C. Z. Wu and W. B. Cheng, Ultraviolet photodetectors based on selectively grown ZnO nanorod arrays, *Appl. Phys. Lett.*, 2009, **94**(20), 203106.
47. F. Yi, Q. Liao, X. Yan, Z. Bai, Z. Wang and X. Chen, Simple fabrication of a ZnO nanorod array UV detector with a high performance, *Phys. E*, 2014, **61**, 180–184.
48. F. Yi, Y. Huang, Z. Zhang, Q. Zhang and Y. Zhang, Photoluminescence and highly selective photoresponse of ZnO nanorod arrays, *Opt. Mater.*, 2013, **35**(8), 1532–1537.
49. W. Lin, X. Yan, X. Zhang, Z. Qin, Z. Zhang, Z. Bai, Y. Lei and Y. Zhang, The comparison of ZnO nanowire detectors working under two wavelengths of ultraviolet, *Solid State Commun.*, 2011, **151**(24), 1860–1863.
50. S. Liu, Q. Liao, S. Lu, X. Zhang, Z. Zhang, G. Zhang and Y. Zhang, Triboelectricity-assisted transfer of graphene for flexible optoelectronic applications, *Nano Res.*, 2016, **9**(4), 899–907.
51. S. Liu, Q. Liao, S. Lu, Z. Zhang, G. Zhang and Y. Zhang, Strain Modulation in Graphene/ZnO Nanorod Film Schottky Junction for Enhanced Photosensing Performance, *Adv. Funct. Mater.*, 2016, **26**(9), 1347–1353.
52. B. O'Regan and M. Grätzel, A low-cost, high-efficiency solar cell based on dye-sensitized colloidal TiO$_2$ films, *Nature*, 1991, **353**(6346), 737–740.
53. K. Keis, E. Magnusson, H. Lindström, S.-E. Lindquist and A. Hagfeldt, A 5% efficient photoelectrochemical solar cell based on nanostructured ZnO electrodes, *Sol. Energy Mater. Sol. Cells*, 2002, **73**(1), 51–58.

54. Y. Tachibana, K. Hara, S. Takano, K. Sayama and H. Arakawa, Investigations on anodic photocurrent loss processes in dye sensitized solar cells: comparison between nanocrystalline SnO_2 and TiO_2 films, *Chem. Phys. Lett.*, 2002, **364**(3–4), 297–302.

55. R. Katoh, A. Furube, T. Yoshihara, K. Hara, G. Fujihashi, S. Takano, S. Murata, H. Arakawa and M. Tachiya, Efficiencies of Electron Injection from Excited N3 Dye into Nanocrystalline Semiconductor (ZrO_2, TiO_2, ZnO, Nb_2O_5, SnO_2, In_2O_3) Films, *J. Phys. Chem. B*, 2004, **108**(15), 4818–4822.

56. K. Sayama, H. Sugihara and H. Arakawa, Photoelectrochemical Properties of a Porous Nb_2O_5 Electrode Sensitized by a Ruthenium Dye, *Chem. Mater.*, 1998, **10**(12), 3825–3832.

57. M. Quintana, T. Edvinsson, A. Hagfeldt and G. Boschloo, Comparison of Dye-Sensitized ZnO and TiO_2 Solar Cells: Studies of Charge Transport and Carrier Lifetime, *J. Phys. Chem. C*, 2007, **111**(2), 1035–1041.

58. C. Bauer, G. Boschloo, E. Mukhtar and A. Hagfeldt, Electron Injection and Recombination in $Ru(dcbpy)_2(NCS)_2$ Sensitized Nanostructured ZnO, *J. Phys. Chem. B*, 2001, **105**(24), 5585–5588.

59. T. P. Chou, Q. Zhang and G. Cao, Effects of Dye Loading Conditions on the Energy Conversion Efficiency of ZnO and TiO_2 Dye-Sensitized Solar Cells, *J. Phys. Chem. C*, 2007, **111**(50), 18804–18811.

60. M. Law, L. E. Greene, A. Radenovic, T. Kuykendall, J. Liphardt and P. Yang, ZnO-Al2O3 and ZnO-TiO_2 core-shell nanowire dye-sensitized solar cells, *J. Phys. Chem. B*, 2006, **110**(45), 22652–22663.

61. Z. Qin, Y. Huang, J. Qi, L. Qu and Y. Zhang, Improvement of the performance and stability of the ZnO nanoparticulate film electrode by surface modification for dye-sensitized solar cells, *Colloids Surf., A*, 2011, **386**(1–3), 179–184.

62. Z. Qin, Y. Huang, J. Qi, Q. Liao, W. Wang and Y. Zhang, Surface destruction and performance reduction of the ZnO nanowire arrays electrode in dye sensitization process, *Mater. Lett.*, 2011, **65**(23–24), 3506–3508.

63. Z. Qin, Y. Huang, Q. Liao, Z. Zhang, X. Zhang and Y. Zhang, Stability improvement of the ZnO nanowire array electrode modified with Al_2O_3 and SiO_2 for dye-sensitized solar cells, *Mater. Lett.*, 2012, **70**, 177–180.

64. G. Zhang, Q. Liao, Z. Qin, Z. Zhang, X. Zhang, P. Li, Q. Wang, S. Liu and Y. Zhang, Fast sensitization process of ZnO-nanorod-array electrodes by electrophoresis for dye-sensitized solar cells, *RSC Adv.*, 2014, **4**(74), 39332–39336.

65. J. Burschka, N. Pellet, S.-J. Moon, R. Humphry-Baker, P. Gao, M. K. Nazeeruddin and M. Gratzel, Sequential deposition as a route to high-performance perovskite-sensitized solar cells, *Nature*, 2013, **499**(7458), 316–319.

66. J.-H. Im, C.-R. Lee, J.-W. Lee, S.-W. Park and N.-G. Park, 6.5% efficient perovskite quantum-dot-sensitized solar cell, *Nanoscale*, 2011, **3**(10), 4088–4093.

67. H.-S. Kim, C.-R. Lee, J.-H. Im, K.-B. Lee, T. Moehl, A. Marchioro, S.-J. Moon, R. Humphry-Baker, J.-H. Yum, J. E. Moser, M. Grätzel and N.-G. Park, Lead Iodide Perovskite Sensitized All-Solid-State Submicron Thin Film Mesoscopic Solar Cell with Efficiency Exceeding 9%, *Sci. Rep.*, 2012, **2**, 591.
68. M. Liu, M. B. Johnston and H. J. Snaith, Efficient planar heterojunction perovskite solar cells by vapour deposition, *Nature*, 2013, **501**(7467), 395–398.
69. W. S. Yang, J. H. Noh, N. J. Jeon, Y. C. Kim, S. Ryu, J. Seo and S. I. Seok, High-performance photovoltaic perovskite layers fabricated through intramolecular exchange, *Science*, 2015, **348**(6240), 1234–1237.
70. J. Min, Z.-G. Zhang, Y. Hou, C. O. Ramirez Quiroz, T. Przybilla, C. Bronnbauer, F. Guo, K. Forberich, H. Azimi and T. Ameri, Interface engineering of perovskite hybrid solar cells with solution-processed perylene–diimide heterojunctions toward high performance, *Chem. Mater.*, 2014, **27**(1), 227–234.
71. D.-Y. Son, K.-H. Bae, H.-S. Kim and N.-G. Park, Effects of seed layer on growth of ZnO nanorod and performance of perovskite solar cell, *J. Phys. Chem. C*, 2015, **119**(19), 10321–10328.
72. X. Xu, H. Zhang, J. Shi, J. Dong, Y. Luo, D. Li and Q. Meng, Highly efficient planar perovskite solar cells with a TiO_2/ ZnO electron transport bilayer, *J. Mater. Chem. A*, 2015, **3**(38), 19288–19293.
73. D. Liu, J. Yang and T. L. Kelly, Compact layer free perovskite solar cells with 13.5% efficiency, *J. Am. Chem. Soc.*, 2014, **136**(49), 17116–17122.
74. K. Mahmood, B. S. Swain and A. Amassian, 16.1% Efficient Hysteresis-Free Mesostructured Perovskite Solar Cells Based on Synergistically Improved ZnO Nanorod Arrays, *Adv. Energy Mater.*, 2015, **5**(17), 1500568.
75. H. Si, Q. Liao, Z. Zhang, Y. Li, X. Yang, G. Zhang, Z. Kang and Y. Zhang, An innovative design of perovskite solar cells with Al_2O_3 inserting at ZnO/perovskite interface for improving the performance and stability, *Nano Energy*, 2016, **22**, 223–231.

CHAPTER 6

Photoelectrochemical Devices

ZHUO KANG, ZHIMING BAI AND YUE ZHANG*

University of Science and Technology Beijing, Beijing, China
*Email: yuezhang@ustb.edu.cn

6.1 Introduction

In recent years, with the sustained growth of global energy needs, research for seeking new energy materials has attracted increasing attention. Hydrogen, as a secondary energy material, has many merits such as being clean, efficient, safe, storable and transportable. Moreover, the calorific value of hydrogen is very high and its overall combustion product is water, which is harmless to the environment. Also, the heat release of hydrogen is 2.7 times that of gasoline at the same weight. Thus, hydrogen is considered to be one of the most ideal green energy sources this century. At present, the main ways of producing hydrogen are steam reforming and cracking petroleum gas. Notably, the raw materials are mainly fossil fuels, which when consumed release carbon dioxide making matters worse from an environmental perspective.[1]

In nature, green plants convert solar energy into chemical energy and store the energy in sugar through photosynthesis.[2,3] Inspired by this idea, solar water splitting, the so-called artificial photosynthesis, uses a water decomposition reaction to produce hydrogen and reserve the solar energy.[4,5] In such processes, multi-junction configurations that use *p*-type and *n*-type semiconductors with different bandgaps, the integration of a photovoltaic device with a photoelectrochemical (PEC) cell and the integration of a photovoltaic device with an electrolyser device are the prominent approaches for the development of PEC water splitting.[6] With the PEC water splitting technique, hydrogen can be yielded with solar illumination during

Nanoscience & Nanotechnology Series No. 43
ZnO Nanostructures: Fabrication and Applications
By Yue Zhang
© Yue Zhang 2017
Published by the Royal Society of Chemistry, www.rsc.org

the day, and subsequently adopted for electricity generation using hydrogen fuel cells instead of solar cells at night. Therefore, no carbon dioxide release, as well as less energy consumption, can be realised when compared with hydrogen production from the steam reforming of fossil fuels and cracking of petroleum gas.

6.2   PEC Principles

Typically, in a PEC system, the semiconductor PEC cell is composed of three parts: light trapping material modified transparent conductive substrate as photoanode and photocathode, and the electrolyte between the two photo-electrodes. In the PEC process, the photons with energy larger than the bandgap of semiconductors are absorbed by electrons in the valence band, thereby generating electron–hole pairs. The photo-generated electrons and holes are separated under the electric field, which is formed in the space charge region between the semiconductor and electrolyte, thus realising the conversion from the solar energy to electric energy. The photo-generated holes have strong oxidizing properties and subsequently realise oxidisation reactions at the photoanode surface, while the photo-generated electrons transfer through the external circuit to the photocathode surface to realise reduction reactions. Figure 6.1 shows a schematic diagram of the PEC water splitting mechanism. Under illumination, after the separation of photo-generated electron–hole pairs, oxygen and hydrogen molecules were yielded at anode and cathode, respectively. When compared with photocatalytic protocols, PEC technology with the assistance of an applied bias significantly reduces the recombination rate and thus improves the solar-to-hydrogen conversion efficiency.

The hydrogen production process of a PEC cell can be divided into four steps: (1) the light absorption and the charge carrier generating; (2) the separation and transportation of the carriers inside photoelectrodes; (3) the extraction of carrier and the generation of electrochemical product at the solid–liquid interface; (4) the control of electrochemical products in the

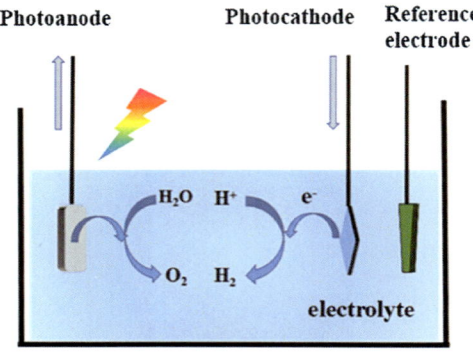

Figure 6.1   Schematic diagram of the energy band structure of a PEC cell.

electrolyte. In the presence of an external electric field at the interface of the semiconductor, the separation efficiency of the photo-generated carriers is greatly elevated. The subsequent electron transport process is closely related to the morphology and structure of photoelectrodes. The energy loss exists in every step of the above mentioned process; for example, the recombination, the trapping, the slow reaction originating from the over potential as well as the interruption of charge transportation. Therefore, the key to increasing the PEC energy conversion efficiency is to improve the PEC performance of the photoelectrode materials, and to simultaneously minimise the unnecessary energy loss process. Consequently, it is important to select a suitable semiconductor for photoelectrode modification. If the semiconductor bandgap is too wide, it is not beneficial for the visible light absorption in consideration of the solar utilisation efficiency. However, for the narrow bandgap semiconductors, it is difficult for photo-generated carriers to perform their redox characteristics, which will intensify the energy loss of the whole PEC process. Through calculation, the ideal semiconductor bandgap for water splitting should be between 1.8–2.2 eV.[3]

The most important parameters of the PEC system that can be used to reflect the PEC efficiency are solar-to-hydrogen (STH) efficiency (η) and incident-photo-to-current efficiency (IPCE).

The STH efficiency is a general benchmark that measures the overall efficiency of a water splitting system when exposed to AM 1.5G illumination under unbiased conditions, which can also be expressed as a fraction of produced chemical energy to the illuminated solar energy. The produced chemical energy can be equal to the product of the current density and the deviation between the 1.23 V and applied bias, and the P_{light} represents the incident light intensity. The formula can be expressed as follows:

$$\eta = \frac{I \times (1.23 - V)}{P_{light}} \tag{6.1}$$

IPCE describes the efficiency of the collected photocurrent per incident photo flux as a function of the illuminated wavelength. For a PEC system, the IPCE measures the utilisation rate of the incident light, which can also be termed as 'electrons out per photons in'. IPCE is calculated as the ratio of the photocurrent measured by the chronoamperometry experiment in three electrode system *vs.* the rate of incident photons from monochromatic light source at various wavelengths.

$$IPCE = \frac{1240 \times I}{\lambda \times P_{light}} \tag{6.2}$$

6.3 PEC Performance Optimisation

The bandgap and energy band position of ZnO are like those of TiO$_2$, and a low valence band position enables a powerful oxidisation capacity of photo-induced holes.[7] Besides, the 1D ZnO nanomaterials usually have a rough surface, which is beneficial for light trapping. The charge transfer efficiency inside single crystalline 1D ZnO nanomaterials is usually several orders of magnitude greater than that in polycrystalline materials. Moreover, the electron mobility of ZnO is dozens of times than that of TiO_2.[8,9] As a result, 1D ZnO nanostructures have attracted widespread attention in the research field of PEC cells.[10] Even so, its wide bandgap restricts the light absorption to the UV region, which only counts for about 5% of the whole solar energy. Therefore, for ZnO nanostructure based photoanodes, extending light absorption in the visible region is the key scientific issue we have to face. Until now, various approaches toward this such as doping,[11,12] noble metal modification,[13–15] quantum dots sensitisation[16,17] and heterostructure construction with narrow bandgap semiconductors[18–22] have been proposed. In addition, countless strategies in terms of charge carrier transfer and separation,[23–25] photoelectrode surface chemical reactions[18,26] and photostability[18,20] were also realised for PEC performance improvement based on ZnO nanostructures.

6.3.1   Light Absorption

As discussed above, the absorption in the visible light region of ZnO materials is poor, which greatly affects solar energy conversion from its origin. The common methods to solve such an issue is to cooperate with novel light absorption materials including noble metals with the surface plasma resonance (SPR) effect, and the narrow bandgap materials to form buried junctions with ZnO.

Recently, plasmonic nanostructures of noble metals, especially Au, have been demonstrated to be promising in photocatalytic and other solar energy conversion applications. In many published works, their tunable interactions with light in the visible and infrared regions as well as strong light scattering abilities and localised (SPR) effect were highlighted.[27–30] Additionally, when compared with inorganic photosensitisers, noble metals usually have ideal corrosion resistance capability during photoreactions.[31]

Au nanoparticles were obtained as follows: 10 mL of chloroauric acid ($HAuCl_4$, 0.01%) solution was adjusted to pH 7 using 0.1 M NaOH solution, and 1 mL 1% polyvinyl alcohol (PVA, $[C_2H_4O]_n$), and 0.5 mL methanol (CH_3OH, 99.5%) were added. Then the ZnO nanorod (NR) arrays were immersed in the mixed solution. After 10 min UV irradiation, the samples were calcined at 350 °C in N_2 atmosphere for 30 min. During this process, Au^{3+} from the $HAuCl_4$ was reduced to neutral Au NPs. The detailed reaction process under UV illumination is described as follows:

$$ZnO \rightarrow ZnO(e + h) \tag{6.3}$$

$$ZnO(e + h) \rightarrow e^- + h^+ \tag{6.4}$$

$$AuCl_4^- + 3e^- \rightarrow Au + 4Cl^- \tag{6.5}$$

With a thin Al_2O_3 coating layer through atom layer deposition (ALD) method, the as-synthesised $ZnO/Au/Al_2O_3$ nanostructures were also applied for PEC hydrogen evolution.[15] The SEM images of well-aligned pristine ZnO NRs with smooth surface and the sample decorated with uniformly distributed Au nanoparticles are shown in Figure 6.2(a) and (b). The ZnO NRs still maintain their vertically-oriented characteristics after Au nanoparticle

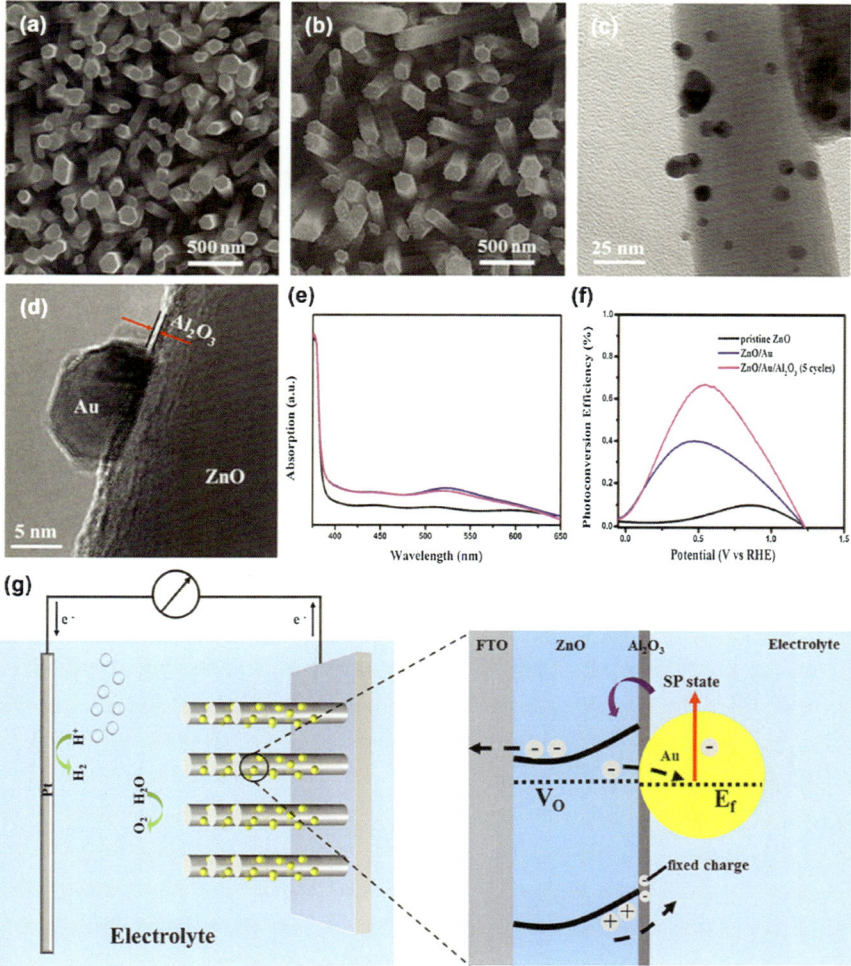

Figure 6.2   (a) Top view SEM image of pristine ZnO; (b) top view SEM image of ZnO/Au; (c) a low-magnification TEM image; (d) a HRTEM image of $ZnO/Au/Al_2O_3$ (5 cycles); (e) UV-vis absorption spectra of pristine ZnO, ZnO/Au and $ZnO/Au/Al_2O_3$ (5 cycles); (f) photoconversion efficiency of the PEC cell with three different photoanodes as a function of the applied potential (*vs.* RHE); (g) scheme for the mechanism of PEC water splitting and energy band diagram of $ZnO/Au/Al_2O_3$ photoanode.
Reproduced with permission from Macmillan Publishers Ltd: *Sci. Rep.* (ref. 15), copyright (2016).

modification. The detailed morphology of Au nanoparticles can be more clearly identified in Figure 6.2(c) and (d). The UV-vis adsorption spectra were performed ranging from 370–650 nm in Figure 6.2(e), and all samples exhibited a huge absorption in UV region. Notably, ZnO/Au and ZnO/Au/Al$_2$O$_3$ samples showed an obvious absorption peak around 525 nm that corresponded well to the localised SPR effect of Au NPs. This demonstrated that the solar energy in the visible light region can be efficiently utilised with the introduction of plasmonic Au nanoparticles. Under visible light illumination, the hot electrons in Au induced by the SPR effect transferred to ZnO to contribute to the enhanced photocurrent. Thus, the STH efficiency of the ZnO/Au/Al$_2$O$_3$ photoanode was calculated to be 0.67% at 0.55 V *vs.* RHE bias (Figure 6.2(f)). Such a performance was much better than that of pristine ZnO photoanode (0.1% at 0.88 V *vs.* RHE bias).

Similar photoanode structures with the assistance of Au nanoparticles for visible light harvesting have been successfully conducted for enhanced PEC performance. In the ZnO/ZnS/Au photoanode,[13] the surface plasmons in Au NPs were photoexcited under visible light illumination, which subsequently led to the hot electron injection from the surface plasmon states (SP states) to the conduction band of ZnO.[14] Therefore, the photocurrent density was improved in the visible region, and finally resulted in a STH efficiency of 0.21%, which was 3.5 times that of a pristine ZnO photoanode. These results indicate that plasmonic metal/ZnO heterostructure photoanodes have great potential for sufficiently utilisation of solar energy in PEC water splitting as well as other related photovoltaic fields.

Another effective way for visible light absorption improvement is to cooperate with narrow bandgap semiconductors. Among various candidates, CdS is regarded as one of the most commonly used quantum dots to modify ZnO nanostructure based photoanodes. Due to the appreciable bandgap of 2.4 eV and suitable energy band positions, the appropriate type-II band alignment between CdS and ZnO can be formed, which favourably facilitates the separation of photogenerated electron–hole pairs and results in improved PEC performance.[22,32–34]

The synthesis of CdS on ZnO is usually realised by a chemical bath deposition method.[19] 0.1234 g cadmium nitrate tetrahydrate and 0.03 g thioacetamide are separately dissolved in 10 mL deionised water. These solutions are then mixed together where the ZnO nanostructures sink into it. The entire reaction is carried out in a 40 °C water bath for 15 min and the acquired CdS/ZnO samples are further rinsed with deionised water for adsorption impurity removal and subsequently dried with nitrogen gas. Based on this, the stair-like type-II band alignment between ZnFe$_2$O$_4$(ZFO), CdS and ZnO was realised to enhance photoanode performance. After illumination, the photo-induced holes inside CdS were favourable for migratation to the ZFO nanoparticles, and subsequently took part in the oxidisation reactions at the photoanode surface. The optimised STH efficiency reached 4.4%, which was claimed to be more than 10 times that of pristine ZnO photoanode (Figure 6.3).

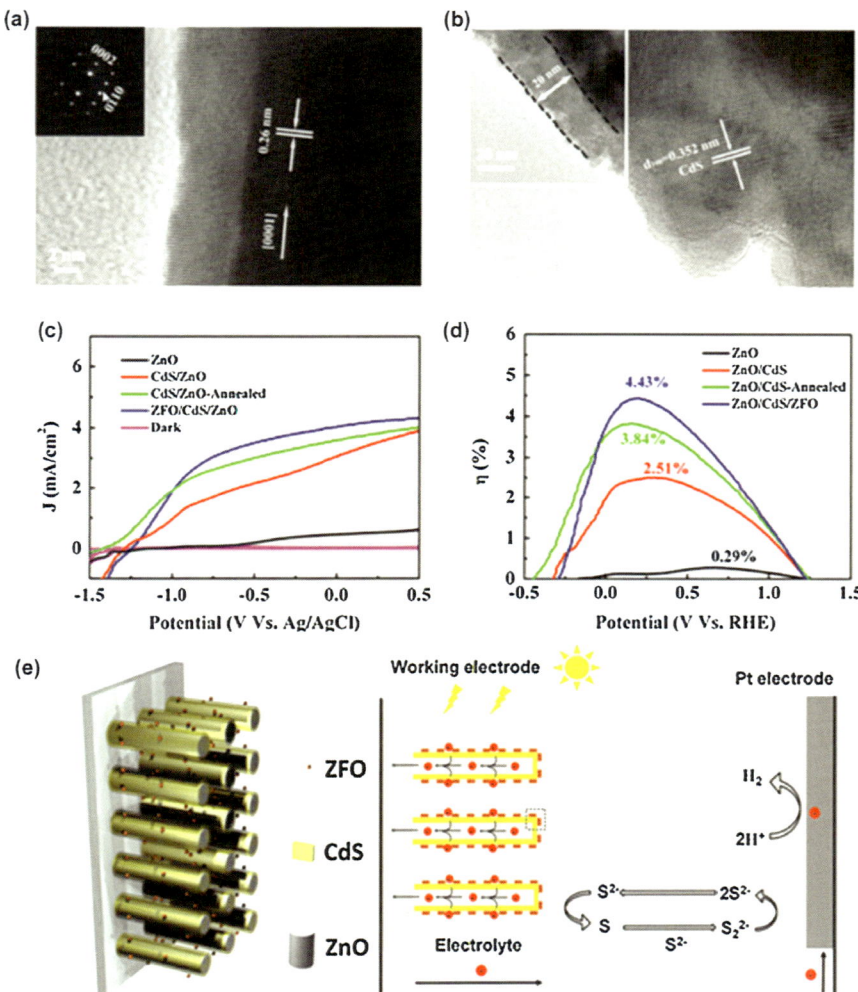

Figure 6.3 (a) HRTEM image of ZnO NR with the inset showing the SAED; (b) HRTEM image of ZnO/CdS core/shell structure; (c) *J–V* characteristics of fabricated ZnO based photoanodes in the dark and under light illumination (70 mW cm^{-2}); (d) STH efficiency of fabricated ZnO based photoanodes; (e) schematic of the ZnO nanorod arrays/CdS/ZFO photoanode and its working mechanism.

Reprinted from Nano Energy, 24, S. Cao, X. Yan, Z. Kang, Q. Liang, X. Liao and Y. Zhang, Band alignment engineering for improved performance and stability of ZnFe$_2$O$_4$ modified CdS/ZnO nanostructured photoanode for PEC water splitting, 25–31. Copyright (2016), with permission from Elsevier.[19]

Besides, the preparation of CdS nanoparticles was also realised through a successive ionic layer adsorption reaction approach.[18] To quantify the light absorption ability of the as-synthesised 3D ZnO NWA-CdS structure, the

IPCE measurements were carried out as a function of incident light wavelength as shown in Figure 6.4(c). The CdS-sensitised samples performed greater IPCE in both the UV and visible region (especially from 400–600 nm) compared with pristine ZnO nanostructures. Together with the UV-vis absorption spectra in Figure 6.4(d), again, a great enhancement in visible light absorption originating from CdS was sufficiently demonstrated. However, in this work, the contribution to the light absorption from the unique 3D branched ZnO nanostructure should not be neglected. The 3D spatial ZnO skeleton is beneficial for loading CdS with a larger quantity and simultaneously increases the solid/liquid interface area, which subsequently contributes to the high light-trapping ability and excellent photo-generated collection capability, respectively. The fabrication processes of the 3D nanostructure photoanode are shown in Figure 6.5. For such photoanodes, the STH efficiency was as high as 3.1%, highlighting the advantages of both 3D branched ZnO structures and CdS. Besides, homogeneous or heterogeneous spatial nanostructures have also been widely adopted in the study of solar energy conversion devices due to their large surface area, strong carrier collection ability and ability to easily form compounds with other functional nanomaterials.[35–38]

Favourable visible light adsorption is a guarantee of outstanding PEC performance. In addition to CdS as visible light sensitiser, another ternary chalcogenide semiconductor Zinc indiumsulfide ($ZnIn_2S_4$) with a narrow bandgap was applied in the ZnO based PEC water splitting system as well.[20] The $ZnIn_2S_4$ electrocatalyst nanosheets were synthesised on reduced graphene oxide (RGO) through a facial solvothermal method, and were subsequently grafted onto ZnO NAs. The morphology for $ZnIn_2S_4$ overcoated on rGO/ZnO is shown in Figure 6.6(b), and its contribution to visible light absorption was demonstrated by UV-vis diffuse reflectance spectra test, as shown in Figure 6.6(c). In the ZnO NAs/RGO/$ZnIn_2S_4$ composite photoanode, ZnO and rGO acted as the matrix for charge transfer and $ZnIn_2S_4$ loading while the $ZnIn_2S_4$ functioned as effective visible light absorption. The formed type-II band alignment between $ZnIn_2S_4$ and ZnO facilitated the photogenerated charge separation and transfer. Moreover, in terms of stability, $ZnIn_2S_4$ was also demonstrated to have considerably superior photostability in aqueous solution compared with CdS.[39,40]

Similarly, other narrow bandgap semiconductors such as Cu_2O,[21] CdSe,[41] and graphitic-C_3N_4[42] were also intensively studied and their dominant working mechanism basically relied on the light adsorption extending to the visible light region, as well as the buried heterojunctions between the two different bandgap semiconductor materials.

6.3.2 Charge Separation Efficiency

Under illumination, electron–hole pairs are generated inside semiconductors. Then the effective separation of such photo-induced charge carriers becomes the second issue that should be taken into consideration.

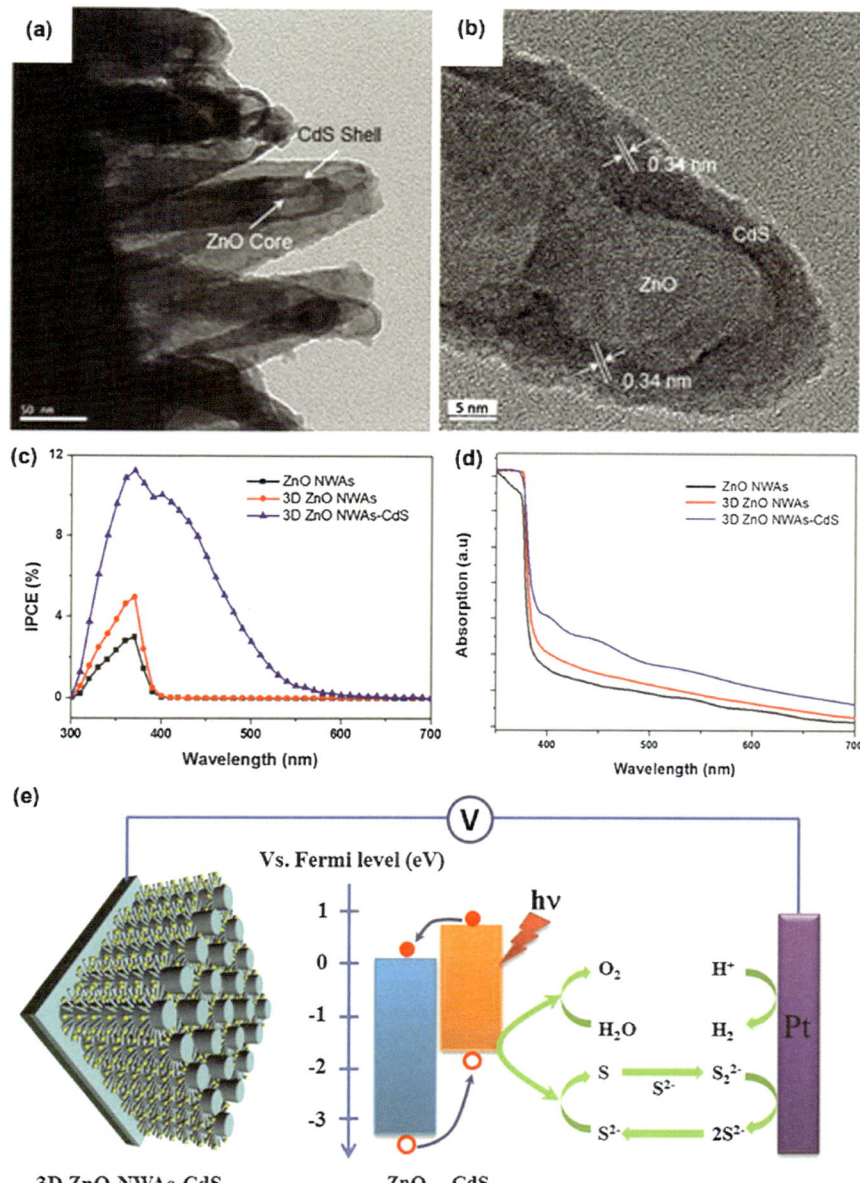

Figure 6.4 (a) TEM and (b) HRTEM images of 3D ZnO nanowire arrays-CdS; (c) IPCE plots of ZnO nanowire arrays, 3D ZnO nanowire arrays-CdS, and 3D ZnO nanowire arrays-CdS in the region of 300–700 nm; (d) UV-vis adsorption spectra of ZnO nanowire arrays, 3D ZnO nanowire arrays, and 3D nanowire arrays-CdS; (e) schematic diagram of the PEC reaction mechanism of 3D ZnO nanowire arrays-CdS photoanode.
Reproduced with permission from John Wiley and Sons. Copyright © 2015 WILEY-VCH Verlag GmbH & Co. KGaA, Weinheim.[18]

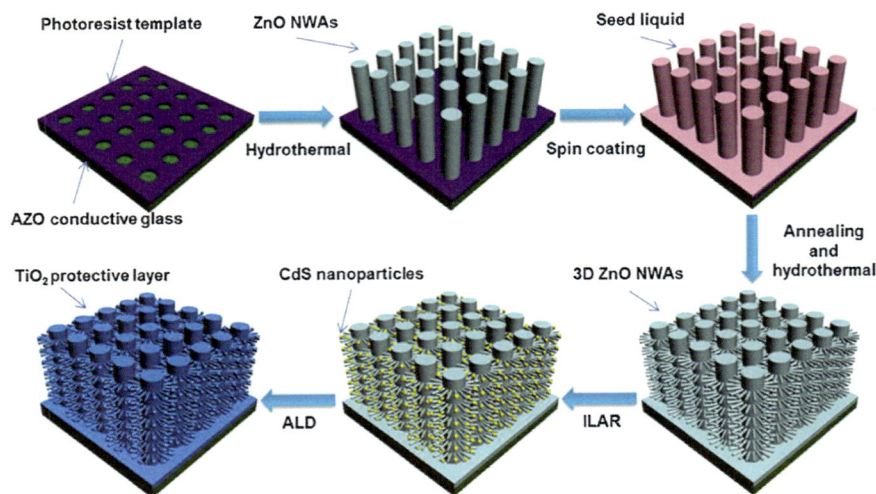

Figure 6.5 Schematic diagram of the processes for the fabrication of 3D ZnO
nanowire array–CdS photoanodes.
Reprinted with permission from John Wiley and Sons. Copyright © 2015
WILEY-VCH Verlag GmbH & Co. KGaA, Weinheim.[18]

Various protocols were proposed in terms of this, for example, incorporating
high electron mobility materials to create electron transport channels for the
separation efficiency enhancement, and introducing passivation layers to
suppress the recombination at the interface.

Graphene, due to its high electron mobility (up to $15\,000\ cm^2\,V^{-1}\,s^{-1}$),[43]
has attracted much attention and is used in numerous research fields. When
incorporating many oxygenic functional groups into graphene through several procedures, they become RGO, which possesses very high specific surface area and many unique characteristics due to the abundant functional
groups. The *in situ* growth of semiconductors on RGO provides stronger
interfacial contact, which greatly facilitates electron transfer and charge
separation, as well as the charge transfer across the solid–liquid interface.
These advantages verify the significance of RGO adoption.

RGO was introduced between ZnO nanowires arrays and $ZnIn_2S_4$ for
photoanode construction.[20] It provided largely improved surface area, and
simultaneously accelerated the PEC process through decreasing the energy
barrier of interfacial electrochemical reactions. In such type-II band alignment structures (Figure 6.7), the photo-generated charge carriers inside ZnO
and $ZnIn_2S_4$ are effectively separated. The photo-generated electrons in the
conduction band (CB) of $ZnIn_2S_4$ can more easily transfer to the CB of ZnO
through the RGO media layer and finally migrate to the photocathode to
participate in water reduction for hydrogen yield with the assistance of an
applied bias, while the photo-generated holes flow to the photoanode surface to be eliminated during water oxidisation. Thus, the formed type-II
band alignment and the RGO sheets contributed synergistically to the over

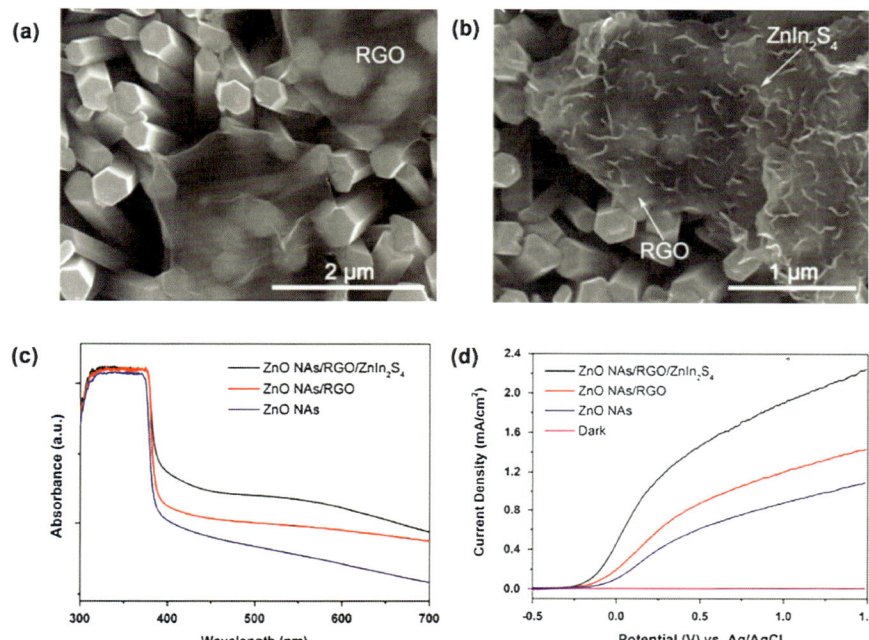

Figure 6.6 (a) SEM images of ZnO/RGO; (b) SEM images of ZnO/RGO/ZnIn$_2$S$_4$; (c) UV-vis diffuse reflectance spectra of ZnO NAs, ZnO NAs/RGO, and ZnO NAs/RGO/ZnIn$_2$S$_4$; (d) variation of current density *vs.* applied potential. Reprinted from Nano Energy, 14, Z. Bai, X. Yan, Z. Kang, Y. Hu, X. Zhang and Y. Zhang, Photoelectrochemical performance enhancement of ZnO photoanodes from ZnIn$_2$S$_4$ nanosheets coating, 392–400. Copyright (2015), with permission from Elsevier.[20]

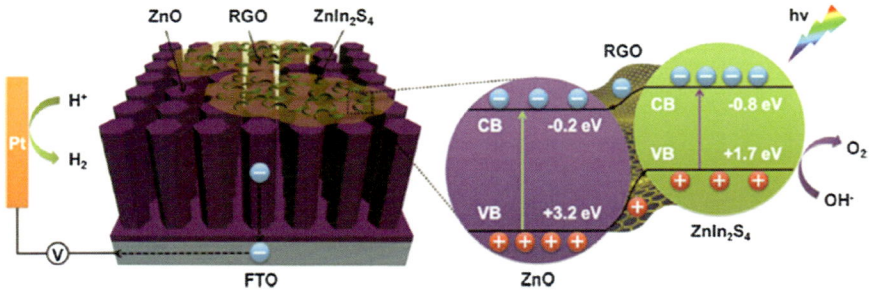

Figure 6.7 Schematic of the energy band structure of the ZnO nanowire arrays/RGO/ZnIn$_2$S$_4$ heterojunction and proposed mechanism for the improvement of the PEC performance.
Reprinted from Nano Energy, 14, Z. Bai, X. Yan, Z. Kang, Y. Hu, X. Zhang and Y. Zhang, Photoelectrochemical performance enhancement of ZnO photoanodes from ZnIn$_2$S$_4$ nanosheets coating, 392–400. Copyright (2015), with permission from Elsevier.[20]

200% enhancement of STH efficiency compared with pristine ZnO nano-structures. Interestingly, the RGO layer can not only be inserted on the top of ZnO nanostructures to facilitate charge transfer and subsequent charge separation, but also be located under ZnO nanostructures. Direct synthesis of ZnO nanowires arrays on the surface of the RGO layer was also realised.[44] The PEC performance elevation verified the enhanced charge extraction property, which was favourable for transport of photo-generated electrons inside ZnO to the external circuit.

Another promising method to effectively enhance PEC performance involves the insertion of a passivation layer to suppress the photo-generated charge carrier recombination at the interface. With the development of the ALD technique, such thin and high-quality passivation layers are usually acquired by the ALD method. Among various passivation materials, Al_2O_3, as a high-κ dielectric, shows great potential in electrochemical energy conversion devices.[45–47] For the construction of PEC photoanodes, Al_2O_3 coatings have been successively employed in Al_2O_3/TiO_2,[45] Al_2O_3/Fe_2O_3,[46] and Al_2O_3/WO_3[48] photoanodes as passivation layers.

In $ZnO/Au/Al_2O_3$ photoanodes,[15] Al_2O_3 coating layers were realised with 5, 10 and 20 cycles at a growth rate of ~1 Å per cycle. Argon (99.998%) at 25 sccm was adopted as a carrier gas. The reactor pressure and temperature were maintained at 0.2 Torr and 150 °C, respectively. Trimethylaluminum ($Al(CH_3)_3$) and DI water were separately pulsed into the chamber with a pulsing time of 20 ms and 15 ms followed by 20s N_2 purging. After that, the chamber was naturally cooled down under N_2 flow. The photocurrent decay indicated the possible block for electron transmission with such an insulation over layer (Figure 6.8(a)). However, in the PL test of Figure 6.8(b), the huge decrease in deep level emission intensity for the $ZnO/Au/Al_2O_3$ compared with ZnO/Au demonstrated that the recombination rate of the photogenerated electron hole pairs was successively reduced with such an

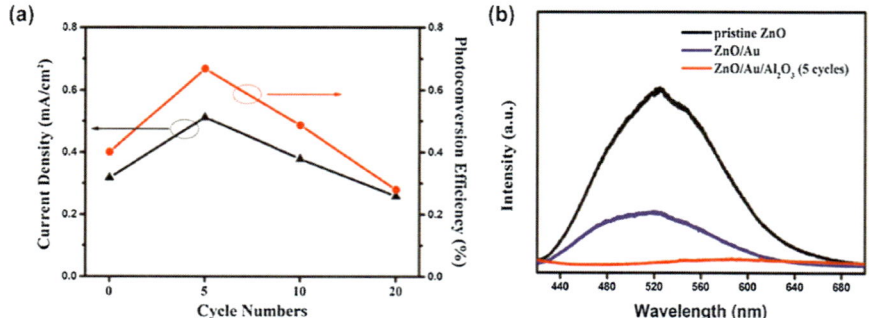

Figure 6.8 (a) Plot of the obtained photocurrent density and photoconversion efficiency as function of the $ZnO/Au/Al_2O_3$ photoanodes with various cycles at a scan rate of 10 mV s^{-1} under 40 mW cm^{-2} light illumination; (b) photoluminescence (PL) spectra of the different photoanodes. Reproduced with permission from Macmillan Publishers Ltd: *Sci. Rep.* (ref. 15), copyright (2016).

Al_2O_3 passivation layer. Finally, the photoanode passivated with 0.5 nm (5 cycle ALD) Al_2O_3 layer acquired optimal STH efficiency due to the effective passivation of the surface states on ZnO nanostructures.

Besides Al_2O_3, ZnS was also utilised in the ZnO based PEC system for the purpose of passivation.[13] A facile chemical conversion method was adopted to realise a ZnS coating layer outside ZnO nanostructures. The ZnO nanorods arrays were immersed in 0.2 M thiacetamide (TAA) aqueous solution and the whole sulfidation process was kept at 95 °C for 15 min. After that, the samples were washed with deionised water and ethanol repeatedly to remove impurities and then dried at room temperature. The TEM images in Figure 6.9(a) clearly show that the ZnS was uniformly coated on the surface of ZnO. Due to the presence of ZnS, the surface trap states of ZnO such as oxygen vacancies and the absorbing of oxygen were significantly reduced, providing higher separation efficiency. Moreover, the lattice distortion at the ZnO/ZnS interface resulted in a compressed bandgap of ZnO. This meant

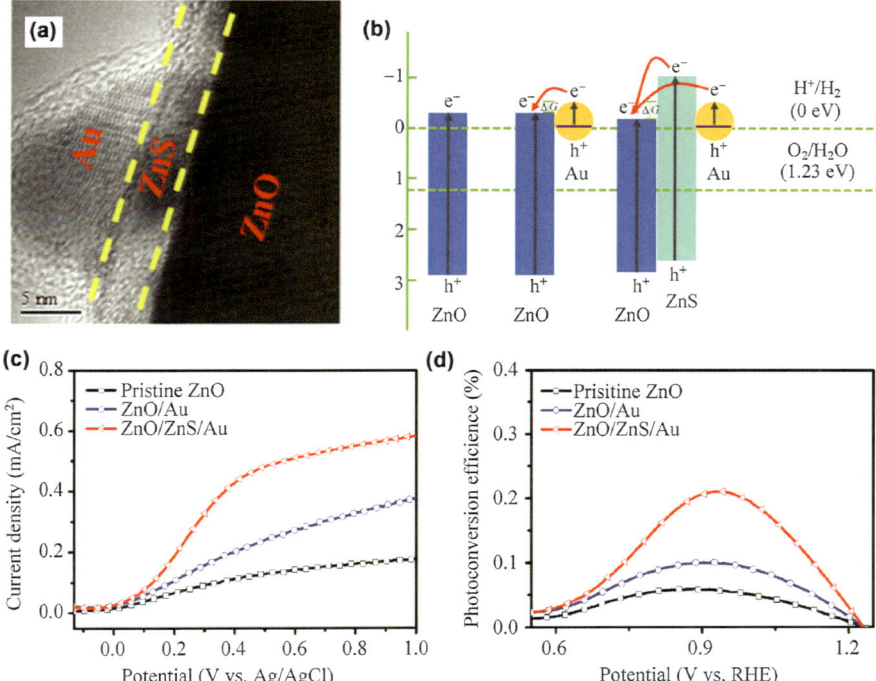

Figure 6.9   (a) TEM images of the ZnO/ZnS/Au; (b) the energy band structure of the ZnO, ZnO/Au and ZnO/ZnS/Au photoanodes; (c) linear sweep voltammetry measurements of ZnO, ZnO/Au and ZnO/ZnS/Au photoanodes with a 20 mV s^{-1} scan rate under illumination; (d) the acquired STH efficiency of fabricated photoanodes.

Reproduced from Nano Res., Design of sandwich-structured ZnO/ZnS/Au photoanode for enhanced efficiency of photoelectrochemical water splitting, 8(9), 2015, 2891–2900, Y. Liu, Y. Gu, X. Yan, Z. Kang, S. Lu, Y. Sun and Y. Zhang (© Tsinghua University Press and Springer-Verlag Berlin Heidelberg 2015) With permission from Springer.[13]

that the free energy change (ΔG) for hot electron injection from Au to ZnO increased, encouraging the tunnelling effect across ZnS and finally improving the utilisation efficiency of visible light.

6.3.3 Photo-stability

When considering the practical applications of the PEC system, the photo-stability is as important as efficiency. Usually, photo-corrosion greatly affects the stability of the PEC system, so protection layers outside photoelectrodes are widely adopted. The responsibilities of such coating layers are not only restricted to passivating surface states for suppressing photogenerated charge carriers, but also include providing ideal chemical stability to resist photocorrosion.

As we know, even though the CdS discussed in the last section is a good choice for light absorption, its photo-stability is always unsatisfactory. The poor stability results from the strong oxidizing conditions in some of the PEC systems, which causes oxidative deactivation of photoanode materials. A series of coating layers were realised to prevent the corrosion of CdS, and the most common candidate is the TiO_2 acquired with ALD. For the 3D ZnO nanowires array/CdS coated with TiO_2,[18] the photocurrent maintained 80% of the original value. In comparison, the sample without TiO_2 protection performed a sharp decay to 45% of the original value (Figure 6.10(b)). The improved current stability can be attributed to the strong anti-corrosion ability of TiO_2.[49] Since TiO_2 has a lower valence band position, the recombination of photogenerated charge carriers may happen inside the CdS nanoparticles and the photogenerated electrons in TiO_2 would consume some oxidizing holes in CdS, which effectively prevents the precipitation of solid sulfur on CdS, thereby reducing the photocorrosion reaction. Even so, the decreased original photocurrent leads to reduced STH efficiency of the PEC cell due to the low conductivity and such inappropriate band alignment between CdS and TiO_2 (Figure 6.10(a)). As we can see, the conflict between efficiency and stability is prominent. So, a comprehensive evaluation of both sides is strongly suggested to balance such contrary influences.

Besides TiO_2, $ZnFe_2O_4$(ZFO) was also adopted to address the photo-stability of ZnO/CdS based photoanodes.[19] The ZFO was acquired through urea-assisted auto-combustion synthesis. 0.3 mmol L^{-1} $Zn(NO_3)_2 \cdot 6H_2O$ and 0.6 mmol L^{-1} $Fe(NO_3)_3 \cdot 9H_2O$ solution were mixed together to keep a Zn/Fe molar ratio of 1:2, then 2 mmol L^{-1} NH_2CONH_2 was added and further stirred for 10 min. The ZFO modified ZnO NAs were prepared by spin-coating ZFO precursor at 800 rpm for 6 s and 3000 rpm for 30 s on ZnO NAs. After heating at 350 °C for 30 min in air, the nanocomposite was subsequently annealed at 600 °C for another 5 h in an Ar atmosphere to eliminate possible organic residues and to stabilise the ZnO/CdS/ZFO microstructures. The results in Figure 6.10(c) show that the photocurrent density of the ZnO/CdS photoanode decreased because of the photocorrosion of the CdS. However, for the samples with a decoration of ZFO, the photocorrosion was obviously

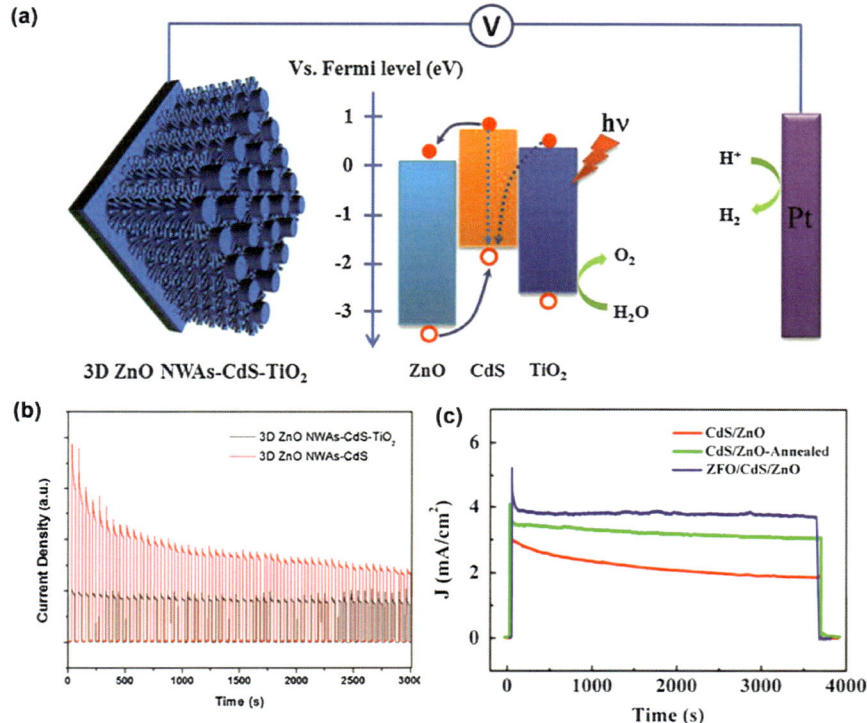

Figure 6.10 (a) The schematic diagram of the PEC reaction mechanism of 3D ZnO nanowire array/CdS/TiO$_2$; (b) transient current density *vs.* time of 3D ZnO nanowire array/CdS and 3D ZnO ZnO nanowire array/CdS/TiO$_2$; (c) stability of different photoanodes under continuous illumination. (a) and (b) reproduced with permission from John Wiley and Sons. Copyright © 2015 WILEY-VCH Verlag GmbH & Co. KGaA, Weinheim;[18] (c) reprinted from Nano Energy, 24, S. Cao, X. Yan, Z. Kang, Q. Liang, X. Liao and Y. Zhang, Band alignment engineering for improved performance and stability of ZnFe$_2$O$_4$ modified CdS/ZnO nano-structured photoanode for PEC water splitting, 25–31. Copyright (2016), with permission from Elsevier.[19]

suppressed. The slightly elevated photocurrent and stability improvement were attributed to the introduction of ZFO, which facilitated the charge carrier separation and extracted photo-generated holes in CdS to migrate to the surface of ZFO. The decreased photo-generated holes at CdS surface hindered its photocorrosion. Thus, the stability of the ZnO/CdS based PEC photoanode was improved through introduction of ZFO.

Many scientific and technical challenges remain in the field of solar water splitting. Further improving the solar utilisation efficiency and suppressing the recombination rate, as well as prolonging the stability duration are critical bottleneck issues. It is highly expected to see a breakthrough in such an artificial photosynthesis system in the foreseeable future.

References

1. A. Kudo and Y. Miseki, Heterogeneous photocatalyst materials for water splitting, *Chem. Soc. Rev.*, 2009, **38**(1), 253–278.
2. F. A. Frame, T. K. Townsend, R. L. Chamousis, E. M. Sabio, T. Dittrich, N. D. Browning and F. E. Osterloh, Photocatalytic Water Oxidation with Nonsensitized IrO_2 Nanocrystals under Visible and UV Light, *J. Am. Chem. Soc.*, 2011, **133**(19), 7264–7267.
3. Y. Tachibana, L. Vayssieres and J. R. Durrant, Artificial photosynthesis for solar water-splitting, *Nat. Photonics*, 2012, **6**(8), 511–518.
4. Y. Li and J. Z. Zhang, Hydrogen generation from photoelectrochemical water splitting based on nanomaterials, *Laser Photonics Rev.*, 2010, **4**(4), 517–528.
5. J. Z. Zhang, Metal oxide nanomaterials for solar hydrogen generation from photoelectrochemical water splitting, *MRS Bull.*, 2011, **36**(1), 48–55.
6. K. Zhang, M. Ma, P. Li, D. H. Wang and J. H. Park, Water Splitting Progress in Tandem Devices: Moving Photolysis beyond Electrolysis, *Adv. Energy Mater.*, 2016, **6**(15), 1600602.
7. Y. Y. Bu, Z. Y. Chen, W. B. Li and B. R. Hou, Highly Efficient Photocatalytic Performance of Graphene-ZnO Quasi-Shell-Core Composite Material, *ACS Appl. Mater. Interfaces*, 2013, **5**(23), 12361–12368.
8. L. E. Greene, M. Law, B. D. Yuhas and P. D. Yang, $ZnO-TiO_2$ core-shell nanorod/P3HT solar cells, *J. Phys. Chem. C*, 2007, **111**(50), 18451–18456.
9. E. Hendry, M. Koeberg, B. O'Regan and M. Bonn, Local field effects on electron transport in nanostructured TiO_2 revealed by terahertz spectroscopy, *Nano Lett.*, 2006, **6**(4), 755–759.
10. Z. Bai, X. Yan, X. Chen, H. Liu, Y. Shen and Y. Zhang, ZnO nanowire array ultraviolet photodetectors with self-powered properties, *Curr. Appl. Phys.*, 2013, **13**(1), 165–169.
11. Y. G. Lin, Y. K. Hsu, Y. C. Chen, L. C. Chen, S. Y. Chen and K. H. Chen, Visible-light-driven photocatalytic carbon-doped porous ZnO nanoarchitectures for solar water-splitting, *Nanoscale*, 2012, **4**(20), 6515–6519.
12. X. Yang, A. Wolcott, G. Wang, A. Sobo, R. C. Fitzmorris, F. Qian, J. Z. Zhang and Y. Li, Nitrogen-Doped ZnO Nanowire Arrays for Photoelectrochemical Water Splitting, *Nano Lett.*, 2009, **9**(6), 2331–2336.
13. Y. Liu, Y. Gu, X. Yan, Z. Kang, S. Lu, Y. Sun and Y. Zhang, Design of sandwich-structured ZnO/ZnS/Au photoanode for enhanced efficiency of photoelectrochemical water splitting, *Nano Res.*, 2015, **8**(9), 2891–2900.
14. Z. Kang, X. Yan, Y. Wang, Y. Zhao, Z. Bai, Y. Liu, K. Zhao, S. Cao and Y. Zhang, Self-powered photoelectrochemical biosensing platform based on Au NPs@ZnO nanorods array, *Nano Res.*, 2015, **9**(2), 344–352.
15. Y. Liu, X. Yan, Z. Kang, Y. Li, Y. Shen, Y. Sun, L. Wang and Y. Zhang, Synergistic Effect of Surface Plasmonic particles and Surface Passivation layer on ZnO Nanorods Array for Improved Photoelectrochemical Water Splitting, *Sci. Rep.*, 2016, **6**, 29907.

16. N. Chouhan, C. Yeh, S. Hu, R. Liu, W. Chang and K. Chen, Photocatalytic CdSe QDs-decorated ZnO nanotubes: an effective photoelectrode for splitting water, *Chem. Commun.*, 2011, **47**(12), 3493–3495.

17. G. Wang, X. Yang, F. Qian, J. Z. Zhang and Y. Li, Double-sided CdS and CdSe quantum dot co-sensitized ZnO nanowire arrays for photoelectrochemical hydrogen generation, *Nano Lett.*, 2010, **10**(3), 1088–1092.

18. Z. Bai, X. Yan, Y. Li, Z. Kang, S. Cao and Y. Zhang, 3D-Branched ZnO/CdS Nanowire Arrays for Solar Water Splitting and the Service Safety Research, *Adv. Energy Mater.*, 2016, **6**(3), 1501459.

19. S. Cao, X. Yan, Z. Kang, Q. Liang, X. Liao and Y. Zhang, Band alignment engineering for improved performance and stability of $ZnFe_2O_4$ modified CdS/ZnO nanostructured photoanode for PEC water splitting, *Nano Energy*, 2016, **24**, 25–31.

20. Z. Bai, X. Yan, Z. Kang, Y. Hu, X. Zhang and Y. Zhang, Photoelectrochemical performance enhancement of ZnO photoanodes from $ZnIn_2S_4$ nanosheets coating, *Nano Energy*, 2015, **14**, 392–400.

21. Z. Kang, X. Yan, Y. Wang, Z. Bai, Y. Liu, Z. Zhang, P. Lin, X. Zhang, H. Yuan, X. Zhang and Y. Zhang, Electronic structure engineering of Cu_2O film/ZnO nanorods array all-oxide p-n heterostructure for enhanced photoelectrochemical property and self-powered biosensing application, *Sci. Rep.*, 2015, **5**, 7882.

22. K. Zhao, X. Yan, Y. Gu, Z. Kang, Z. Bai, S. Cao, Y. Liu, X. Zhang and Y. Zhang, Self-Powered Photoelectrochemical Biosensor Based on CdS/RGO/ZnO Nanowire Array Heterostructure, *Small*, 2016, **12**(2), 245–251.

23. S. W. Boettcher, E. L. Warren, M. C. Putnam, E. A. Santori, D. Turner-Evans, M. D. Kelzenberg, M. G. Walter, J. R. McKone, B. S. Brunschwig, H. A. Atwater and N. S. Lewis, Photoelectrochemical hydrogen evolution using Si microwire arrays, *J. Am. Chem. Soc.*, 2011, **133**(5), 1216–1219.

24. Y. Lin, S. Zhou, S. W. Sheehan and D. Wang, Nanonet-based hematite heteronanostructures for efficient solar water splitting, *J. Am. Chem. Soc.*, 2011, **133**(8), 2398–2401.

25. J. Shi and X. Wang, Hierarchical TiO_2–Si nanowire architecture with photoelectrochemical activity under visible light illumination, *Energy Environ. Sci.*, 2012, **5**(7), 7918–7922.

26. X. Sun, K. Maeda, M. Le Faucheur, K. Teramura and K. Domen, Preparation of $(Ga_{1-x}Zn_x)(N_{1-x}O_x)$ solid-solution from $ZnGa_2O_4$ and ZnO as a photo-catalyst for overall water splitting under visible light, *Appl. Catal., A*, 2007, **327**(1), 114–121.

27. Y. C. Pu, G. Wang, K. D. Chang, Y. Ling, Y. K. Lin, B. C. Fitzmorris, C. M. Liu, X. Lu, Y. Tong, J. Z. Zhang, Y. J. Hsu and Y. Li, Au nanostructure-decorated TiO_2 nanowires exhibiting photoactivity across entire UV-visible region for photoelectrochemical water splitting, *Nano Lett.*, 2013, **13**(8), 3817–3823.

28. Z. Zhang, L. Zhang, M. N. Hedhili, H. Zhang and P. Wang, Plasmonic Gold Nanocrystals Coupled with Photonic Crystal Seamlessly on TiO_2

Nanotube Photoelectrodes for Efficient Visible Light Photoelectrochemical Water Splitting, *Nano Lett.*, 2013, **13**(1), 14–20.

29. J. Azevedo, L. Steier, P. Dias, M. Stefik, C. T. Sousa, J. P. Araujo, A. Mendes, M. Graetzel and S. D. Tilley, On the stability enhancement of cuprous oxide water splitting photocathodes by low temperature steam annealing, *Energy Environ. Sci.*, 2014, **7**(12), 4044–4052.

30. X. Zhang, Y. Liu, S.-T. Lee, S. Yang and Z. Kang, Coupling surface plasmon resonance of gold nanoparticles with slow-photon-effect of TiO_2 photonic crystals for synergistically enhanced photoelectrochemical water splitting, *Energy Environ. Sci.*, 2014, **7**(4), 1409–1419.

31. S. C. Warren and E. Thimsen, Plasmonic solar water splitting, *Energy Environ. Sci.*, 2012, **5**(1), 5133–5146.

32. Y. Wang, F. Wang and J. He, Controlled fabrication and photocatalytic properties of a three-dimensional ZnO nanowire/reduced graphene oxide/CdS heterostructure on carbon cloth, *Nanoscale*, 2013, **5**(22), 11291–11297.

33. C. Li, T. Ahmed, M. Ma, T. Edvinsson and J. Zhu, A facile approach to ZnO/CdS nanoarrays and their photocatalytic and photoelectrochemical properties, *Appl. Catal., B*, 2013, **138**, 175–183.

34. Z. Liu, Y. Wang, B. Wang, Y. Li, Z. Liu, J. Han, K. Guo, Y. Li, T. Cui, L. Han, C. Liu and G. Li, PEC electrode of ZnO nanorods sensitized by CdS with different size and its photoelectric properties, *Int. J. Hydrogen Energy*, 2013, **38**(25), 10226–10234.

35. Y. Hou, F. Zuo, A. Dagg and P. Y. Feng, A Three-Dimensional Branched Cobalt-Doped alpha-Fe_2O_3 Nanorod/$MgFe_2O_4$ Heterojunction Array as a Flexible Photoanode for Efficient Photoelectrochemical Water Oxidation, *Angew. Chem., Int. Ed.*, 2013, **52**(4), 1248–1252.

36. X. Zhang, Y. Liu and Z. H. Kang, 3D Branched ZnO Nanowire Arrays Decorated with Plasmonic Au Nanoparticles for High-Performance Photoelectrochemical Water Splitting, *ACS Appl. Mater. Interfaces*, 2014, **6**(6), 4480–4489.

37. H. M. Cheng, W. H. Chiu, C. H. Lee, S. Y. Tsai and W. F. Hsieh, Formation of Branched ZnO Nanowires from Solvothermal Method and Dye-Sensitized Solar Cells Applications, *J. Phys. Chem. C*, 2008, **112**(42), 16359–16364.

38. H. Dai, Y. Zhou, Q. Liu, Z. D. Li, C. X. Bao, T. Yu and Z. G. Zhou, Controllable growth of dendritic ZnO nanowire arrays on a stainless steel mesh towards the fabrication of large area, flexible dye-sensitized solar cells, *Nanoscale*, 2012, **4**(17), 5454–5460.

39. M. Li, J. Su and L. Guo, Preparation and characterization of $ZnIn_2S_4$ thin films deposited by spray pyrolysis for hydrogen production, *Int. J. Hydrogen Energy*, 2008, **33**(12), 2891–2896.

40. Y. Yu, G. Chen, G. Wang and Z. Lv, Visible-light-driven $ZnIn_2S_4$/$CdIn_2S_4$ composite photocatalyst with enhanced performance for photocatalytic H_2 evolution, *Int. J. Hydrogen Energy*, 2013, **38**(38), 1278–1285.

41. M. A. Holmes, T. K. Townsend and F. E. Osterloh, Quantum confinement controlled photocatalytic water splitting by suspended CdSe nanocrystals, *Chem. Commun.*, 2012, **48**(3), 371–373.
42. X. Wang, K. Maeda, A. Thomas, K. Takanabe, G. Xin, J. M. Carlsson, K. Domen and M. Antonietti, A metal-free polymeric photocatalyst for hydrogen production from water under visible light, *Nat. Mater.*, 2009, **8**(1), 76–80.
43. K. S. Novoselov, Z. Jiang, Y. Zhang, S. V. Morozov, H. L. Stormer, U. Zeitler, J. C. Maan, G. S. Boebinger, P. Kim and A. K. Geim, Room-Temperature Quantum Hall Effect in Graphene, *Science*, 2007, **315**(5817), 1379.
44. Z. Kang, Y. Gu, X. Yan, Z. Bai, Y. Liu, S. Liu, X. Zhang, Z. Zhang, X. Zhang and Y. Zhang, Enhanced photoelectrochemical property of ZnO nanorods array synthesized on reduced graphene oxide for self-powered biosensing application, *Biosens. Bioelectron.*, 2015, **64**, 499–504.
45. Y. J. Hwang, C. Hahn, B. Liu and P. D. Yang, Photoelectrochemical Properties of TiO_2 Nanowire Arrays: A Study of the Dependence on Length and Atomic Layer Deposition Coating, *ACS Nano*, 2012, **6**(6), 5060–5069.
46. F. Le Formal, N. Tetreault, M. Cornuz, T. Moehl, M. Gratzel and K. Sivula, Passivating surface states on water splitting hematite photoanodes with alumina overlayers, *Chem. Sci.*, 2011, **2**(4), 737–743.
47. C. Lin, F. Y. Tsai, M. H. Lee, C. H. Lee, T. C. Tien, L. P. Wang and S. Y. Tsai, Enhanced performance of dye-sensitized solar cells by an Al_2O_3 charge-recombination barrier formed by low-temperature atomic layer deposition, *J. Mater. Chem.*, 2009, **19**(19), 2999–3003.
48. W. Kim, T. Tachikawa, D. Monllor-Satoca, H. I. Kim, T. Majima and W. Choi, Promoting water photooxidation on transparent WO_3 thin films using an alumina overlayer, *Energy Environ. Sci.*, 2013, **6**(12), 3732–3739.
49. Y. W. Chen, J. D. Prange, S. Duhnen, Y. Park, M. Gunji, C. E. Chidsey and P. C. McIntyre, Atomic layer-deposited tunnel oxide stabilizes silicon photoanodes for water oxidation, *Nat. Mater.*, 2011, **10**(7), 539–544.

CHAPTER 7

Biosensing Devices

YU SONG, ZHUO KANG AND YUE ZHANG*

University of Science and Technology Beijing, Beijing, China
*Email: yuezhang@ustb.edu.cn

7.1 Electrochemical Biosensors

An electrochemical biosensor, or three electrode detection system, is used as a classic protocol in understanding oxidation–reduction reactions in biological and chemical progress. Usually, the detection system consists of three electrodes, electrolyte solution and measurement circuits. The three electrodes are called working electrodes, auxiliary electrodes and reference electrodes, respectively. These three electrodes contribute to input and output of electrical power of the detection system, and the boundary of the working electrodes is the place for target reactions. At the boundary of the working electrodes and electrolyte solution, the target reaction is observed, recorded and analysed. In order not to be involved in the target reaction, materials of the working electrodes are usually required to be chemically resistant, efficiently conductive and easily machinable. The auxiliary electrodes are used to construct circuits with the working electrodes and the outside measurement circuits, and the reference electrodes are used to calibrate the potential of the working electrodes.

The three electrode detection system can be used to investigate biochemical reactions in detail by modifying the working electrodes with appropriate BREs and specific enzymes.[1] When compared to a negative value of static isoelectric points of enzymes including glucose oxidase (GOx), uricase and lactic oxidase (LOD), the static isoelectric point of ZnO is positive and can adsorb these enzymes naturally. In addition, considering the ZnO

Nanoscience & Nanotechnology Series No. 43
ZnO Nanostructures: Fabrication and Applications
By Yue Zhang
© Yue Zhang 2017
Published by the Royal Society of Chemistry, www.rsc.org

structural merits of biocompatibility, piezoelectricity, chemical resistance and easy fabrication, ZnO are suitable for use as potential BREs modified on the surface of the electrodes to form biosensors with specific biomolecular targets. ZnO nanofibers with diameters ranging from 195–350 nm are proposed to modify a gold electrode.[2] These ZnO nanofibres are synthesised by the electrospinning method, a popular means widely applied for producing polymer fibres. Following transfer of ZnO nanofibres onto the gold electrodes, GOx molecules are functionalised and immobilised to form a GOx/ZnO/Au biosensing electrode. Besides physically modifying electrodes using ZnO nanostructures, directly growing ZnO on electrodes is deemed to be more efficient for taking advantage of the electrical properties of ZnO nanostructures. A novel biosensing electrode is constructed by directly growing ZnO nanotubes on the surface of Au electrodes through chemical etching, an electrochemical fabrication technique. A cross-linking method is adopted to immobilise GOx following ZnO nanotube deposition on Au electrodes to form a GOx/ZnO/Au electrode. This combination enables the as-synthesised biosensor to be systematically designed and developed.[3] A novel glucose electrode is developed by combining ZnO nanorods with Au/Pt alloy through a series of steps of chemosynthesis. Then GOx and lectin are immobilised on the as-synthesised hybrid electrodes to form a glucose target electrode with a detection range of 1.8 μM to 5.15 mM and a detection limit of 0.6 μM.[4] ZnO NRs are functionalised with uricase on the Au electrode surface to form a reagent-less uric acid biosensor. This uric acid biosensing electrode possesses thermal stability, anti-interference ability and a low apparent Michaelis–Menten value of 0.238 mM.[5] LOD is immobilised on ZnO nanorods in favour of glutaraldehyde as a cross linker to construct a potentiometric electrochemical sensor for lactic acid detection in a concentration range from 0.1 μM to 1 mM.[6] In short, enzyme/ZnO electrodes have been greatly developed and many kinds of enzyme electrodes have also been fabricated for specific detection purposes.

There are two main methods to integrate ZnO onto working electrodes: (1) transferring and (2) directly growing. ZnO structures are fabricated by using the methods mentioned in Chapter 2. Here, we use ZnO NRs as an example. Before modifying electrodes, all electrodes are required to be polished to get rid of contaminants and oxidation layers. In the methods of transferring ZnO onto surfaces of Au electrodes, these as-synthesised ZnO NRs are dispersed into phosphate buffer saline (PBS) solution. One drop of PBS solution containing ZnO NRs is dropped on the top of the Au electrodes and dried. This step is repeated several times until the thickness of the ZnO NRs layer reaches 1 mm. As for growing ZnO NRs on Au electrodes, the Au electrodes are placed horizontally into reaction solution. After full reaction as mentioned in the ZnO fabrication method, a layer of ZnO NRs products with a thickness of around 1 mm is obtained on the top of the Au electrodes.[7] Besides ZnO NRs, other ZnO morphologies, ZnO nanotetrapods[8–10] and ZnO nanowires,[11,12] are used to modify Au electrodes. In addition, since gold nanoparticles have merits for catalysis and immobilisation, Au nanoparticles

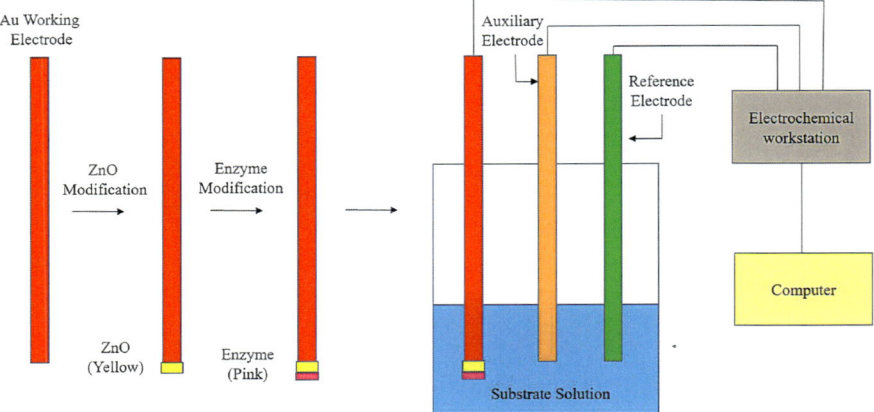

Figure 7.1 Illustration of electrochemical biosensor fabrication. After sequent modification of ZnO nanostructures (yellow) and enzymes (pink) on the top of Au electrodes, the as-fabricated biosensor is used for specific substrates detection in solution. All signals are collected by the electrochemical workstation and processed by computer programs.

and ZnO nanorods are integrated to modify Au electrodes[13,14] for further development of the performance of sensing.

Following the step of ZnO/Au electrode construction for biosensing is enzyme immobilisation. Enzyme solution with an active concentration, such as GOx, uricase, and LOD, is dropped on the fabricated ZnO/Au electrodes, followed by a drop of Nafion solution. After incubation overnight at 4 °C, Nafion film can further enhance immobilisation of enzymes on ZnO and ZnO on Au electrodes, making these fabricated biosensors ready for detection. The general electrochemical electrode process is shown in Figure 7.1. The three electrodes are all submerged into detected solution (working electrodes in red, auxiliary electrodes in orange and reference electrodes in green). On the top of the working electrodes, enzyme and ZnO are immobilised (ZnO in yellow, and enzymes in pink).

Glucose is an important monosaccharide in metabolism. For living cells, it is a key energy source as well as a pivotal product during metabolism. In the human body, glucose serves as a major energy source through glycolysis. However, a high level of glucose concentration in human blood is a marker of diabetes mellitus, indicating a metabolic disorder caused by malfunction of the pancreas. This condition is suffered among people over 50-years-old. To monitor health conditions, it is helpful to develop powerful glucose sensors for glucose concentration detection. Considering that GOx is an enzyme specific to glucose, it can be used to construct glucose biosensors. The reaction that GOx catalyses is:

$$\text{glucose} + O_2 \rightarrow \text{gluconic acid} + H_2O_2 \tag{7.1}$$

$$H_2O_2 \rightarrow O_2 + 2H^+ + 2e^- \tag{7.2}$$

Based on the biosensor process described above, an as-fabricated Nafion/ GOx/ZnO/Au electrode is validated for glucose detection by using cyclic voltammogram (CV) curves. From the CV curve, ZnO immobilisation and GOx adsorption are determined to be reliable and stable. Successive addition of various glucose concentration ranging from 4 μM to 10 mM is used to test the sensitivity of the electrodes. The results show that a detection limit of 4 μM and a sensitivity of 25.3 μA mM^{-1} cm^{-2}.[8] A novel GOx/ZnO/Au electrode is further developed by introducing gold nanoparticles to increase enzyme adsorption and signal transmitting. When Au nanoparticles are combined with ZnO nanorods, the superficial surface greatly increases, providing more places for reaction.[14]

In Figure 7.2(a), the series of CV curves of GOx/ZnO electrodes in 3 mM glucose solution are collected at various scan rates of 20, 40, 60, 80 and 100 mV s^{-1} (the arrowhead indicates the increase direction). The inset presents the relationship between the peak current and the square roots of scan rates, indicating that the typical electrochemical behaviour of a surface diffusion process is dominant during the reaction of GOx and glucose on the surface of ZnO/Au electrodes. Figure 7.2(b) presents the results of the GOx/ZnO/Au electrodes used for successive detection of various concentrations of glucose ranging from 4 μM to 10 mM.

As per healthily individual conditions, the concentration of uric acid, as a product during purine metabolism, maintains a balance between uptake and excretion. However, if malefaction of the kidney causes a delay in excretion of uric acid, the accumulation of uric acid inside the body will lead to a decrease in the pH value of body fluids, breaking the balance of acid and alkali and leading to gout. Thus, fast detection of uric acid concentration in body fluids is another crucial clue for health monitoring. The following equations give the uricase catalysing reaction:

$$\text{uric acid} + O_2 \rightarrow \text{allantoin} + CO_2 + H_2O_2 \tag{7.3}$$

$$H_2O_2 \rightarrow O_2 + 2H^+ + 2e^- \tag{7.4}$$

An efficient Nafion/Uricase/ZnO/Au electrode for uric acid detection has been proposed.[9] Field-emission scanning electron microscopy (FESEM), energy-dispersive X-ray spectroscopy (EDX), X-ray diffraction and Raman spectroscopy, respectively, are used to study the structure of ZnO nanotetrapods. Further CV curves conclude that the as-fabricated uric acid electrode possesses a detection limit of 0.8 μM, a sensitivity of 80.0 μA mM^{-1} cm^{-2} and a fast-current response time of around 9 s. Au electrodes with ZnO micro/ nanowires are fabricated to increase sensitivity to 89.74 μA mM^{-1} cm^{-2}. Like Lei's work, FESEM, EDX, X-ray diffraction, and Raman spectroscopy are applied to investigate the structure of ZnO micro/nanowires to confirm their capability of being transducers for adsorbing uricase. The detection range is from 0.1 mM to 0.59 mM.[11] In Figure 7.3, the fabricated uricase/ZnO nanowires/Au electrode is used to detect uric acid in a mixture containing glucose, lactic acid and uric acid. This result shows that the fabricated uric acid

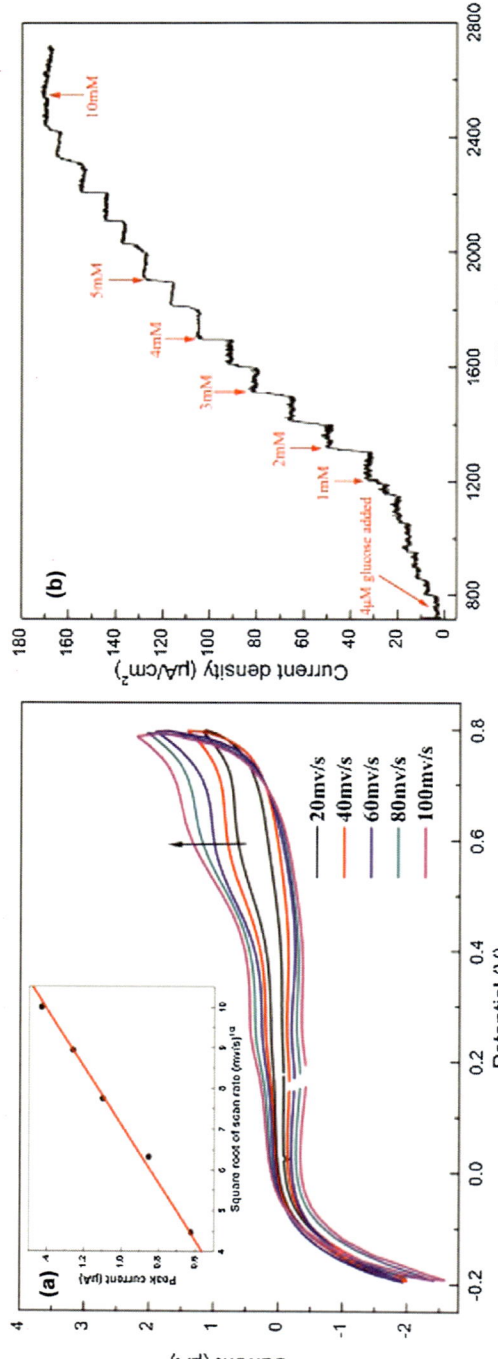

Figure 7.2 (a) CV curves of the GOx/ZnO/Au electrode in 3 mM glucose at different scan rates. The inset shows the peak current as a linear function of the scanning rates; (b) real-time response current of the ZnO/Au biosensor for successive additions of glucose to the stirring air-saturated PBS.
Reprinted from Colloids Surf., A, 361(1–3), Y. Lei, X. Yan, N. Luo, Y. Song and Y. Zhang, ZnO nanotetrapod network as the adsorption layer for the improvement of glucose detection *via* multiterminal electron-exchange, 169–173. Copyright (2010), with permission from Elsevier.[8]

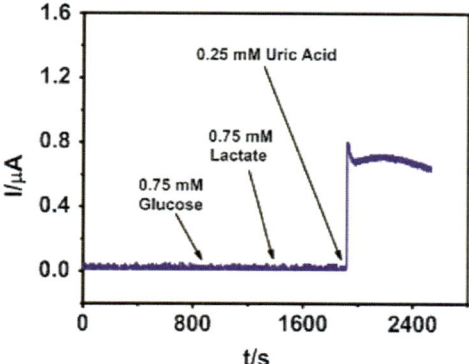

Figure 7.3 The selectivity of the uric acid biosensor under the interference of glucose and lactate.
Reproduced from Microchim. Acta, Highly sensitive uric acid biosensor based on individual zinc oxide micro/nanowires, 180(9), 2013, 759–766, Y. Zhao, (© Springer-Verlag Wien 2013) With permission from Springer.[11]

electrode only responds to uric acid and not to the other components, confirming its excellent performance in sensitivity and selectivity.

Lactic acid is a waste product of the metabolism after consuming energy inside the body. Like uric acid, the concentration of lactic acid is maintained at a certain level in healthy individuals. However, accumulation of lactic acid decreases the pH value of body fluids, putting healthy cells under huge pressure to survive. Therefore, it is necessary to monitor the concentration of lactic acid in the body. For the detection of L-lactic acid, the catalytic reaction of lactate oxidase (LOD) can be described as:

$$\text{L-lactic acid} + O_2 \rightarrow \text{pyruvate} + H_2O_2 \tag{7.5}$$

$$H_2O_2 \rightarrow O_2 + 2H^+ + 2e^- \tag{7.6}$$

A novel amperometric biosensor constructed by Nafion/LOD/ZnO/Au was fabricated for lactic acid detection.[10] By analysing the results obtained from the CV curve, an efficient lactic acid electrode with a high sensitivity of 28.0 μA mM^{-1} cm^{-2} has been validated. Concentrations of lactic acid ranging from 1.2–120 μM are successively added into testing solution and the linear relationship of concentration and current density is in the range of 3.6 μM to 0.6 mM. In addition, this as-fabricated enzyme electrode has a low apparent Michaelis-Menten constant of 0.58 mM. This electrode model is further developed by replacing ZnO nanotetropods with ZnO nanowires because ZnO nanowires possess the advantage of being BREs.[12] Immobilisation of LOD onto the 500-nm length ZnO nanowires is demonstrated by fluorescence characterisation. As per the analysis of CV curves, a sensitivity of 15.6 μA mM^{-1} cm^{-2} and a detection range from 12 μm to 1.2 mM with a detection limit of 12 μm are obtained. ZnO NRs integrated with gold

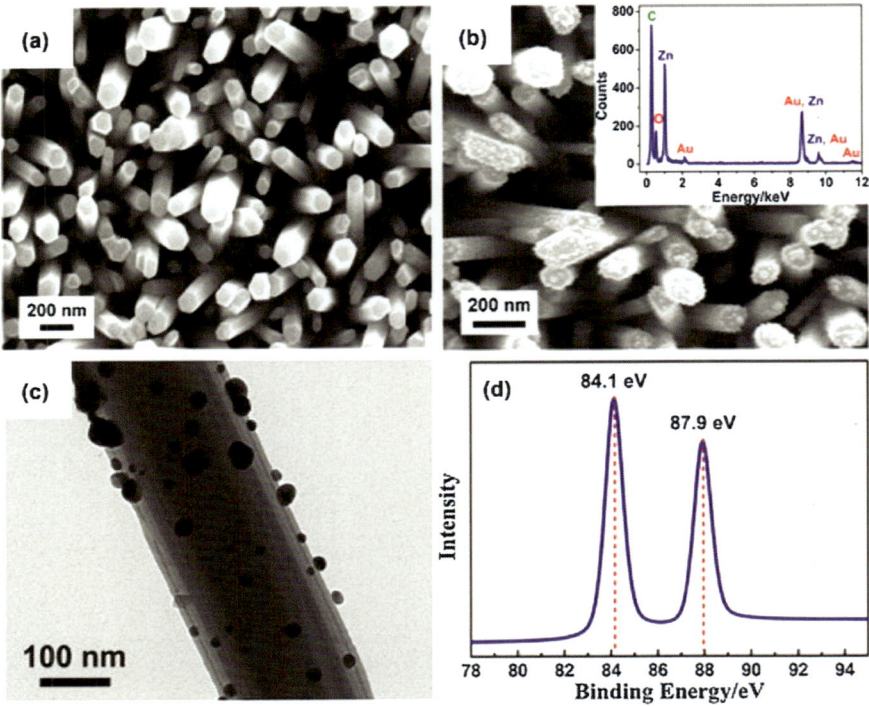

Figure 7.4 (a) FE-SEM image of ZnO NR arrays grown on fluorine-doped tin oxide (FTO) glass; FE-SEM image (b) and EDS pattern (inset of (b)) of Au NPs/ZnO NR arrays on FTO; (c) TEM image of Au NPs/ZnO NR; (d) XPS spectrum of Au 4f region.
Reprinted from Colloids Surf. B, 126, Y. Zhao, X. Fang, Y. Gu, X. Yan, Z. Kang, X. Zheng, P. Lin, L. Zhao and Y. Zhang, Gold nanoparticles coated zinc oxide nanorods as the matrix for enhanced L-lactate sensing, 476–480. Copyright (2015), with permission from Elsevier.[13]

nanoparticles (Au NPs) are proposed to improve the detection performance of the former LOD/ZnONR arrays/Au NPs electrodes.[13] Figure 7.4(a)–(c) shows the morphology of synthesised ZnO NR arrays and Au NPs/ZnO NR arrays with diameters of 100–200 nm, and Au NPs with average diameters of 10–18 nm are homogeneously absorbed onto the ZnO NRs. The EDS pattern in the inset of Figure 7.4(b) demonstrates the existence of Au element except for Zn and O elements. The XPS spectrum of the Au 4f region is shown in Figure 7.4(d), and the peaks of 84.1 and 87.9 eV are ascribed to metallic Au. This brand-new combination increases the sensitivity to $24.56 \ \mu A \ mM^{-1} \ cm^{-2}$ and reaches a smaller Michaelis-Menten constant of 1.58 mM compared to that of pure ZnO nanowires. A detection limit of 6 μM and a wider linear range of 10 μM to 0.6 mM for L-lactate detection are obtained.

Electrochemical biosensors, or three electrode detection systems, are flexible for forming various biological substrate detection devices just by modifying ZnO/Au working electrodes with specific substrate recognition molecules, with their powerful performance in wide ranges of concentration detection, stability, reliability and low cost. However, there are still some trivial disadvantages that need to be improved for future application. First, the detection range, especially at low concentrations has a lot of room for improvement. Second, considering the usage of reference electrodes, the size of this detection system cannot be minimised on a small scale. Third, the constructed three electrode detection system needs quite a longer response time to detect substrates and recovery time to reset from the last detection. When these issues are solved in the future, the three electrode detection system will have great potential in industrial applications.

7.2 FET-based Biosensors

The field effect transistor (FET) is an emerging electronic device widely applied in electronic and electrical engineering. The development of FET technologies boosts prosperities in integrated circuits industry vastly, bringing huge convenience to daily life and tremendous profits to commercial business. With the advantages of large input impedance, low noise, thermal stability and anti-radio ability, FETs can realise various kinds of engineering properties in amplification circuits. The concept of FET takes advantage of major carriers for signal amplification. In the model of FETs, the output source-drain current signal of FETs is controlled by a gate voltage applied on the gate area. There are two main categories of FETs: junction FETs (JFETs) and metal-oxide-semiconductor FETs (MOSFETs). When it comes to easy integration with semiconductor materials, MOSFETs are usually a priority in electrical engineering research and programs and hence in this chapter we mainly discuss MOSFET based biosensors, or FET based biosensors. MOSFETs are divided into two types: *n*-type FETs and *p*-type FETs. In *n*-type FETs, positive ions doped in the gate area push positive vacancies far away from the gate area, leaving only negative electrons near the gate area to form an electron channel, whereas it is the opposite case in *p*-type FETs.

Bending ZnO nanowires can generate a piezoelectric effect, and this property is utilised to design a piezoelectric field effect transistor (PE-FET) to measure nanoforce/pressure when the source to drain current changes.[15] In addition, ZnO nanowires, as per their electrical properties, can be deemed to be *n*-type FETs. Undoped ZnO nanowires are very sensitive to surrounding environment, especially in a pH sensitive environment. Two Pt electrodes are placed on a substrate full of ZnO nanorods/nanowires to construct a humidity sensor. This sensor is very sensitive to the change of humidity, ranging from 12–97%. The impedance of the sensor decreases with the increase in humidity.[16] Furthermore, a ZnO nanowire-based humidity sensor is designed by randomly bridging a ZnO nanowire across two electrodes. ZnO: Ga/SiO_2/Si as the template is used to grow ZnO nanowires. The impedance of the fabricated humidity sensor also decreases with increasing of humidity.[17]

The biochemical environment, to some extent, is a more sophisticated pH environment. During metabolism, many biochemical products are consumed and generated, involving lots of gene transcription, enzyme catalysis and energy consumption, accompanied by large-scale pH changes. If as-fabricated pH ZnO sensors are going to be used in detecting biochemical signals, an upgrade will be required to modify simple ZnO nanowires. In this chapter, we mainly focus on enzyme biosensor fabrication, especially in uricase/ZnO nanowire FET based biosensors. A uric acid sensor is fabricated based on an uricase-coated single ZnO nanowire FET biosensor. The as-synthesised biosensor is very sensitive at low concentrations ranging from 1 pM to 50 μM, and its response time turns out to be in the order of milliseconds.[18] In addition, a comprehensive ZnO NR based FET biosensor is constructed to selectively detect the concentration of glucose, cholesterol and urea, contributing to the development of the organs-on-chips concept.[19]

Ultra-long ZnO nanowires are prepared as per the fabrication methods described in Chapter 2. The as-synthesised ZnO nanowires are dispersed by ethanol on a silicon substrate with a thin oxide layer.[18] Oxygen plasma (0.3 Torr, 25 W power for 60 s) is applied on the surface of the ZnO nanowires to remove contaminants and add hydroxyl groups, followed by a treatment of 2% ethanol solution of 3-aminopropyltriethoxysilane (APTES). After silanisation and rinsing twice with ethanol, the ZnO nanowires undergo incubation at 120 °C for 10 min under N_2. After preparation of Ti/Au metal layers as the source and drain electrodes, a modified ZnO nanowire is placed onto the two electrodes, immobilised by silver paste carefully. In addition, polymethyl methacrylate (PMMA) passivates the electrodes to reduce the leak current, excluding influences of other effects except the conductance changes of the ZnO nanowires. Finally, 5 μL uricase (5 units per mL) is dropped on the ZnO, and incubated on the ZnO nanowire device for another 40 min in saturated glutaraldehyde vapour. After rinsing with $0.01 \times PBS$ and deionised water for another 15 min, the fabricated ZnO nanowire biosensors are ready for uric acid detection. The structure and procedure of the fabricated uricase/ZnO nanowire FET biosensor is depicted in Figure 7.5. The ZnO nanowire (yellow) is placed on the top of two Ti/Au source and drain electrodes (red), fixed by silver paste (silver), and modified by uricase (pink).

Beside uric acid biosensors, ZnO FET biosensors can also be used for other substrate molecule detection, including protein, antibodies, glucose, lactic acid, *etc.* A 3D bio-FET sensor was developed by immobilising prostate specific antigen/1-antichymotrypsin (PSA-ACT) antibody as a probe on the biosensor to detect PSA-ACT complex antigen, which is a biomarker for the diagnosis of prostate cancer.[20] ZnO FET biosensors can also be applied in neuroscience by integrating with a specific designed BRE, which can immobilise on both ZnO and target neuropeptide, Orexin A, which is an indicator of fatigue and cognitive performance in blood and saliva.[21] ZnO FET biosensors are also used for heart disease diagnosis. The sensing areas are modified using APTES, glutaraldhyde and cardiac Troponin I (cTnl) monoclonal antibody sequentially to form a bio-receptor for cTnl detection,

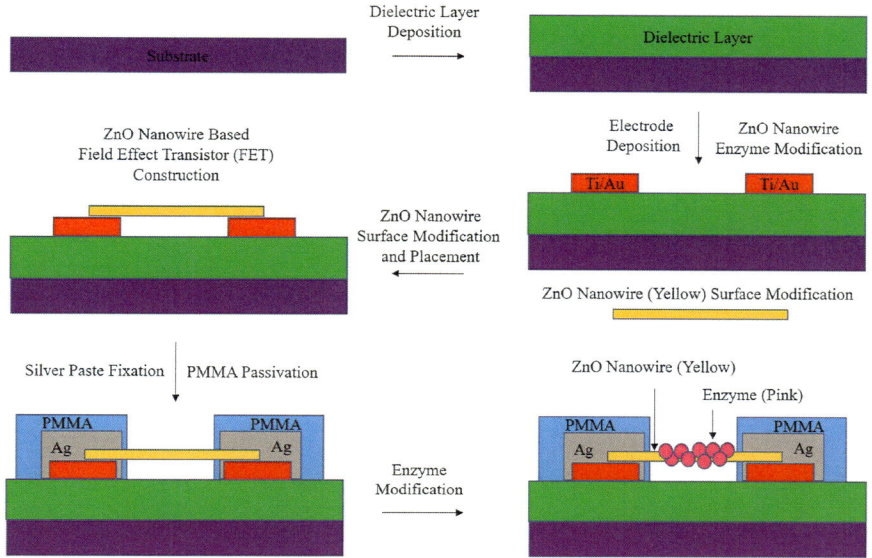

Figure 7.5 Illustration of FET-based biosensor fabrication. After deposition of Ti/Au electrodes on the dielectric layer, ZnO nanowires (yellow) are placed on the electrodes. Silver paste is used to fix the position of the ZnO nanowires and PMMA is used to passivate the electrodes which is followed by enzyme addition. The silver paste is connected to the outside measurement circuits, and all signals are processed by computer programs.

the concentration of which in blood can be used as an indicator of heart attack.[22] In addition, ZnO FET biosensors are proposed for vitamin detection. A specific BRE is designed for riboflavin, which is important for body growth and red blood cell production, and an immobilised as-designed DNA aptamer to ZnO FET.[23] Furthermore, a high-performance cholesterol sensor is fabricated by immobilised cholesterol oxidase (ChOx) enzyme on ZnO FET biosensors.[24] In short, ZnO FET based biosensors are very flexible to being updated to various kinds of species sensing by using specific species recognition 'receptors' to modify sensing areas and hence they have great potential in industrial production and technology upgrades.

The as-fabricated uricase/ZnO nanowire/FET biosensor is used to detect various concentrations of uric acid in solution. Before detection, bovine saline albumin (BSA) is used to block excessive binding areas and maintain an enzyme-friendly environment. When the signal collected from BSA solution returns to the baseline, the biosensor is ready for uric acid detection.[18] As shown in Figure 7.6(a), a successive concentration of uricase ranging from 1 pM to 50 μM is tested. The conductance of the biosensor increases with the increment of the concentration. The same trend can be also observed in the relationship of the source-drain current and the concentration, as shown in Figure 7.6(b). As per the data, it is demonstrated that

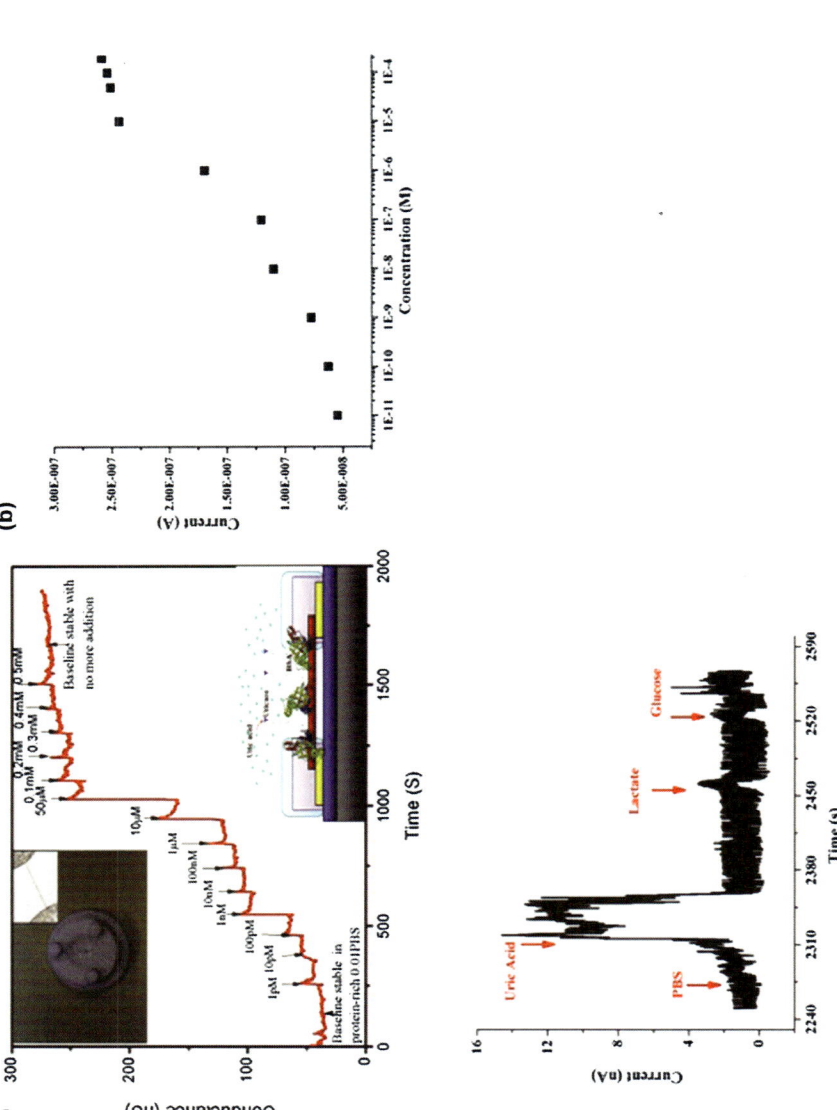

Figure 7.6 (a) Conductance of the device *vs.* time following the addition of uric acid in the buffer solution. The upper inset picture is the home-made reaction cell for sensing, and the optical image of silver paste immobilised ultra-long ZnO nanowire; the lower inset is the configuration of the device during active sensing measurements; (b) plot of stable current *vs.* different concentrations of uric acid; (c) effect of potentially interfering substances on sensor response upon adding 300 μM lactate and 300 mM glucose solution. Reprinted from Sens. Actuators, B, 176, X. Liu, P. Lin, X. Yan, Z. Kang, Y. Zhao, Y. Lei, C. Li, H. Du and Y. Zhang, Enzyme-coated single ZnO single nanowire FET biosensor for detection of uric acid, 22–27. Copyright (2013), with permission from Elsevier.[18]

the uric acid biosensor is very sensitive in low concentration detection, and this can be attributed to the amplification effects of ZnO FETs. The weak change of potential surrounding the gate areas can be amplified to a macroscale current signal. In addition, reliability experiments are also implemented to confirm that the uric acid biosensor is exclusive for other species, as shown in Figure 7.6(c). 300 μM lactate and 300 mM glucose are used to stimulate the biosensor, but the biosensor does not respond to this stimulation, indicating that the biosensor is specific to uric acid.

There are many kinds of ZnO structures, including ZnO nanowires, nanorods, nanotetrapods, belts, nanocombs, *etc.*, which can be used to construct state-of-the-art biosensors that display various electrical, physical and biological properties. ZnO FET biosensors, based on information summarised above, are very feasible for updating to meet the requirements for different kinds of species detection. Natural BREs as well as artificially designed BREs are encouraged to modify sensing areas and these modification processes are usually easy to be implemented based on current fabrication protocols. These merits make ZnO FET biosensors promising both in experimental research and industrial application, especially in individual medication. However, there are still some challenges in ZnO FET biosensor applications. (1) The fabrication of ZnO FET needs to be done in a clean room and to avoid impairment, the fabricated biosensors needs to be stored in a biological-friendly environment. (2) The packaging technology for the fabricated biosensors is still another issue that needs to be discussed in future. Although these challenges prevent ZnO FET biosensors being applied immediately in current commercial marketing, these enzyme/ZnO/FET biosensors, with the development of relevant technologies, will show great potential in industrial production and individual medication.

7.3 HEMT-based Biosensors

In the 1970s, the Bell lab discovered that there is high electron mobility in the heterojunction structure between pure GaAs and *n*-type AlGaAs because the electron donor AlGaAs layer is separated from GaAs and no electron scattering occurs during electron moving, compared to the mobility of electron in FETs prevented by ion purities scattering in conductive carriers. Then novel FETs based on the heterojunction structure were constructed, such as the high electron mobility transistor (HEMT). Usually, the molecular beam epitaxy (MBE) is used to fabricate HEMTs, causing discontinuity of the energy band due to discontinuity between the conductance band and valence band, or the wideness of the forbidden band. This change of energy forms an electron trap near the boundary in the GaAs layer to restrain the mobility of electrons in the z direction while promoting the mobility in the x–y plane. When compared to relatively low electron mobility in traditional FETs in which donors and receptors of major carriers are in the same transmission channel, high electron mobility is obtained in HEMTs because donors from *n*-type AlGaAs form a two-dimensional electron gas channel in the GaAs layer in which there are no ion impurities scattering during the mobility of electrons. Compared to the source-drain current signals of traditional FETs controlled by the density of carriers, those of HEMTs are more sensitive to stimulation due to the higher density of electron carriers in the two-dimensional electron gas channel.

Like ZnO FET biosensors, if the gate areas can be modified by ZnO transducers and proper recognition targets, HEMTs can be updated to various specific biosensors for different detection requirements. The gate region of AlGaN/GaN based HEMTs is modified with ZnO nanorods and GOx to detect the concentration of glucose in solution.[25] Further development of this model leads to construction of lactic acid biosensor by modifying the sensing areas of ZnO nanorods with LOD.[26] Since ZnO nanotetrapods (T-ZnOs) and ZnO nanowires show better electrochemical properties in signal transmitting, these two morphologies of ZnO are promoted to modify the sensing areas of HEMTs. In particular the T-ZnOs nanostructure firstly discovered in Zhang's lab; this structure has multiple electron channels and therefore allows electron transferring with high efficiency. In order to take advantage of the T-ZnOs structure, the sensing areas of AlGaAs/GaAs based HEMTs is modified with T-ZnO nanostructures and uricase to construct a uric acid biosensor, with a detection limit of 0.2 nM and a response time of 1 s.[27] To further develop this biosensor model, the AlGaAs/GaAs HEMT is modified with ZnO nanowires and LOD to construct a lactic acid biosensor with a detection limit of 0.03 nM and a response time of less than 1 second.[28] In order to improve the detection performance of this lactate biosensor model, ZnO nanowires used as a BRE are creatively doped with In for better efficiency in LOD adsorption and signal transmitting.[29]

Following this, we will set an example of making enzyme/ZnO/HEMTs biosensors by using the protocol of uric acid biosensor fabrication to describe how to fabricate the biosensors. HEMTs are fabricated first, followed by the gate areas modification for substrate molecular detection. After modification and incubation, the as-fabricated enzyme/ZnO/HEMTs biosensors are ready for detection of corresponding substrates in solution.

The HEMT layer structure is constructed by using the GEN-II MBE system. From bottom to top, there is a 3 inches thick semi-insulate GaAs substrate, a 1 μm GaAs layer, a 30 Å spacing AlGaAs layer (the portion of Al $X_{Al} = 0.3$), a 220 Å Si doped AlGaAs layer (the portion of Al $X_{Al} = 0.3$, the Si doped concentration $= 1.4 \times 10^{18}$ cm^{-3}), and a 50 Å Si doped GaAs cap layer (the Si doped concentration $= 5 \times 10^{18}$ cm^{-3}).[27] During HEMT layer fabrication, the working temperature is 580 °C, and the flow rates of Ga, As and Al in $Al_{0.3}Ga_{0.7}As$ are 1.1×10^{-7}, 1.0×10^{-5} and 2.5×10^{-8} Torr, respectively. The growth rates of the GaAs layer and $Al_{0.26}Ga_{0.7}As$ layer are 2.6 and 3.71 Å s^{-1}. Only if these fabrication parameters are very strictly controlled, can the spontaneous polarisation boundary between GaAs and AlGaAs be generated for the two-dimensional electron gas. The overall parameters of the as-synthesised HEMT layer structure are: when $T = 300$ K, the mobility $= 7035$ cm^2 v^{-1} s^{-1}, and the density of the 2D electron cloud $= 1.79 \times 10^{-11}$ cm^{-2}; when $T = 77$ K, the mobility $= 145\,700$ cm^2 v^{-1} s^{-1}, and the density of the 2D electron cloud $= 2.77 \times 10^{-11}$ cm^{-2}. Then two Ni/GeAu/Ni/Au (Ni: 50 nm, GeAu: 204 nm, Ni: 10 nm, Au: 50 nm) electrodes with a size of 2 mm $\times$ 5 mm and a SiO$_2$ insulation layer (200 nm) are deposited on the cap layers by using two masks. The electrodes are maintained at 400 °C for 20 s, making metal atoms disperse to the 2D electron gas channel to form ohmic contact. The structure layers and procedure for the HEMT based biosensor are shown in Figure 7.7. The ZnO nanostructures used for gate area modification are labelled in yellow and the enzyme adsorbed on the ZnO nanostructures is labelled in pink.

T-ZnOs dispersed in an ethanol suspension are dropped on the as-synthesised HEMT gate area at room temperature. Uricase is then added to these transferred T-ZnOs after ethanol evaporation and immobilised following Nafion solution addition. The fabricated devices can be stored at 4 °C before detection. As for the pure ZnO nanowires and in-doped ZnO nanowires used in HEMT based lactic acid biosensors, they can also be transferred to the HEMT gate area by dispersal in ethanol solution and drying.[28,29] Then specific LOD is introduced to these ZnO nanowires and immobilised by ZnO adsorption and Nafion film. These LOD/ZnO/In-doped ZnO nanowires/HEMT biosensors are also stored at 4 °C before detection.

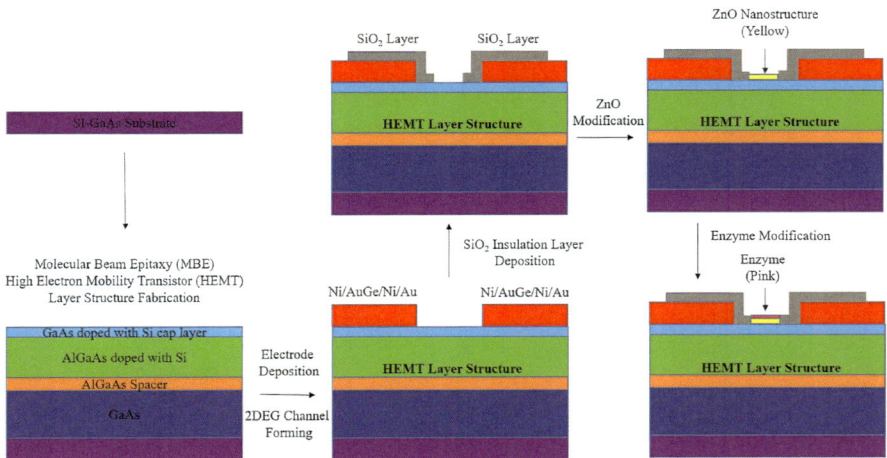

Figure 7.7 Illustration of HEMT-based biosensor fabrication. After layer-by-layer structure fabrication of HEMTs on a substrate by MBE deposition, two Ni/AuGe/Ni/Au electrodes are deposited on the cap layer. Proper annealing allows diffusion of these metal atoms to form a two-dimensional electron gas channel at the boundary area. Then two SiO$_2$ layers are used to insulate the electrodes and sensing areas. ZnO nanostructures and enzymes are sequentially transferred to the sensing areas to form the detection area, followed by addition of Nafion film. All biological signals are converted to electrical signals and transmitted through the two electrodes, and processed by computer programs.

When compared to the wideness of the forbidden band of AlGaN/GaN used in other work, which is 3.44 eV, the wideness of the forbidden band of AlGaAs/GaAs is 1.42 eV. The smaller wideness of the forbidden band allows AlGaAs/GaAs to be an ideal potential semiconductor material for biosensor construction because the change of surface ambience is more easily transmitted to the gate area.

Before detecting uric acid in solution, the sensors are warmed to 37 °C for 30 min, and PBS with pH = 6.9 is used to humidify the as-fabricated uricase/T-ZnOs/HEMT biosensor.[27] Alternatively, adding PBS solution and 2 nM uric acid solution to the biosensor can be used to validate the reliability and stability of the fabricated biosensors, as shown in Figure 7.8(a). Then concentrations of uric acid in PBS ranging from 0.2 nM to 2 mM are applied to the biosensors to verify the sensitivity. The signals are collected and amplified by an electrochemical workstation and the recorded results are analysed using Origin engineering software. Results show that a detection limit of 0.2 nM and a fast response time of around 1 s are obtained, as shown in Figure 7.8(b), indicating that the fabricated biosensors are qualified for uric acid detection.

When compared to amperometic-based lactic acid sensors, no fixed reference electrodes are required in HEMT based lactic acid sensors.[28] This simplification allows for a flexible application of the lactic acid sensors for lactic acid detection. After validation of the PBS solution, successive concentrations of lactic acid ranging from 0.03 nM to 300 mM are continuously introduced to the sensors. The recorded results show that ZnO-nanowires sensors have a better detection performance with a lower detection limit of 0.03 nM and a faster response time of less than 1 s, when compared to T-ZnO uric acid biosensors. A further update of the ZnO nanowires with In-doping greatly improves the electrochemical properties of the ZnO nanowires, enhancing the density of 2D electron gas, and optimising the performance of the sensing efficiency of the gate area.[29] The In-doped ZnO nanowire lactic

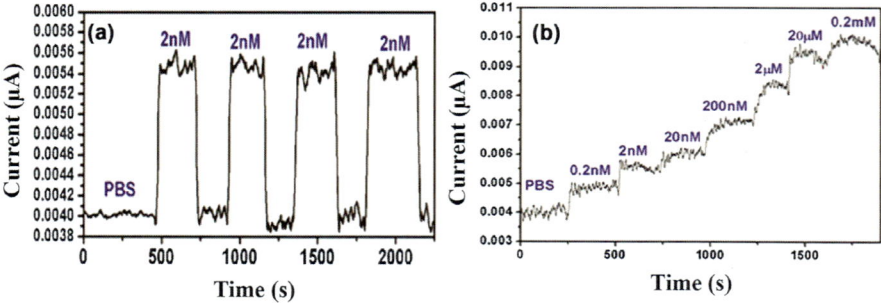

Figure 7.8 (a) Repeatability of uricase/T-ZnOs/HEMTs biosensor in detection. Amperometic response of one uricase/T-ZnOs/HEMT biosensor to 2 nM UA four times at a bias voltage of 0.5 V; (b) amperometric response of uricase/T-ZnOs/HEMT biosensor to a series of concentrations of UA. Reproduced from ref. 27 with permission from AIP Publishing.

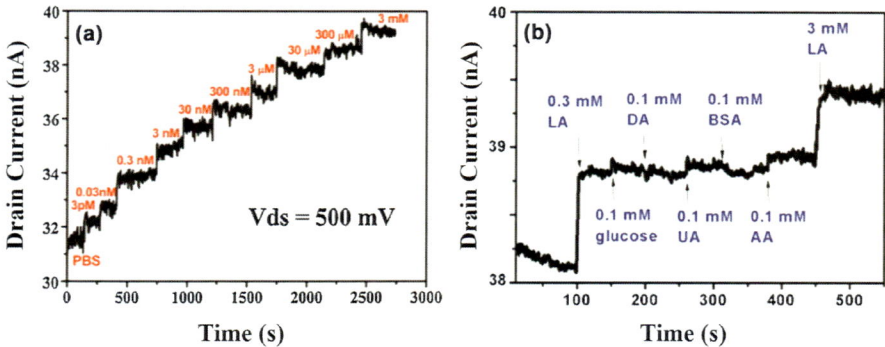

Figure 7.9 (a) Plot of drain current *vs.* time with successive exposure to higher lactic acid concentrations, ranging from 3 pM to 3 mM in PBS with a pH value of 7.4; (b) selectivity measurement of the In-doped ZnO-modified AlGaAs/GaAs HEMT lactic acid biosensor.

Reprinted from Sens. Actuators, B, 212, S. Ma, X. Zhang, Q. Liao, H. Liu, Y. Huang, Y. Song, Y. Zhao and Y. Zhang, Enzymatic lactic acid sensing by In-doped ZnO nanowires functionalized AlGaAs/GaAs high electron mobility transistor, 41–46. Copyright (2015), with permission from Elsevier.[29]

acid sensors possess a lower detection limit of 3 pM than pure ZnO nanowire sensors, as shown in Figure 7.9(a). Since the In-doped ZnO nanowires can further enhance the density of the 2D electron gas, the as-fabricated sensors show higher efficiency in selecting the response to lactic acid when exposed to a mixture of lactic acid, uric acid, glucose and BSA at 0.1 mM level, as shown in Figure 7.9(b), and are deemed to be promising for future industrial applications and individual medication.

As an emerging prototype of biosensors fabrication, enzymes/ZnO/HEMT biosensors have attracted much attention due to their excellent performance for the low concentration detection of substrates. Like FET-based bio-sensors, HEMT-based sensors can be upgraded to meet different substrate detection requirements by immobilising specific enzymes at the gate areas. This flexibility allows HEMT-based biosensors to be widely applied in a variety of circumstances. However, there is still much to improve for HEMT-based biosensors. HEMT is a very sophisticated semiconductor device and this restrains its wide application in industrial applications. Like FET-based biosensors, the packaging issue is another barrier preventing commercialisation of HEMT-based biosensors. Although these challenges are not going to be met immediately, more advanced technologies will be integrated into current fabrication protocols, giving enzyme/ZnO nanostructures/HEMT-based biosensors the potential for applications in real-time health monitoring and individual medication.

In this chapter, the topics of enzyme/ZnO biosensors were reviewed in detail and divided into three main subjects: three electrode measurement systems with enzyme/ZnO modification, enzyme/ZnO based FET biosensors and enzyme/ZnO based HEMT biosensors. The electrochemical biosensor, or the three electrode detection system, is a classic and conventional means for biosensors fabrication. Due to its excellent performance in terms of biocompatibility, conductance and physical adsorption, ZnO is proposed to be an efficient BRE in many enzyme electrode designs. After modification of specific enzymes onto the ZnO structure, the as-fabricated electrodes are ready to detect the corresponding substances, like glucose, uric acid and lactic acid. In order to further minimise the three electrode recording system into a small scale with a large detection limit, enzyme/ZnO/FET biosensors are proposed considering the piezoelectricity of ZnO structure materials. This integration of enzyme and ZnO promotes the development of micro-biosensor technologies. The ZnO nanowire based FET lactic acid biosensors successfully detect low concentrations of lactic acid in solution. HEMTs, as one kind of advanced FETs, can greatly increase the sensitivity of former FET based biosensors due to its electron channels generating from two-dimensional electron gas. On the sensing gate areas of ZnO based HEMTs, uricase and LOD are used to form corresponding substrate detection biosensors, respectively. These biosensors as discussed above are products of current biosensing technologies as well as milestones for future corresponding technology development. Although some trivial drawbacks are currently restraining the wide application of these prototypes of biosensors, the excellent properties of these biosensors have already provided a blueprint for future applications in industrial production and individual medication.

References

1. Y. Zhang, Z. Kang, X. Yan and Q. Liao, ZnO nanostructures in enzyme biosensors, *Sci. China Mater.*, 2015, **58**(1), 60–76.
2. M. Ahamd, C. Pan, Z. Luo and J. Zhu, A single ZnO nanofiber-based highly sensitive amperometric glucose biosensor, *J. Phys. Chem. C*, 2010, **114**(20), 9308–9313.
3. T. Kong, Y. Chen, Y. Ye, K. Zhang, Z. Wang and X. Wang, An amperometric glucose biosensor based on the immobilization of glucose oxidase on the ZnO nanotubes, *Sens. Actuators, B*, 2009, **138**(1), 344–350.
4. J. Zhang, C. Wang, S. Chen, D. Yuan and X. Zhong, Amperometric glucose biosensor based on glucose oxidase–lectin biospecific interaction, *Enzyme Microb. Technol.*, 2013, **52**(3), 134–140.
5. F. Zhang, X. Wang, S. Ai, Z. Sun, Q. Wan, Z. Zhu, Y. Xian, L. Jin and K. Yamamoto, Immobilization of uricase on ZnO nanorods for a reagentless uric acid biosensor, *Anal. Chim. Acta*, 2004, **519**(2), 155–160.
6. Z. Ibupoto, S. Shah, K. Khun and M. Willander, Electrochemical L-lactic acid sensor based on immobilized ZnO nanorods with lactate oxidase, *Sensors*, 2012, **12**(3), 2456–2466.
7. Y. Lei, X. Yan, J. Zhao, X. Liu, Y. Song, N. Luo and Y. Zhang, Improved glucose electrochemical biosensor by appropriate immobilization of nano-ZnO, *Colloids Surf., B*, 2011, **82**(1), 168–172.

8. Y. Lei, X. Yan, N. Luo, Y. Song and Y. Zhang, ZnO nanotetrapod network as the adsorption layer for the improvement of glucose detection via multiterminal electron-exchange, *Colloids Surf., A*, 2010, **361**(1–3), 169–173.

9. Y. Lei, X. Liu, X. Yan, Y. Song, Z. Kang, N. Luo and Y. Zhang, Multicenter uric acid biosensor based on tetrapod-shaped ZnO nanostructures, *J. Nanosci. Nanotechnol.*, 2012, **12**(1), 513–518.

10. Y. Lei, N. Luo, X. Yan, Y. Zhao, G. Zhang and Y. Zhang, A highly sensitive electrodchemcial biosensor based on zinc oxide nanotetrapods for L-lactic acid detection, *Nanoscale*, 2012, **4**, 3438–3443.

11. Y. Zhao, X. Yan, Z. Kang, P. Lin, X. Fang, Y. Lei, S. Ma and Y. Zhang, Highly sensitive uric acid biosensor based on individual zinc oxide micro/nanowires, *Microchim. Acta*, 2013, **180**(9/10), 759–766.

12. Y. Zhao, X. Yan, Z. Kang, X. Fang, X. Zheng, L. Zhao, H. Du and Y. Zhang, Zinc oxide nanowire-based electrochemical biosensor for L-lactic acid amperometric detection, *J. Nanopart. Res.*, 2014, **16**(5), 2398.

13. Y. Zhao, X. Fang, Y. Gu, X. Yan, Z. Kang, X. Zheng, P. Lin, L. Zhao and Y. Zhang, Gold nanoparticles coated zinc oxide nanorods as the matrix for enhanced L-lactic sensing, *Colloids Surf., B*, 2015, **126**, 476–480.

14. Y. Zhao, X. Fang, X. Yan, X. Zhang, Z. Kang, G. Zhang and Y. Zhang, Nanorod arrays composed of zinc oxide modified with gold nanoparticles and glucose oxidase for enzymatic sensing of glucose, *Microchim. Acta*, 2015, **182**(3/4), 605–610.

15. X. Wang, J. Zhou, J. Song, J. Liu, N. Xu and Z. Wang, Piezoelectric field effect transistor and nanoforce sensor based on a single ZnO nanowire, *Nano Lett.*, 2006, **6**(12), 2768–2772.

16. Y. Zhang, K. Yu, D. Jiang, Z. Zhu, H. Geng and L. Luo, Zinc oxide nanorod and nanowire for humidity sensor, *Appl. Surf. Sci.*, 2005, **242**, 212–217.

17. S. Chang, S. Chang, C. Lu, M. Li, C. Hsu, Y. Chiou, T. Hsueh and I. Chen, A ZnO nanowire-based humidity sensor, *Superlattices Microstruct.*, 2010, **47**(6), 772–778.

18. X. Liu, P. Lin, X. Yan, Z. Kang, Y. Zhao, Y. Lei, C. Li, H. Du and Y. Zhang, Enzyme-coated single ZnO nanowire FET biosensor for detection of uric acid, *Sens. Actuators, B*, 2013, **176**, 22–27.

19. R. Ahmad, N. Tripathy, J. Park and Y. Hahn, A comprehensive biosensor integrated with a ZnO nanorod FET array for selective detection of glucose, cholesterol and urea, *Chem. Commun.*, 2015, **51**, 11968–11971.

20. B. Kim, I. Sohn, D. Lee, G. Han, W. Lee, H. Jung and N. Lee, Ultrarapid and ultrasensitive electrical detection of proteins in a three-dimensional biosensor with high capture efficiency, *Nanoscale*, 2015, **7**(21), 9844–9851.

21. J. Hagen, W. Lyon, Y. Chushak, M. Tomczak, R. Naik, M. Stone and N. Kelley-Loughnane, Detection of orexin A neuropeptide in biological fluids using a zinc oxide field effect transistor, *ACS Chem. Neurosci.*, 2013, **4**(3), 444–453.

22. M. Fathil, M. Md Arshad, A. Ruslinda, S. Gopinath, M. Nuzaihan, R. Adzhri, U. Hashim and H. Lam, Substrate-gate coupling in ZnO-FET biosensor for cardiac troponin I detection, *Sens. Actuators, B*, 2017, **242**, 1142–1154.

23. J. Hagen, S. Kim, N. Kelley-Loughnane, R. Naik and M. Stone, Selective vapor phase sensing of small molecules using biofunctionalized field effect transistors, *Proc. SPIE*, 2011, **8018**, 80180B.

24. R. Ahmad, N. Tripathy and Y. Hahn, High-performance cholesterol sensor based on the solution-gated field effect transistor fabricated with ZnO nanorods, *Biosens. Bioelectron.*, 2013, **45**, 281–286.

25. B. Kang, H. Wang, F. Ren, S. Pearton, T. Morey, D. Dennis, J. Johnson, P. Rajagopal, J. Roberts, E. Piner and K. Linthicum, Enzymatic glucose detection using ZnO nanorods on the gate region of AlGaN/GaN high electron mobility transistors, *Appl. Phys. Lett.*, 2007, **91**(25), 252103.

26. B. Chu, B. Kang, F. Ren, C. Chang, Y. Wang, S. Pearton, A. Glushakov, D. Dennis, J. Johnson, P. Rajagopal, J. Roberts, E. Piner and K. Linthicum, Enzyme-based lactic acid detection using AlGaN/GaN high electron mobility transistors with ZnO nanorods grown on the gate region, *Appl. Phys. Lett.*, 2008, **93**(4), 042114.

27. Y. Song, X. Zhang, X. Yan, Q. Liao, Z. Wang and Y. Zhang, An enzymatic biosensor based on three-dimensional ZnO nanotetrapods spatial net modified AlGaAs/GaAs high electron mobility transistors, *Appl. Phys. Lett.*, 2014, **105**(21), 213703.

28. S. Ma, Q. Liao, H. Liu, Y. Song, P. Li, Y. Huang and Y. Zhang, An excellent enzymatic lactic acid biosensor with ZnO nanowires-gated AlGaAs/GaAs high electron mobility transistor, *Nanoscale*, 2012, **4**, 6415–6418.

29. S. Ma, X. Zhang, Q. Liao, H. Liu, Y. Huang, Y. Song, Y. Zhao and Y. Zhang, Enzymatic lactic acid sensing by In-doped ZnO nanowires functionalized AlGaAs/GaAs high electron mobility transistor, *Sens. Actuators, B*, 2015, **212**, 41–46.

CHAPTER 8

Self-powered Devices

ZHENG ZHANG, ZHIMING BAI AND YUE ZHANG*

University of Science and Technology Beijing, Beijing, China
*Email: yuezhang@ustb.edu.cn

8.1 ZnO Nanostructure-based Self-powered Photodetectors

The photovoltaic effect and photoconductive effect are two typical mechanisms for conventional photodetectors (PD). In conventional photodetectors, an external power supplement is essential for transmitting the separated photogenerated carriers into the circuit. By utilising the advanced electrical transmission properties and optical gain effect of low-dimensional ZnO nanostructures, a new type of self-powered (SP) PD was proposed in 2010. A SP PD based on the photovoltaic effect can operate at zero bias without external power. When ZnO absorbs photons with larger energy than its bandgap, electron–hole pairs can be generated. Then, the photogenerated carriers are separated by the built-in electric field at the interface, forming a photocurrent. Thus, in SP PDs, a stable and strong built-in electric field is important for its performance. As per the type of contact junctions, SP PDs can be divided into three categories: (1) *p–n* junction type, (2) Schottky junction type and (3) solid–liquid junction type.

8.1.1 Schottky Junction SP Photodetectors

Photovoltaic effects can usually be observed in Schottky junction-based devices, which can provide energy for themselves in applications as photodetectors. This feature makes it possible to detect light irradiation without

Nanoscience & Nanotechnology Series No. 43
ZnO Nanostructures: Fabrication and Applications
By Yue Zhang
Published by the Royal Society of Chemistry, www.rsc.org

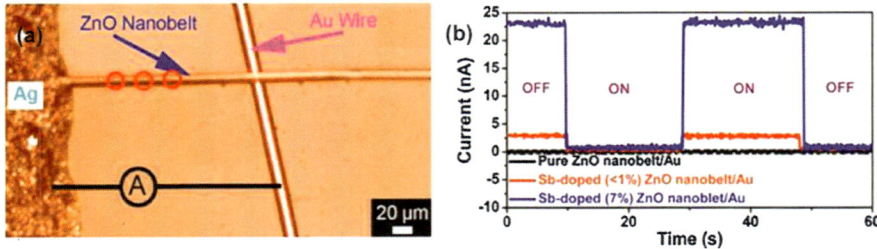

Figure 8.1 (a) Schematic diagram of a SP PD based on a single Sb-doped ZnO nanobelt; (b) the photoresponse as a function of time at zero bias. Reproduced from ref. 1 with permission from AIP Publishing.

an external power source. In 2010, a cross-structured SP PD was fabricated using a single ZnO nanobelt and Au microwires, as shown in Figure 8.1. It had a photosensitivity of about 22, and a response time of less than 100 ms at zero bias. The photocurrent of the detector was improved with increasing Sb doping concentration in the ZnO belt. The high donor impurity density due to Sb doping in ZnO resulted in an obvious increase of carrier concentration and mobility. Thus, more photogenerated electron–hole pairs in this area will be effectively separated, which can also result in the enhancement of UV photocurrent.[1]

In a Schottky junction-based SP PD, a larger junction contact area could improve the device performance effectively. The nanowire networks are promising building blocks for future photodetectors. UV photodetectors based on ZnO micro/nanowire networks with Pt contacts have been fabricated on glass substrates. The turn-on voltage was estimated to be about 1.5 V and the rectification ratio at ± 5 was about 70 (Figure 8.2). As per the thermionic emission model, the ideal factor and Schottky barrier height were 16 and 1.1 eV, respectively, indicating good Schottky contact between ZnO and Pt. Because the conductance of the device is dominated by a Schottky barrier and a nanowire–nanowire contact barrier, the network photodetector has a dark current at the 10^{-10} ampere scale. The photodetectors exhibited a high photosensitivity (5×10^3) for 365 nm UV light with a fast recovery time of 0.2 s at zero bias. The light-induced modulation of the Schottky barrier and nanowire–nanowire contact barrier were used to account for the results. It was also observed that the photodetector had a high on–off ratio of 800 without external bias. The photovoltaic effect was proposed to explain the SP working mechanism.[2]

UV photodetectors based on integrated ZnO nanowires were also studied for the application of SP PDs. Here, a ZnO nanowire array UV PD with Pt Schottky contacts was fabricated on a glass substrate (Figure 8.3). Under UV light, this PD showed a high on–off ratio of 892 at 30 V bias. The rise time was estimated to be about 70 s; meanwhile the decay process was excellently fitted with a biexponential relaxation equation. The first decay process takes a short time, in which photogenerated electrons in the conduction band

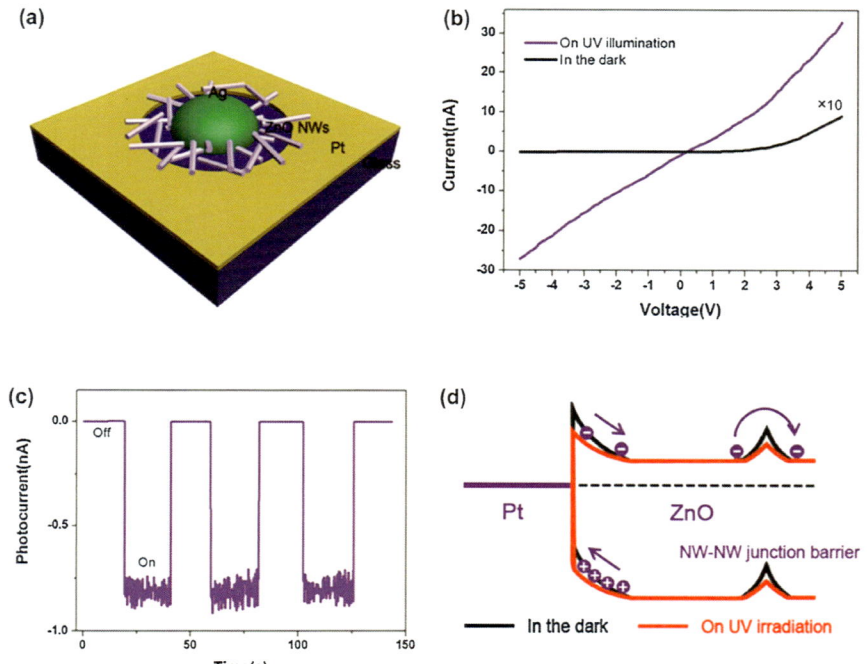

Figure 8.2 (a) Schematic of the ZnO nanowire network PD; (b) *I–V* characteristics of the ZnO nanowire network photodetector in the dark and upon 365 nm UV illumination; (c) the photocurrent response curve of the photodetector under UV illumination being turned on and off at zero bias; (d) schematic of the carrier generation, Schottky barrier and nanowire-nanowire contact barrier for electron transfer in the ZnO nanowire network PDs.

Reprinted from Prog. Nat. Sci., 24(1), Z. Bai, X. Yan, X. Chen, K. Zhao, P. Lin and Y. Zhang, High sensitivity, fast speed and self-powered ultraviolet photodetectors based on ZnO micro/nanowire networks, 1–5. Copyright (2014), with permission from Elsevier.[2]

recombine with photogenerated holes in the valence band. The second decay process is longer, which was related to the re-adsorption of oxygen molecules on the surfaces of ZnO nanowires. Besides, it was observed that the photocurrent not only decayed when the UV light was off, but also decayed on steady UV light illumination. After 14 min illumination, the photocurrent decreased to 77% of its maximum value. This phenomenon is called anomalous photoconductivity, which goes through two-electron relaxation processes: electron trapping by the adsorbed O_2 molecules and recombination with the positively charged deep level defect states. Interestingly, it was also found that this PD had a responsivity of 475 at zero bias. By fitting the *I–V* curve in the dark, it was found that there was a difference of 30 meV between the Schottky barrier heights at the two sides of the PD. The two-electron flowed in opposite directions resulting in the formation of

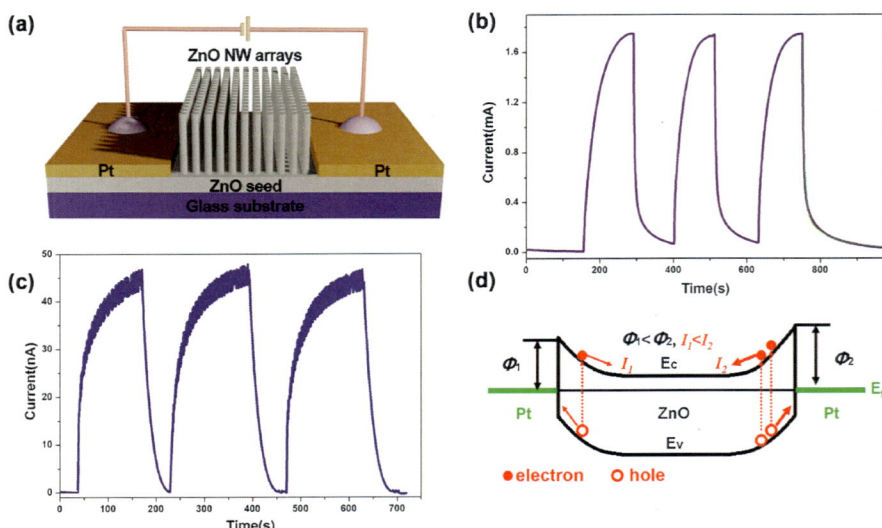

Figure 8.3   (a) The schematic diagram of the ZnO nanowire array UV PD; (b) the current
response curve of the device upon UV light illumination being turned on
and off at 30V bias; (c) the current response curve of the ZnO nanowire array
UV PD upon UV light (365 nm) illumination being turned on and off at
0 V bias; (d) bandgap diagram of the device upon UV light illumination
at 0 V bias.

Reprinted from Curr. Appl. Phys., 13(1), Z. Bai, X. Yan, X. Chen, H. Liu,
Y. Shen and Y. Zhang, ZnO nanowire array ultraviolet photodetectors with
self-powered properties, 165–169. Copyright (2013), with permission from
Elsevier.[3]

photocurrent.[3] This work suggested that ZnO nanowire arrays are promising
materials for fabricating low-cost SP UV PDs.

The effect of ambient temperature on the performance of SP PDs has been
investigated for applications in environmental monitoring. An SP UV detector was fabricated based on selectively grown ZnO nanowires with Al/Pt
interdigitated electrodes. Without applied bias, the PD showed good UV
light selectivity with a quick rise time of about 80 ms, as shown in Figure 8.4.
Upon heating, the increased amount of adsorbed oxygen at the interface
enhanced the Schottky barrier height, which led to a maximum sensitivity of
3.1×10^4 at 340 K, which was 82% higher than that obtained at room temperature. At low temperature, the absorption quantity of oxygen increases,
resulting in a larger interface state density and then a higher interface band
bending. Therefore, when the temperature was below the turning point
(340 K), the Schottky barrier height was proportional to temperature, which
was ascribed to interface oxygen adsorption. As the temperature increased,
the Schottky barrier height gradually decreases due to the higher energies of
the electrons at higher temperature. By using the thermionic emission–
diffusion theory, photocurrents at different temperatures were simulated,
which were in line with the experimental data.[4]

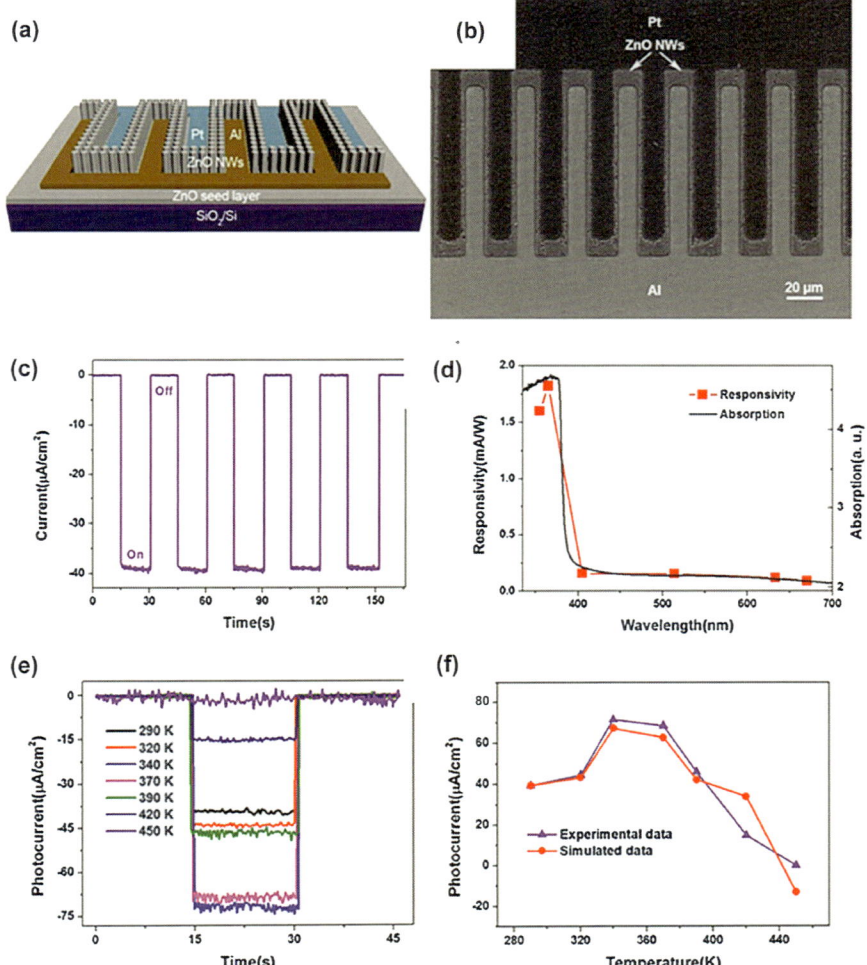

Figure 8.4 The self-powered ultraviolet photodetector constructed by Al/ZnO nanowire array/Pt. (a) The structure diagram of the device; (b) the SEM morphology of the device; (c) photocurrent response at zero bias; (d) IPCE curve; (e) photocurrent response at zero bias with different temperatures; (f) the relationship between the photocurrent and the temperature.

Reproduced from ref. 4 with permission from The Royal Society of Chemistry.

8.1.2 *p–n* Junction Self-powered Photodetectors

Junctions between *p-* and *n*-type semiconductors provide a strong driving force to separate photo-generated electrons and holes. In fact, *p–n* junction-type PDs without any external power source have already been reported by many groups. A single *p–n* homojunction ZnO nanowire has been fabricated as a SP detector.[5] It shows the obvious rectifying characteristics of a diode and good UV sensing performance. Under zero bias, the short circuit photocurrent and the open circuit voltage were 1 μA and 0.2 V. The rise time and the recovery time were calculated to be 30 ms and 50 ms, respectively, which were smaller than those of reported photoconductive UV detectors based on ZnO nanowires. However, a single ZnO nanowire/GaN film heterojunction was fabricated to act as a SP visible-blind UV PD.[6] This UV detector can be driven by UV light with a short circuit current density of 5×10^4 mA cm^{-2}, an open circuit voltage of 2.7 V, and an output power of 1.1 μW. The response time of the ZnO/GaN heterojunction was much smaller (rise time about 20 μs, decay time about 219 μs) than traditional ZnO detectors. A selective multi-wavelength PD was constructed *via* combining the SP device with a CdSe nanowire based device. The integrated device not only demonstrated UV and red light photosensing properties, but also could act as an optically and logic gate controlled by UV light.

A single ZnO nanowire/Si film heterostructure PD has been fabricated, which showed an ultrafast response time of 7.4 ms and a high sensitivity of 2×10^4 for UV light and 5×10^3 for visible light. Under 325 nm laser illumination, the output power of the SP PD reached 1.7 nW. Under 514 nm laser illumination, the output power was 0.5 nW. In addition, the photocurrent at zero bias and open-circuit voltage exhibited square root and logarithmical dependences on incident light power, respectively.[7] When compared with the common *p–n* heterojunction-type devices, the fabricated devices with double heterojunctions based on a single *n*-type ZnO microwire and a *p*-type Si film exhibited excellent electrical performance. The ideal factor of the heterojunction was calculated to be 3, and the rectification ratio was about 300 at ±5 V, indicating its ideal rectification behaviour. At zero bias, the fabricated device can deliver a photocurrent of 71 nA, a high photosensitivity of about 3170 under UV light (0.58 mW cm^{-2}) illumination and a fast rising and falling time of both less than 0.3 s. The device of the double heterojunctions of a *n*-ZnO microwire/*p*-Si fillm might be a good candidate for potential applications in photodetection, sensors and photocatalysis.[8]

In recent years, there has been a growing interest in inorganic–organic hybrid systems because they integrate the high electron mobility of inorganic semiconductors and the solution processability of polymers. A new UV PD based on a single ZnO tetrapod and PEDOT: PSS heterostructure has been constructed and investigated. At zero bias, the detector showed an on–off ratio of about 1100, a rise time of about 3.5 s and a decay time of about 4.5 s when 325 nm UV (0.16 mW) illuminated the *p–n* heterojunction. The SP properties were driven by the photovoltaic effect, with a short-circuit current of ~1.1 nA, an open-circuit voltage of ~0.2 V and a fill factor of ~25%.[9]

However, the photocurrent of SP PDs based on single ZnO nanowires can only reach the microamp level, because of the small junction contact area. A structure of *n*-ZnO/*p*-NiO core–shell nanowire arrays was proposed to increase the *p–n* junction contact area.[10] The peak responsivity of the device at zero bias was about 0.493 mA W^{-1}. It also showed high selectivity to UV light, which was caused by the filter effect of the outer layer of the *p*-NiO shell layer under front illustration conditions. A sandwich-like structure PD composed of one layer of *p*-type polyaniline (PANI) nanowires and two layers of *n*-type ZnO nanorods was designed and fabricated to improve the photocurrent at zero bias.[11] By increasing the layers of the PSS, the photocurrent could reach about 14 µA with a sensitivity of over 10^5 at zero bias. When the irradiation direction of the UV light was changed, the movement direction of electron in the device was also changed.

8.1.3 Solid–Liquid Junction Self-powered Photodetector

A solid–liquid junction SP UV detector was fabricated based on a ZnO nanowire array/H_2O heterojunction, which has a similar structure to a typical dye-sensitised solar cell (DSSC) but without dye adsorption. The working mechanism of the solid–liquid junction SP UV detector was like the photoelectrochemical (PEC) effect. The band bending at the solid–liquid interface acts as a built-in potential, which has Schottky barrier behaviour. The built-in potential can separate photogenerated electron–hole pairs and form a photocurrent. That is to say, the ZnO/H_2O heterojunction can operate in photovoltaic mode without external power sources. When compared with *p–n* junction type and Schottky type SP UV detectors, the solid–liquid junctions have a much larger short circuit current in the milliampere range because of the enhance junction contact area. They also have some advantages, including low-cost, simple manufacturing process and being composed of abundant and non-toxic raw materials.[12–14]

In 2011, an UV PD based on TiO_2/water solid–liquid heterojunction was reported.[15] It showed a maximum responsivity of about 69.2 mA W^{-1} at a wavelength of 347.5 nm at zero bias. Additionally, the spectral response could be modified and the performance could be enhanced *via* precise design of the active layer, electrolyte and substrate of the device. This study opened the door for more sophisticated commercial photodetection using a semiconductor–liquid heterojunction. Soon afterwards, a PEC cell based on a nanocrystalline TiO_2 film was designed, which had the same structure as a DSSC without dye adsorption, exhibiting trap-to-trap electron transport.[16] Upon UV light illumination (33 mW cm^{-2}), the device showed a photocurrent density of about 0.55 mA cm^{-2} at zero bias and an open circuit voltage of 0.6 V. It had a rise time of 80 ms and decay time of 30 ms, a sensitivity of about 2699, and good photoresponse linearity in a light power range of 25 µW cm^{-2} to 33 mW cm^{-2}. A PEC cell with unsensitised ZnO nanostrawberry aggregates/TiO_2 film was fabricated as a SP UV PD, which exhibited a high on–off ratio of 37 900, a fast rise time of 22 ms and a decay time of 9 ms for short circuit current signal.[17] The ZnO nanostrawberry aggregates/TiO_2 film has better light time response performance than the TiO_2 film under low light intensity.

ZnO has a similar band energy structure to TiO_2 and a larger electron mobility than TiO_2. Therefore, ZnO is a promising alternative to TiO_2 for the use in PEC type SP UV detectors. It was reported that a SP UV detector was built based on a ZnO nanowire array photoanode and H_2O electrolyte.[18] The monocrystalline ZnO nanowire provides a direct pathway for the photo-generated electrons and the solid–liquid interface provided the driving force for the separation of photogenerated electron–hole pairs. Upon illumination of 1.25 mW cm^{-2} of UV light (365 nm), this PEC cell has a high photosensing performance, generating a short circuit current of 0.8 μA and an open circuit voltage of 0.5 V. The responsivity peak (22 mA W^{-1}) is centred at *ca.* 385 nm, indicating that it was suitable to be used in the range of 320–400 nm. The current showed linearity with increasing light intensity in a weak light intensity range, and the current gradually saturates in a high light intensity range. The rise time and decay time were both about 100 ms. The high short circuit current, fast response speed, high light selectivity and good photosensitivity linearity combined with low cost and facile manufacturing process make the PEC type SP UV PD a promising potential for practical applications.

8.1.4 Piezotronic Engineering for Self-powered Photodetector Optimisation

It is well known that the working principle for most optoelectronic devices relies on the charge carrier separation/combination process at the interface rather than in bulk. Therefore, the arbitrary regulation of interfacial electronic states could be implemented to modulate the performance or endow the device with novel functionality. So far, multiple approaches have been introduced to realise the precise tailoring of interfacial energetics, but most of these were restricted to complicated fabrication processes and specific device configuration. Due to the lack of inversion symmetry, semiconductor materials with the wurtzite structure such as ZnO, GaN and CdS could generate non-mobile piezoelectric polarisation charges at the interface in the presence of mechanical deformation. By taking advantage of this strain-induced piezopolarisation, charge carrier transport across a Schottky interface or *p–n* junction could be effectively tuned.

As mentioned above, the operation of SPPD relies on the separation of photon-generated electron–hole pairs within the built-in electric field at the *p–n* junction or Schottky contact interface. Therefore, its performance was exquisitely sensitive to the barrier height and strength of internal field. Performance enhancement of the SP UV sensor with strain engineering was first verified based on the PEDOT: PSS/ZnO heterostructure, where the photocurrent at zero voltage bias could be significantly improved with increasing tensile strain.[19] At zero bias, the sensitivity of the device was about 10^3 under the illumination of 325 nm UV light, along with a fast response time of <1 s. Because there is no external bias, this photocurrent originates from the separation of photogenerated carriers within the interfacial built-in electric field at the heterojunction. Due to the intrinsicly coupled semiconducting and piezoelectric properties of ZnO, the strain-induced piezo-potential could tune the energy band profile at the interface and increase/decrease the barrier height. When the $+c$ axis points from PEDOT: PSS to

ZnO and tensile strain was applied, a negative piezopotential was created at the PEDOT: PSS/ZnO interface, which raised the energy of the electrons in ZnO conduction band. Therefore, the energy band difference between the conduction band of ZnO and the lowest unoccupied molecular orbit of PEDOT: PSS would be reduced and the barrier height lowered. This decreased barrier height weakened the strength of the built-in electric field. Therefore, the recombination of photogenerated electron–hole pairs could be enhanced, and the photocurrent decreased. The other route of contact between the PEDOT and ZnO nanowire is when the $+c$ axis points from ZnO to PEDOT: PSS; a positive piezopotential would be established at the interface under tensile strain. The positive potential lowers the conduction band energy of ZnO, therefore the barrier height increases. The increased barrier height will induce the broadening of the space-charge region in ZnO and the strengthening of the built-in electric field. Thus, more effective electron–hole pairs would be generated and could be separated more efficiently, giving rise to the enhanced photocurrent and sensitivity.

The piezopolarisation charges were not only capable of inducing remarkable modulation of band shifting in ZnO but also the adjacent semiconductor which forms heterojunction with it. Recently, a heterostructure of Cu_2O/ZnO has been demonstrated.[20] Under an illumination density of 17.2 mW cm^{-2}, a photoresponse increase of 18.6% was obtained when applying -0.88% compressive strain. Our results also demonstrate the illumination-density-dependent photoresponse enhancement due to the screening effect, with a 2.2% response enhancement per 0.1% strain being obtained under an illumination density of 17.2 mW cm^{-2}, but an enhancement of only 1.2% for an illumination density of 87.8 mW cm^{-2}. This was because more generated free charges carriers could partially screen the piezopotential and thus the modulation ability was weakened. Because of the preferred $+c$ orientation growth of ZnO NRs, a permanent positive piezopotential will be produced at the Cu_2O–ZnO interface when applying compressive strain. This positive piezopotential, just like an externally applied positive bias, produces a sharper and extended built-in field and depletion region in Cu_2O. The enlarged space charge region in Cu_2O means that more excitons were generated and can be separated more effectively, leading to the enhanced photoresponse current.

By using piezotronic effect, the photoresponse of a ZnO/Au Schottky junction based SP UV detector has been greatly improved. A 440% augment of photocurrent was obtained when the device was subjected to 0.580% tensile strain. Meanwhile, the enhanced photoresponse under strain does not influence response time, stability, or repeatability of the device. The underlying mechanism of this enhancement of the device performance under strain could be clarified in light of their energy band realignment under strain. By utilizing a remnant piezopotential, charge transport behaviour at the interface was effectively modulated. A negative piezopotential was produced at the metal–semiconductor interface, repelling electrons away from the interface and consequently further depleting the interface and raising the barrier height. Undoubtedly, the stronger built-in field was favourable for photoexcitation separation and extraction. Though the piezoresistance effect would certainly take place and contribute to the charge transport behavior, it is a symmetrical effect that changes the bulk conductance of semiconductor instead of the interface property. Therefore, piezotronic effect dominates the overall charge transport behavior in this work because of the asymmetrical variation of current under positive and negative bias. In addition, because the polarity of piezopotential depends on the relationship between the direction of the *c* axis of ZnO wire and that of applied strain, when the +*c* axis of ZnO points toward epoxy, a positive

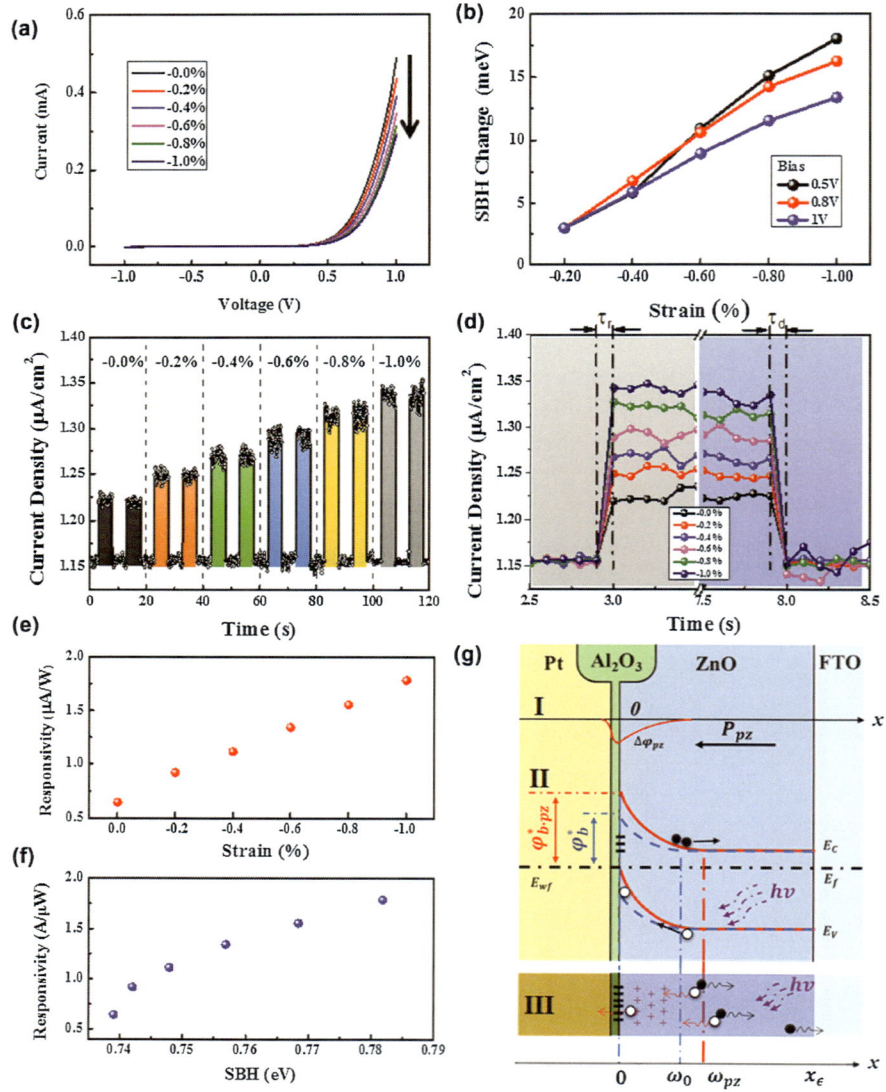

piezopotential appears at the interface under tensile strain, which causes electrons to accumulate at the interface, resulting in a narrower depletion region and lower barrier height. The shrunken and blunter built-in potential suppresses the separation and extraction of photoexcites, and inferior photoresponse properties are thus forecasted.[21]

For a metal-semiconductor PD, the deep depletion region formed in the semiconductor provides strong built-in electric-field to separate electron–hole pairs. Furthermore, a thin insulator layer can be employed between semiconductor and metal electrode forming a metal-insulator-semiconductor contact, where the insulator layer reduces the tunnelling current and improves the reliability and performance. Here a metal-insulator-semiconductor ($Pt/Al_2O_3/ZnO$) based SPPD has been developed. The PD has a sensitive response to light illumination without any external bias. Applying an ultrathin dielectric layer and piezotronic effect are two effective strategies for interface engineering to enhance the photoresponse properties. The dielectric layer can significantly enhance the effective Schottky barrier height. In addition, the Schottky barrier height can be actively modulated by the piezopolarisation-induced built-in electric field variation under compressive strain. Thus, the photoresponse properties of the SPPD were largely improved by the Schottky barrier height enhancement. The responsivity and detectivity of the SPPD were increased 2.77 times and 2.78 times, respectively, under a compressive strain of 1.0%. Figure 8.5 shows a schematic diagram of the band structure under compressive strain and illumination. When a compressive strain was applied vertically on to the Pt electrode along the length direction of ZnO nanowire, a piezoelectric field was generated inside ZnO with a negative polarisation formed at the ZnO/Pt interface. This strain-induced polarisation could modulate the depletion region in ZnO, the insulator layer, and the contact surface of Pt electrode. Because the carrier concentration in Pt was much greater than

---

Figure 8.5 *I–V* characteristics of PDs under strain. (a) *I–V* curves under a series of compressive strains; (b) calculated SBH change under compressive strain at different biases of 0.5 V, 0.8 V and 1 V; (c) the photoresponses at illumination of 100 mW cm^{-2} under different compressive strains; (d) detailed current density profiles of the response and recovery process; (e) the relationship between responsivity and strain; (f) the relationship between responsivity and SBH; (g) schematic diagram of the devices under compressive strain and illumination. I is the potential variation at the interface under a compressive strain. II is the band structure diagram. The blue dashed line is the original energy band structure of the ZnO; the red solid line is the band structure under negative piezo-polarisation; and the black dash dot line is the Fermi level at thermal equilibrium. III illustrates a gradient piezoelectric field formed in ZnO with a negative piezo-polarisation at the interface, which widens the depletion region.
Reprinted from Nano Energy, 9, Z. Zhang, Q. Liao, Y. Yu, X. Wang and Y. Zhang, Enhanced photoresponse of ZnO nanorods-based self-powered photodetector by piezotronic interface engineering, 237–244. Copyright (2014), with permission from Elsevier.[22]

ZnO and Al_2O_3, the influence of piezopotential in Pt electrode occurs at the interface and decays rapidly. Due to the extremely small thickness of the Al_2O_3 layer, the potential reduction across Al_2O_3 was negligible. Figure 8.5(g)-I shows the potential variation at the interface in the presence of piezo-polarisation. In Figure 8.5(g)-II, the blue dash line is the original energy band structure of ZnO, the red solid line is the band structure with negative piezo-polarisation, and the black dash dot line is the Fermi level at thermal equilibrium. Under compressive strain, the depletion region was widened by the negative piezo-polarisation. Such piezo-polarisation enhanced built-in field and widened depletion region were favourable for photogenerated electron–hole pair separation, and thus rapid response and higher sensitivity were obtained.[22]

Until now, several kinds of ZnO-based *p–n* junction PDs have been reported, in which organic–inorganic hybrid PDs have attracted tremendous research interest because of their unique properties combining high flexibility, low cost, and easy fabrication at low temperature of organic polymers with high stability of inorganic materials. Amongst various organic materials, 2,2′,7,7′-tetrakis (*N,N*-di-*p*-methoxyphenyl amine) 9,9′-spirobifluorene (Spiro-MeOTAD) is a non-hazardous solid hole conductor and exhibits efficient hole regeneration capability and high conductivity, and thus may be a good choice in UV PDs. Here, a flexible SP organic–inorganic hybrid heterojunction UV PD fabricated using *n*-type RF-sputtered ZnO films and *p*-type Spiro-MeOTAD was demonstrated. The obtained device has a fast and stable response to UV light illumination at zero bias. Together with responsivity and detectivity, the photocurrent can be increased about 1-fold upon applying a 0.753% tensile strain. However, compressive strain decreases the barrier height. Because the *c* axis of RF-sputtered ZnO thin film grown on ITO/PET substrate is pointing from ITO/PET substrate to ZnO, once the ZnO film suffers a tensile strain, permanent positive piezoelectric polarisation charges will be induced at the ZnO/Spiro-MeOTAD interface. This positive piezopotential at the interface can lower the local conduction band level of ZnO, thus increasing the barrier height from ΔE_0 to ΔE^+. The increased barrier height will strengthen the built-in electric field, and therefore, the separation efficiency of photogenerated electron–hole pairs is enhanced, leading to the improved photocurrent and responsivity. However, when the RF-sputtered ZnO film suffers a compressive strain, the induced negative piezoelectric polarisation charges at the interface will lift the local conduction band level of ZnO, thus lowering the barrier height from ΔE_0 to ΔE^-. Thus, the strength of built-in electric field was weakened; the charge separation efficiency was jeopardised, resulting in decreased photocurrent and responsivity.[23]

When an *n*-type semiconductor comes into contact with the electrolyte, the transfer of electric charge occurs until electronic equilibrium at the interface is reached, thus producing a space–charge region and upward band bending in the semiconductor. Meanwhile, due to the relatively high conductivity and charge concentration of electrolyte, most of the depletion region falls on the semiconductor side. Therefore, when the semiconductor is illuminated by light with an energy greater than its bandgap, this semiconductor–electrolyte interface can be operated by the photovoltaic effect and the built-in potential acts as the driving force to separate photo-generated electron–hole pairs. Here, a SPPD based on a ZnO film/electrolyte solid–liquid heterojunction without any external power supply was constructed. At zero bias voltage, the device shows an on–off ratio larger than 10^2 with response and recovery times in the range of milliseconds. A 48% performance enhancement was achieved with the introduction of 0.15% compressive strain due to the generation of piezo-polarisation charges.

This phenomenon can be interpreted from the energy band diagram of the ZnO–electrolyte heterojunction in the presence of strain. Because of the preferred $+c$ orientation growth of sputtered ZnO, permanent positive piezoelectric polarisation charges were induced near the ZnO–electrolyte interface when applying tensile strain. This positive piezopotential attracts free electrons and lowers the energy of the conduction band in ZnO, thus decreasing the barrier height. The decreased barrier height weakens the strength of built-in electric field and the separation efficiency of excitons was jeopardised, resulting in decreased photocurrent. Furthermore, under these conditions, the photo-generated electrons migrate into the electrolyte more easily, which could facilitate the reduction reaction process. High reproducibility and stability of the response could be demonstrated upon applying a constant strain periodically to ZnO. In contrast, when the ZnO film is compressively strained, negative piezo-polarisation charges deplete electrons in ZnO and increase the Schottky barrier height at the interface of the ZnO–electrolyte. Therefore, more excitons could be generated and the separation efficiency is enhanced, leading to the increased photocurrent. In this case, the valence band bending favours the transfer of holes to the electrolyte for oxidation reaction. The results reported here provide a new approach to tuning the properties of the semiconductor–electrolyte interface and may be extended to photoelectrochemical or photocatalytic processes.[24]

As a new member of the SP nanosystems, SP PDs can be driven by light signals and convert them into electric signals. They have many advantages over traditional UV PDs such as higher sensitivity, larger conversion efficiency and faster response time. Although SP PDs have made great progress in recent years, practical applications set high requirements for the performance of the devices. The photoelectric conversion efficiency needs to be further elevated to ensure sufficiently high output, because the light signals are usually very weak. Stability needs to be improved to make sure they can continuously operate in harsh environments. A protective layer is needed to enhance the acid and alkali resistance of ZnO nanomaterials. The size and weight of the SP devices must to be reduced, especially in applications for biomedicine. If all the problems are solved, SP PDs will have great application potential in light communication, light detection, biosensing and environmental monitoring.

8.2   PEC Biosensing

Recently, biosensing devices based on the PEC mechanism have gradually attracted wide-ranging attention, and are regarded as one of the most promising analytical methods.[25] Such types of sensing device take the photoexcitation process and electrochemical detection in combination to realise advantages both from optical and electrochemical sensors. Since the excitation source (light) and detection signal (photocurrent) are totally separate, the undesired background noise signal can be avoided, thus improving the sensitivity when compared with traditional biosensors.[26–28] The electrochemical biosensing devices also have obvious advantages like ease of construction, low cost, ease of miniaturisation *etc.* Most importantly, solar light can power the device without further applied bias, so the so-called 'self-powered' biosensing devices have been realised.

Taking glutathione (GSH) as an example molecule, ZnO nanostructure-based PEC biosensors are discussed as follows. GSH is a tripeptide that is composed of glutamyl (Glu), cysteine (Cys), glycine (Gly) and endogenous antioxidant, which are widely distributed in the intracellular environment and play indispensable roles in organisms. GSH can be oxidised to glutathione disulfide (GSSG), which is suitable for the oxidation reaction at the photoanode in the PEC system. Based on ZnO nanostructures, various modification protocols have been applied, including electrode structure optimisation, plasmonic metal decoration with SPR effect, $p–n$ heterojunction construction and interface engineering to acquire high performance for the bioanalysis of GSH.

In terms of electrode structure optimisation, the direct synthesis of a ZnO nanorod array was realised on the RGO surface.[29] Under UV illumination, due to the inserting an RGO layer at the bottom of ZnO, the high efficient electron extraction from the ZnO to the external circuit significantly suppressed the recombination rate and facilitated the charge transfer inside photoanodes. Therefore, the RGO/ZnO heterostructure performance improved the PEC properties compared to pristine ZnO. The photo-generated holes accumulated at the ZnO surface and subsequently participated in the oxidisation of GSH. There is a certain relationship between the holes consumed during oxidisation and the electron flow through the external circuits, so the bioanalysis of GSH concentration was realised with solar excitation. With 0 V bias (*vs.* Ag/AgCl), the constructed PEC biosensing photoanode exhibited a detection limit of 2.17 μM ($n = 3$) and a linear range of 10–200 μM ($R^2 = 0.997$). For further optimisation of the electrode structure, the ZnO nanorod array was also synthesised on the surface of 3D graphene foam. Under illumination, the large surface area as well as high electron mobility promoted the PEC properties to a better stage.

The detection limit towards GSH was down to 1.79 μM ($n = 3$) and the linear range was 10–300 μM ($R^2 = 0.991$), demonstrating the superiorities of electrode structure optimisation for PEC performance enhancement.

In terms of plasmonic metal decoration, Au nanoparticles were uniformly deposited on the surface of ZnO nanostructures *via* a UV reduction method (Figure 8.6(a) and (b)).[30] Under solar illumination, the SPR induced hot electrons injected into the conduction band of ZnO to enhance the photocurrent, which effectively took advantage of visible light (Figure 8.6(c)). With 0 V bias (*vs.* Ag/AgCl), the constructed Au@ZnO PEC biosensing photoanodes acquired a detection limit of 3.29 μM ($n = 3$) and a linear range of 20–1000 μM ($R^2 = 0.996$), (Figure 8.6(d) and (e)).

Regarding the *p–n* junction construction and interface engineering, a ZnO/Cu$_2$O heterostructure was fabricated.[31] The thickness of the Cu$_2$O layer, as well as the interface area, were effectively modulated *via* adjusting the electrodeposition duration of *p*-type Cu$_2$O (Figure 8.7(a)–(f)). Moreover, through changing the pH value of the electrode position electrolyte, the charge carrier density inside Cu$_2$O was controlled, which finally engineered the electronic structure at the interface between ZnO and Cu$_2$O. Therefore, the optimal built-in electric field of such a *p–n* junction was acquired, which indicated that the highest photo-generated charge separation efficiency and best PEC performance were acquired (Figure 8.7(g)). The results show that the pH 11 Cu$_2$O/ZnO sample had the best PEC response. Based on this, the detection limit toward GSH dropped to 0.42 μM ($n = 3$) and a wide linear range of 10–1000 μM ($R^2 = 0.991$) was achieved.

Through crosswise comparison of above mentioned PEC biosensors, we can conclude that the PEC biosensing performance toward GSH was gradually improved along with the increased PEC photoanode performance (Figure 8.8). Specifically, the upper limitation of linear range increased with the PEC photoanode property improvement. This is because in a certain time period, the amount of photo-generated holes at the photoanode surface determines the quantity of GSH molecules that are to be oxidised. The number of consumed photo-generated holes corresponds to the amount of export electrons which is the photocurrent intensity. Thus, the acquired PEC biosensing linear ranges are basically positively related to the PEC photoanode properties. However, the detection limit basically decreases with the PEC photoanode property improvement. When the GSH with trace quantities was oxidised at the photoanode surface, the output electric signal corresponds to the photo-generated electrons flow to the external circuit instead of being trapped or recombined. Therefore, the better the PEC photoanode properties are, the lower the detection limit will be. Besides, the diffusion, absorption and immobilisation of GSH molecules will also affect PEC biosensing performance. In general, the PEC biosensing performance depends on both the supplier side that provides photo-generated holes and the consumer side that consumes photo-generated holes.

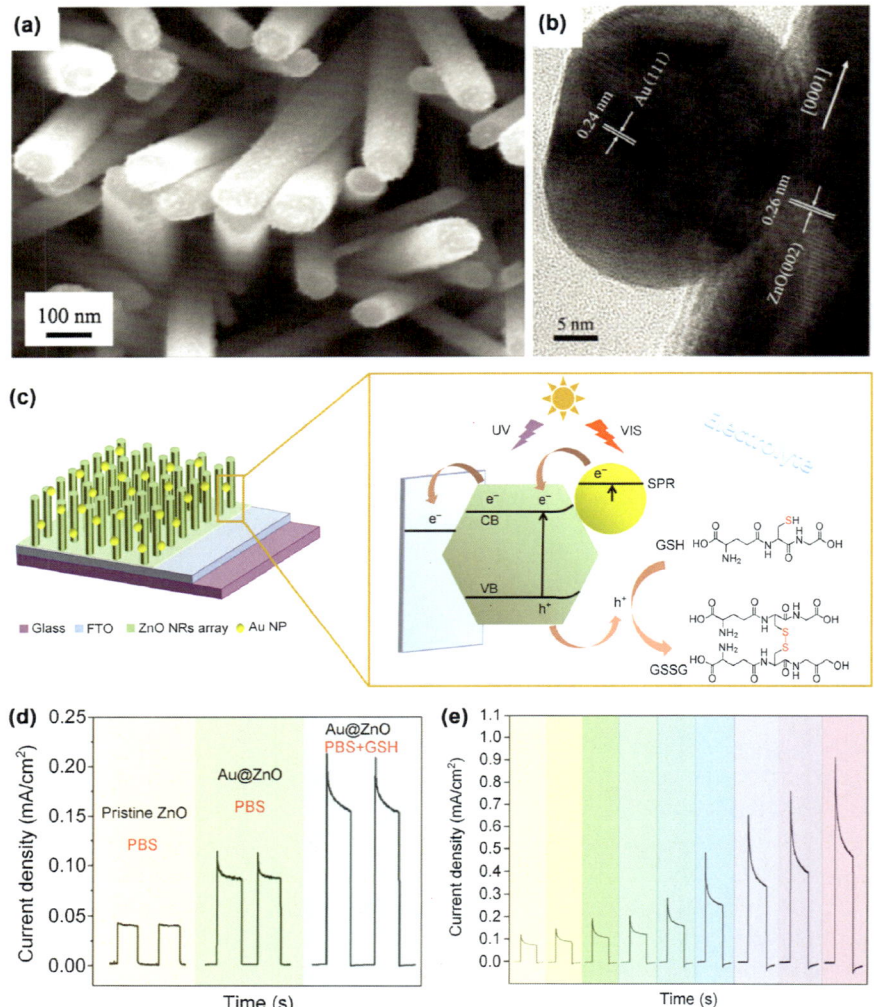

Figure 8.6 (a) SEM images of Au NPs modified ZnO NRs; (b) HRTEM image of the interface between Au nanoparticle and ZnO nanorod; (c) schematic illustration of the PEC biosensing mechanism toward GSH detection based on a Au@ZnO photoanode; (d) photoresponses of pristine ZnO and Au@ZnO in PBS or 100 μM GSH containing PBS with a bias of 0 V *vs.* Ag/AgCl under illumination; (e) photocurrent response of a Au@ZnO based PEC biosensor in 0.1 M PBS in the presence of various GSH concentrations.
Reproduced from Nano Res., Self-powered photoelectrochemical biosensing platform based on Au NPs@ZnO nanorods array, 9(2), 2015, 344–352, Z. Kang, X. Yan, Y. Wang, Y. Zhao, Z. Bai, Y. Liu, K. Zhao, S. Cao and Y. Zhang, (© Tsinghua University Press and Springer-Verlag Berlin Heidelberg 2015) With permission from Springer.[30]

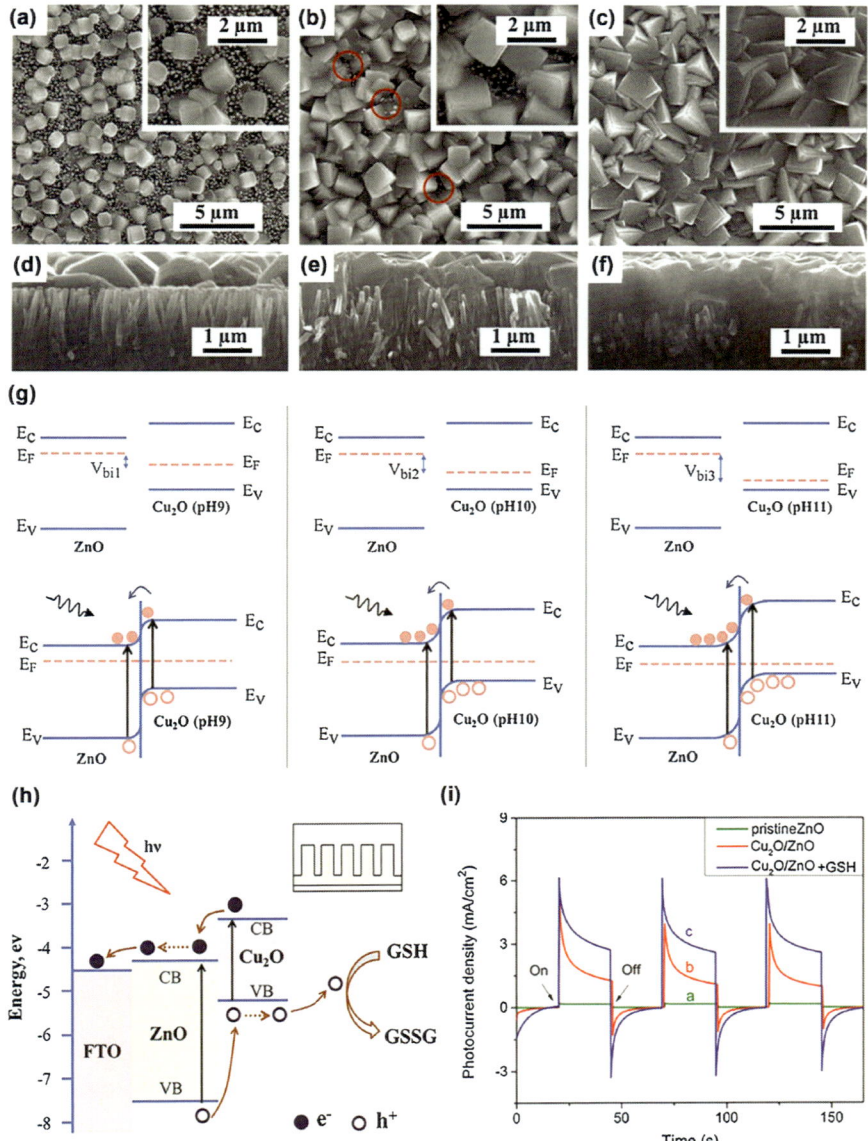

Figure 8.7 SEM images of Cu₂O/ZnO NRs array heterostructures acquired by electrodepositing Cu₂O at pH 11 for 10 min (a) and (b), 20 min (c) and (d) and 30 min (e) and (f); (g) energy band diagrams for separated ZnO and Cu₂O electrodeposited at pH from 9–11, and the Cu₂O/ZnO heterostructure; (h) illustration of the PEC mechanism for biosensing GSH at the Cu₂O/ZnO photoanode; (i) photoresponse of pristine ZnO and Cu₂O/ZnO heterostructure in PBS or GSH.
Reproduced by permission from Macmillan Publishers Ltd: *Sci. Rep.* (ref. 31), copyright (2015).

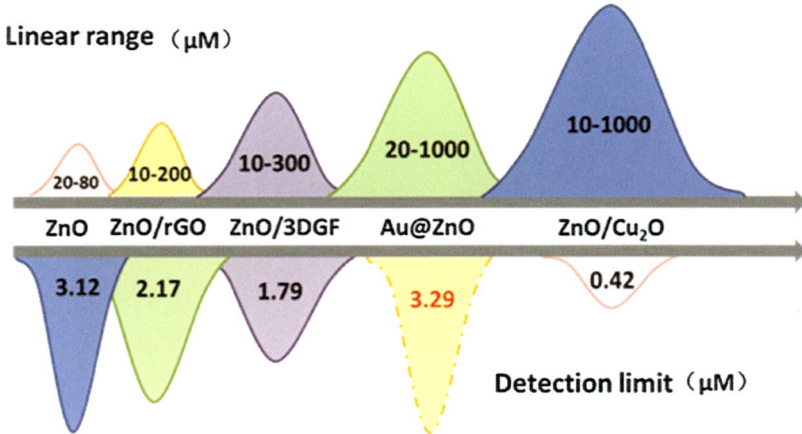

Figure 8.8   Illustration of the relationship between PEC biosensing performance and PEC photoanode properties.

References

1. Y. Yang, W. Guo, J. Qi and J. Zhao, Self-powered ultraviolet photo-detector based on a single sb-doped ZnO nanobelt, *Appl. Phys. Lett.*, 2010, **97**(22), 223113.
2. Z. Bai, X. Yan, X. Chen, K. Zhao, P. Lin and Y. Zhang, High sensitivity, fast speed and self-powered ultraviolet photodetectors based on ZnO micro/nanowire networks, *Prog. Nat. Sci.:Mater.*, 2014, **24**(1), 1–5.
3. Z. Bai, X. Yan, X. Chen, H. Liu, Y. Shen and Y. Zhang, ZnO nanowire array ultraviolet photodetectors with self-powered properties, *Curr. Appl. Phys.*, 2013, **13**(1), 165–169.
4. Z. Bai, X. Chen, X. Yan, X. Zheng, Z. Kang and Y. Zhang, Self-powered ultraviolet photodetectors based on selectively grown ZnO nanowire arrays with thermal tuning performance, *Phys. Chem. Chem. Phys.*, 2014, **16**(20), 9525–9529.
5. H. D. Cho, A. S. Zakirov, S. U. Yuldashev, C. W. Ahn, Y. K. Yeo and T. W. Kang, Photovoltaic device on a single ZnO nanowire p-n homo-junction, *Nanotechnology*, 2012, **23**(11), 115401.
6. Y. Bie, Z. Liao, H. Zhang, G. Li, Y. Ye, Y. Zhou, J. Xu, Z. Qin, L. Dai and D. Yu, Self-powered, ultrafast, visible-blind uv detection and optical lo-gical operation based on ZnO/gan nanoscale p-n junctions, *Adv. Mater.*, 2011, **23**(5), 649–653.
7. Z. Bai, X. Yan, X. Chen, Y. Cui, P. Lin, Y. Shen and Y. Zhang, Ultraviolet and visible photoresponse properties of a ZnO/Si heterojunction at zero bias, *RSC Adv.*, 2013, **3**(39), 17682–17688.
8. J. Qi, X. Hu, Z. Wang, X. Li, W. Liu and Y. Zhang, A self-powered ultra-violet detector based on a single ZnO microwire/p-Si film with double heterojunctions, *Nanoscale*, 2014, **6**(11), 6025–6029.

9. F. Yi, Q. Liao, Y. Huang, Y. Gu and Y. Zhang, Self-powered ultraviolet photodetector based on a single ZnO tetrapod/PEDOT: PSS hetero-structure, *Semicond. Sci. Technol.*, 2013, **28**(10), 2016–2018.

10. P. N. Ni, C. X. Shan, S. P. Wang, X. Y. Liu and D. Z. Shen, Self-powered spectrum-selective photodetectors fabricated from n-ZnO/p-NiO core–shell nanowire arrays, *J. Mater. Chem. C.*, 2013, **1**(29), 4445–4449.

11. S. Yang, J. Gong and Y. Deng, A sandwich-structured ultraviolet photo-detector driven only by opposite heterojunctions, *J. Mater. Chem.*, 2012, **22**(28), 13899–13902.

12. X. Li, C. Gao, H. Duan, B. Lu, Y. Wang, L. Chen, Z. Zhang, X. Pan and E. Xie, High-performance photoelectrochemical-type self-powered uv photodetector using epitaxial TiO_2/SnO_2 branched heterojunction nanostructure, *Small*, 2013, **9**(11), 2005–2011.

13. Y. Xie, L. Wei, Q. Li, Y. Chen, H. Liu, S. Yan, J. Jiao, G. Liu and L. Mei, A high performance quasi-solid-state self-powered uv photodetector based on TiO_2 nanorod arrays, *Nanoscale*, 2014, **6**(15), 9116–9121.

14. Y. Wang, W. Han, B. Zhao, L. Chen, F. Teng, X. Li, C. Gao, J. Zhou and E. Xie, Performance optimization of self-powered ultraviolet detectors based on photoelectrochemical reaction by utilizing dendriform tita-nium dioxide nanowires as photoanode, *Sol. Energy Mater. Sol. Cells*, 2015, **140**, 376–381.

15. W. J. Lee and M. H. Hon, An ultraviolet photo-detector based on TiO_2/water solid-liquid heterojunction, *Appl. Phys. Lett.*, 2011, **99**(25), 251102–251103.

16. X. Li, C. Gao, H. Duan, B. Lu, X. Pan and E. Xie, Nanocrystalline TiO_2 film based photoelectrochemical cell as self-powered uv-photodetector, *Nano Energy*, 2012, **1**(4), 640–645.

17. C. Gao, X. Li, Y. Wang, L. Chen, X. Pan, Z. Zhang and E. Xie, Titanium dioxide coated zinc oxide nanostrawberry aggregates for dye-sensitized solar cell and self-powered uv-photodetector, *J. Power Source*, 2013, **239**(10), 458–465.

18. Q. Li, L. Wei, Y. Xie, K. Zhang, L. Liu, D. Zhu, J. Jiao, Y. Chen, S. Yan and G. Liu, ZnO nanoneedle/H_2O solid-liquid heterojunction-based self-powered ultraviolet detector, *Nanoscale Res. Lett.*, 2013, **8**(1), 1–7.

19. P. Lin, X. Yan, Z. Zhang, Y. Shen, Y. Zhao, Z. Bai and Y. Zhang, Self-powered uv photosensor based on PEDOT: PSS/ZnO micro/nanowire with strain-modulated photoresponse, *ACS Appl. Mater. Interface*, 2013, **5**(9), 3671–3676.

20. P. Lin, X. Chen, X. Yan, Z. Zhang, H. Yuan, P. Li, Y. Zhao and Y. Zhang, Enhanced photoresponse of Cu_2O/ZnO heterojunction with piezo-modulated interface engineering, *Nano Res.*, 2014, **6**(6), 860–868.

21. S. Lu, J. Qi, S. Liu, Z. Zhang, Z. Wang, P. Lin, Q. Liao, Q. Liang and Y. Zhang, Piezotronic interface engineering on ZnO/Au-based schottky junction for enhanced photoresponse of a flexible self-powered uv detector, *ACS Appl. Mater. Interfaces*, 2014, **6**(16), 14116–14122.

22. Z. Zhang, Q. L. Liao, Y. H. Yu, X. D. Wang and Y. Zhang, Enhanced photoresponse of ZnO nanorods-based self-powered photodetector by piezotronic interface engineering, *Nano Energy*, 2014, **9**, 237–244.
23. Y. Shen, X. Yan, H. Si, P. Lin, Y. Liu, Y. Sun and Y. Zhang, Improved photoresponse performance of self-powered ZnO/spiro-MeOTAD heterojunction ultraviolet photodetector by piezo-phototronic effect, *ACS Appl. Mater. Interfaces*, 2016, **8**(9), 6137–6143.
24. P. Lin, X. Yan, Y. Liu, P. Li, S. Lu and Y. Zhang, A tunable ZnO/electrolyte heterojunction for a self-powered photodetector, *Phys. Chem. Chem. Phys.*, 2014, **16**(48), 26697–26700.
25. W.-W. Zhao, J.-J. Xu and H.-Y. Chen, Photoelectrochemical DNA Biosensors, *Chem. Rev.*, 2014, **114**(15), 7421–7441.
26. M. Liang, S. Jia, S. Zhu and L.-H. Guo, Photoelectrochemical sensor for the rapid detection of in situ DNA damage induced by enzyme-catalyzed Fenton reaction, *Environ. Sci. Technol.*, 2008, **42**(2), 635–639.
27. A. Devadoss, P. Sudhagar, C. Terashima, K. Nakata and A. Fujishima, Photoelectrochemical biosensors: New insights into promising photoelectrodes and signal amplification strategies, *J. Photochem. Photobiol., C*, 2015, **24**, 43–63.
28. K. Yan, R. Wang and J. Zhang, A photoelectrochemical biosensor for o-aminophenol based on assembling of CdSe and DNA on TiO$_2$ film electrode, *Biosens. Bioelectron.*, 2014, **53**, 301–304.
29. Z. Kang, Y. Gu, X. Yan, Z. Bai, Y. Liu, S. Liu, X. Zhang, Z. Zhang, X. Zhang and Y. Zhang, Enhanced photoelectrochemical property of ZnO nanorods array synthesized on reduced graphene oxide for self-powered biosensing application, *Biosens. Bioelectron.*, 2015, **64**, 499–504.
30. Z. Kang, X. Yan, Y. Wang, Y. Zhao, Z. Bai, Y. Liu, K. Zhao, S. Cao and Y. Zhang, Self-powered photoelectrochemical biosensing platform based on Au NPs@ZnO nanorods array, *Nano Res.*, 2015, **9**(2), 344–352.
31. Z. Kang, X. Yan, Y. Wang, Z. Bai, Y. Liu, Z. Zhang, P. Lin, X. Zhang, H. Yuan, X. Zhang and Y. Zhang, Electronic structure engineering of Cu$_2$O film/ZnO nanorods array all-oxide p-n heterostructure for enhanced photoelectrochemical property and self-powered biosensing application, *Sci. Rep.*, 2015, **5**, 7882.

CHAPTER 9

Service Behaviours

PEIFENG LI, QI ZHANG AND YUE ZHANG*

University of Science and Technology Beijing, Beijing, China
*Email: yuezhang@ustb.edu.cn

9.1 Introduction

Relative to traditional bulk materials, nanomaterials are used not only as structural units but also as functional units. Like the traditional bulk materials, damage and failure also occur in nanomaterials and nanodevices under the influence of service or external conditions. Similar phenomena have also been observed in the investigation of one-dimensional (1D) ZnO nanomaterials and nanodevices. Typical characterisation includes structural damage, performance degradation, stability degradation, functional failure, short life *etc.* Problems are exposed under some experimental or service conditions such as illumination, currents or electric fields, external pressure or forces, some chemical gases or solvents,[1-8] and so on. These problems seriously obstruct the process of the investigation of 1D ZnO nanomaterials and nanodevices. For the above phenomena, the concepts of nanodamage and nanofailure in nanomaterials and nanodevices have been proposed by our group, and have attracted more and more attention. To guarantee the stability, reliability, safety, and long life of the nanodevices when applied in daily life and other areas, it is necessary to investigate the nanodamage and nanofailure security service parameters and mechanisms of 1D ZnO nano-materials and nanodevices thoroughly under simulated and actual applied conditions before the large-scale industrialization of 1D ZnO nanodevices. In addition, the investigation into nanodamage and nanofailure would not only guide us in improving and promoting the properties of existing

Nanoscience & Nanotechnology Series No. 43
ZnO Nanostructures: Fabrication and Applications
By Yue Zhang
© Yue Zhang 2017
Published by the Royal Society of Chemistry, www.rsc.org

applied nanodevices, but also guide us in developing and designing some new potential nanodevices to satisfy the needs of daily life or industrialisation.

To facilitate investigating the nanodamage and nanofailure phenomena, damage and failure mechanisms, and security service parameters of 1D ZnO nanomaterials and nanodevices, typical nanodamage and nanofailure were classified per their service conditions. The current classifications mainly include electrical, mechanical, electromechanical, and chemical nano-damage and nanofailure.

9.2   Electrical Nanodamage and Nanofailure

The shrinking of nanostructures would generally increase the current density and thus lead to escalated resistive (Joule) heating.[9] Hence, the robustness of nanostructures under electrical fields has become one of the main concerns during device design, for which safety, stability, reliability and durability issues should be seriously considered.[10,11]

Electrical nanodamage and nanofailure is the structural damage or functional failure of nanomaterials and nanodevices under an electron beam or electric field. As a typical semiconductor, the electrical properties of 1D ZnO nanomaterials and nanodevices are important points that cannot be ignored. Until now, electrical nanodamage and nanofailure in 1D ZnO nanomaterials have been reported when investigating their electrical properties.

The obvious fracture phenomenon in ZnO : Sb nanobelts (NBs) had been observed under the irradiation of a 200 kV convergent electron beam in a transmission electron microscope (TEM). Figure 9.1 shows TEM images of the ZnO : Sb NBs under electron beam irradiation at different times. It was found that the damage extent becomes more obvious with increasing time. The investigation indicated that the Sb element led to the instability of the ZnO structure.

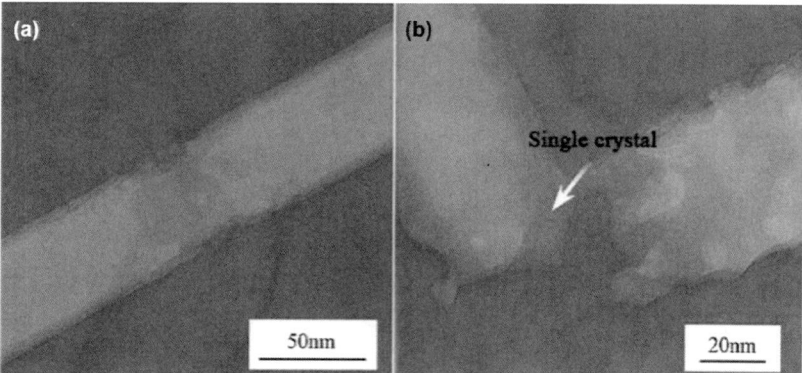

Figure 9.1   Nanodamage of Sb-doped ZnO NBs under irradiation of the TEM convergent electron beam for (a) 15 s; and (b) 30 s.

The pore was also observed in the ZnO nanowires (NWs) and was subjected to a convergent electron beam irradiation in a 300 kV TEM, which is shown in Figure 9.2.[1] The size of the perforated hexagonal pores generated by irradiation can vary with the beam size and the polar surfaces were resistant to the electron beam irradiation damage in contrast with the non-polar ones.

Electrical failure was also observed in the single-crystalline ZnO:In NBs in the current loop when the current density increased to 7.4×10^6 A cm^{-2} using an *in situ* approach.[3] Figure 9.3 shows the transfer curve of the ZnO:In NB device, the *I–V* curve in the breakdown process and the SEM images before and after fracture. The researchers proposed that the electrical failure of ZnO:In NB was due to Joule heating.

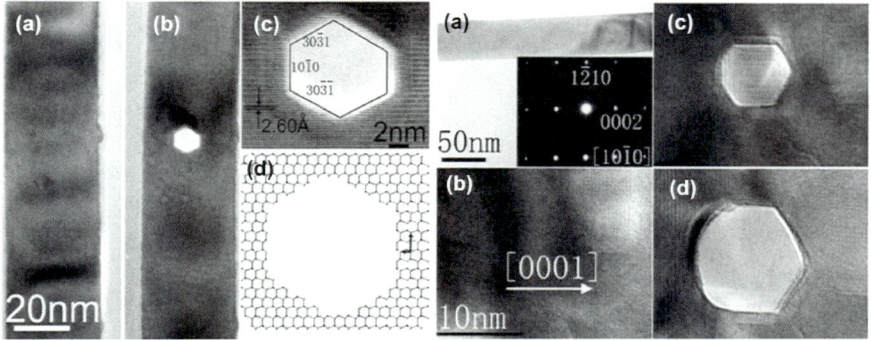

Figure 9.2 Left: TEM image of a ZnO NW (a) before and (b) after convergent electron beam irradiation (30 s); (c) HRTEM image taken near a hexagonal pore; (d) structural model displaying the six edges of a hexagonal pore. Right: (a) TEM image of a typical ZnO NW with (10$\bar{1}$0) top face and SAED pattern; TEM image of a ZnO NW before and after focused beam irradiation for (b) 0 s; (c) 10 s; and (d) 40 s.
Reproduced from ref. 1 with permission of AIP Publishing.

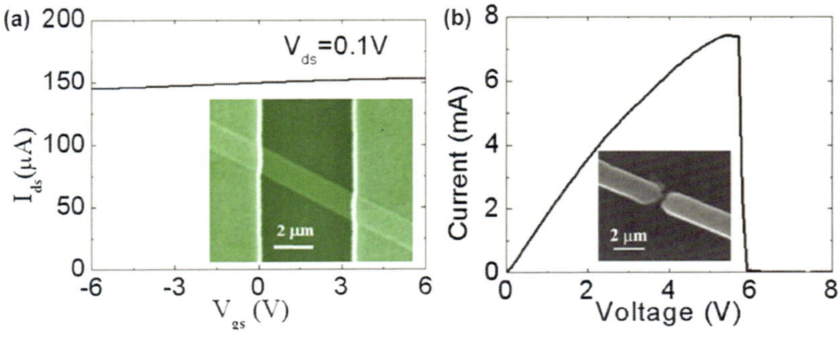

Figure 9.3 (a) Transfer curve of the ZnO:In NB device at $V_{ds} = 0.1$ V. Inset: SEM image of the individual ZnO:In NB bridge between two Ti/Au electrodes on SiO$_2$/Si substrates; (b) *I–V* curve recorded for an individual ZnO:In NB that breaks down at high current. Inset: SEM picture of the ZnO:In NB after current breakdown.
Reproduced from ref. 3 with permission of AIP Publishing.

We have also thoroughly researched electrical nanodamage and nano-failure of 1D ZnO nanomaterials by AFM and *in situ* SEM manipulators in recent years. The electrical nanodamage of single ZnO NWs was measured with a conductive atomic force microscope (C-AFM) tip.[5] Figure 9.4 shows AFM images of the ZnO NW before and after nanodamage and the morphology under different applied biases. The nanodamage threshold voltage was estimated to be 7.0 ± 0.5 V for the 5 nm NWs and the nano-damage became more obvious with increasing time. The thickness along a single NW during the damaging process can be accurately modulated by controlling the evaporation time and applied voltage in selected areas.

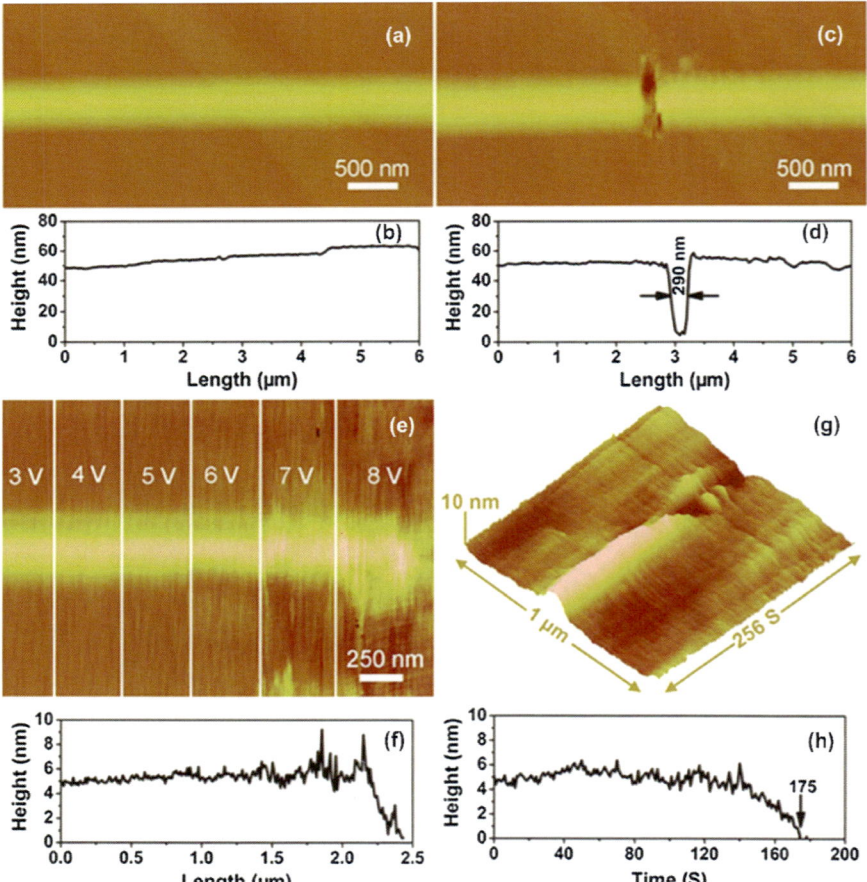

Figure 9.4   Height images and line profiles along the longitudinal direction of a ZnO NW: before (a) and (b) and after (c) and (d) electrical nanodamage; 2D height image (e) and line profile (f) of thinning of ZnO NWs under different applied bias; 3D height image (g) and line profile (h) of thinning of ZnO NWs in the different evaporation time under the controlled bias voltage of 7 V.
Reproduced from ref. 5 with permission of AIP Publishing.

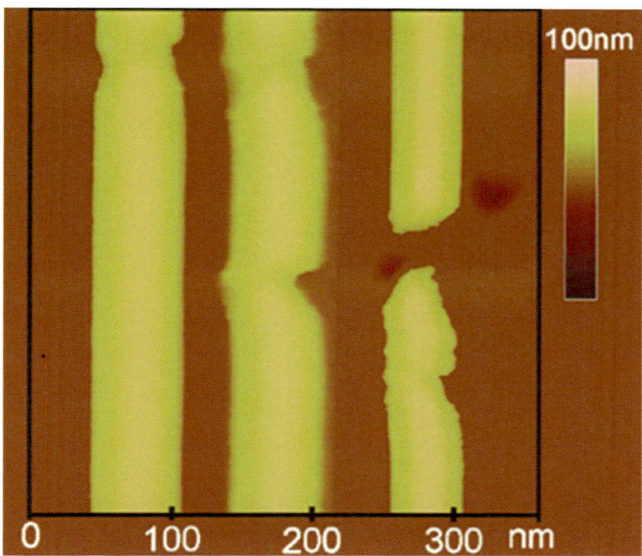

Figure 9.5 AFM images of ZnO : Sb NBs under 5 V by C-AFM tip after 5 s, 15 s and 30 s (from left to right).

The nanodamage mechanism may be due to the Joule heating generated under the transverse electric field which leads to the melting of ZnO nanomaterials.

The behaviour of the ZnO : Sb nanobelts was measured under 5 V by a C-AFM tip after different times and nanodamage and fracture phenomena were also found. The morphology changes of the doped ZnO NB under 5 V as time elapses is shown in Figure 9.5. From the images, we can see that the damage extent increased with time until fracture, which also indicated that the instability of the ZnO structure was affected by Sb.

The stability of the ZnO NWs in a metal-semiconductor-metal structure was also investigated by applying a longitudinal electric field inside a scanning electron microscope (SEM) equipped with manipulators.[8] Figure 9.6 shows SEM images and *I–V* curves of the ZnO NW before and after fracture. The failure of single crystalline ZnO NWs was directly observed when the applied electric field reached the break point, an intensity of about 10^6 V m^{-1}. In the failure process, recrystallisation would occur from single crystal to a polycrystalline pearl-like structure that can be seen in the high-resolution TEM (HRTEM) images in the inset of Figure 9.6(a) and (c). The experimental results indicated that the failure was attributed to a joint effect of high electric field and Joule heating.

In addition, the electrical service behaviour of ZnO NWs with various diameters was investigated using a nano-manipulation technique.[12] Figure 9.7 shows the *I–V* curves, SEM images, threshold voltages, critical current densities and the proposed core–shell model. The threshold voltages of the ZnO NWs increased linearly from 15–60 V with the diameter

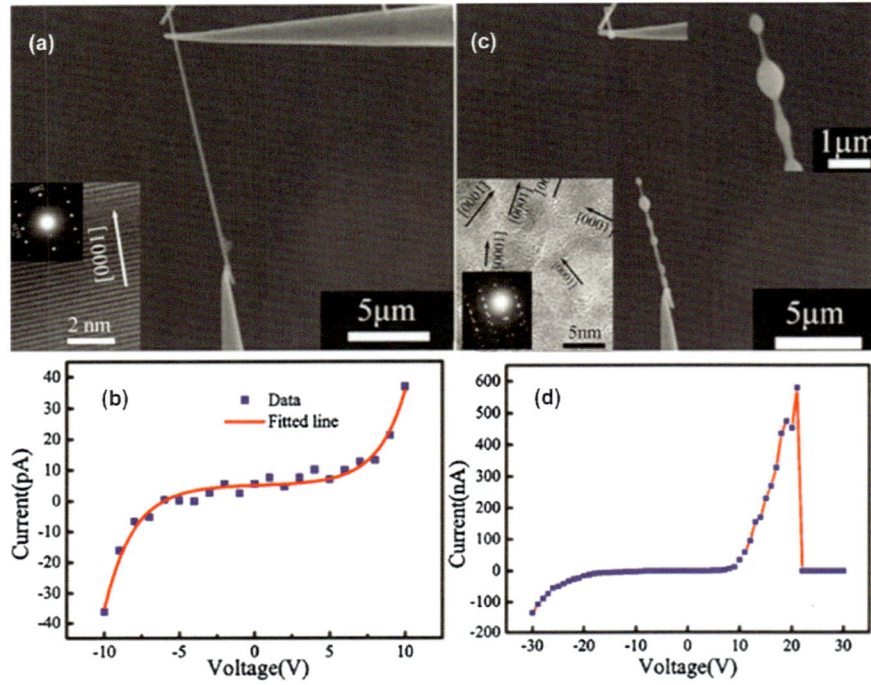

Figure 9.6   SEM images of the single ZnO NW (a) before and (c) after failure; *I–V* curves of the single ZnO NW (b) before and (d) after failure corresponding to the SEM images.
Reproduced from ref. 8 with permission of AIP Publishing.

increasing from 103–807 nm. The critical current densities were distributed from 19.50×10^6 to 56.90×10^6 A m^{-2} and the reciprocal of the critical current density increased linearly with increasing diameter as well. The threshold voltage and critical current density can be used as the evaluation and criterion to judge the work state of the ZnO NWs for building nanodevices. The thermal core–shell model was proposed to explain the electrical nanodamage and nanofailure mechanism of ZnO NWs. The main reason for the nanodamage or nanofailure of the ZnO NWs is attributed to the Joule heat generated on the surface of the MS junction, which melts the ZnO NW once the temperature exceeds the melting point.

Moreover, a redshift of the band edge emission was found in the molten pearl part after the electric-induced nanowire fracture, as illustrated in Figure 9.8(a); currently, strong light scattering from the polycrystalline parts can be observed when compared to the nanowire before the failure process happens (Figure 9.8(b)).[13] In the single crystal, the different dielectric constants along the *c* and *a* axes for the hexagonal ZnO single crystal lead to the confinement of light along the *c* axis (Figure 9.8(c)). Therefore, these achievements on the electrical failure of nanoscale elements are quite necessary for their practical applications.

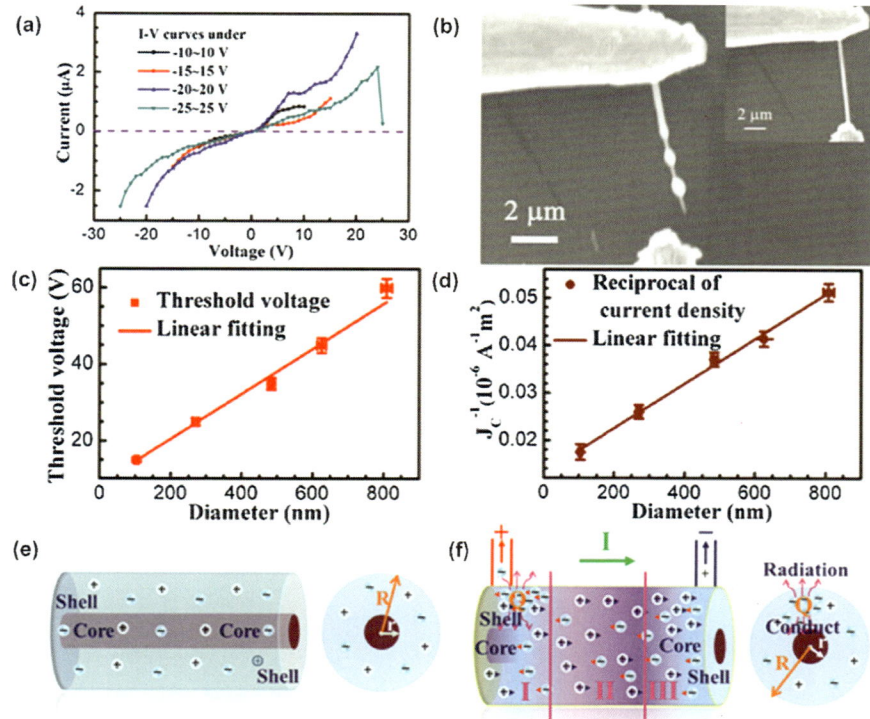

Figure 9.7 (a) *I–V* curves of a single ZnO NW; (b) SEM images of the ZnO NW before (inset) and after failure. Relationship between the (c) threshold voltages, (d) critical current densities of the ZnO NWs and the diameters; the running states of electrons and holes in a ZnO NW with a thermal core–shell model: (e) without and (f) under longitudinal voltage.
Reproduced with permission from ref. 12. Copyright (2014) American Chemical Society.

9.3 Mechanical Nanodamage and Nanofailure

Low-dimensional nanostructures, especially single crystal nanowires, might be the most important components in the miniaturisation of next-generation electromechanical devices. From a practical viewpoint, the robustness of these building blocks which can withstand sufficient external loads is of great significance.[14,15] It is particularly important to conduct quantitative experiments while keeping the experimental conditions constant, because the extremely small dimensions impose a tremendous burden on an experimentalist.[16]

Mechanical nanodamage or nanofailure is the structural damage or functional failure of nanomaterials and nanodevices under external forces, pressures or vibrations. The realisation of 1D ZnO nanodevices would be affected by the external environments which makes the mechanical nanodamage or nanofailure of 1D ZnO inevitable. Excellent mechanical

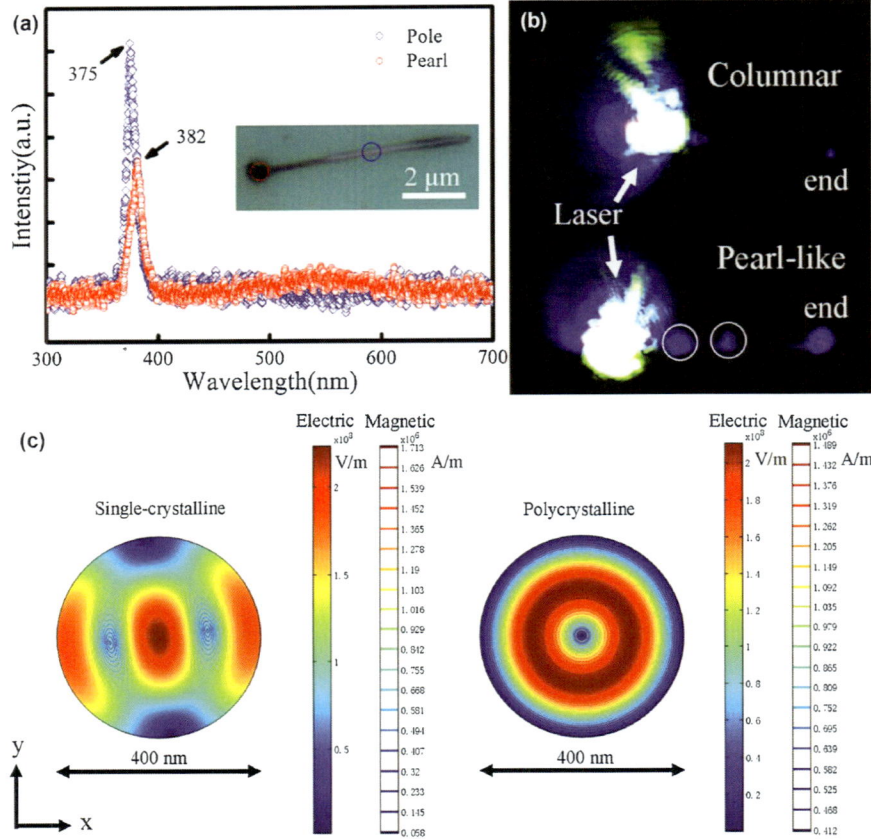

Figure 9.8 (a) Photoluminescence study on the different positions of a pearl-like ZnO NW, showing obvious redshift in the pearl part; (b) wave guiding performance of the ZnO NW before and after fracture; (c) FEM calculation of the crystal structure dependent light confinement abilities. Reproduced with permission from Optical Society of America.[13]

properties are indispensable conditions for the nanodevices either as structural units or functional units, and this attracts lots of attention. Some mechanical nanodamage or nanofailure phenomena of 1D ZnO nanomaterials have been observed by researchers in recent years.

In early research, the nanofracture phenomena of ZnO NBs and nanohelices were observed in the manipulation processes;[15,17] however, the mechanical properties were not measured further. Figure 9.9 shows the fracture morphologies of the ZnO NB and nanohelix.

The mechanical properties of ZnO NWs with diameters from 60–310 nm and a typical length of 2 μm were studied through tensile and bending experiments with a nanomanipulator in SEM.[18] The tensile fracture stresses and strains were about 3.7–5.5 GPa and 5%, respectively; while the bending fracture stresses and strains were about 7.5 GPa and 7.7%, respectively.

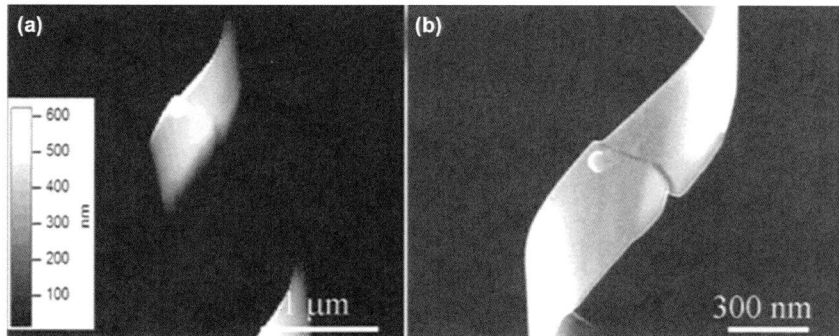

Figure 9.9   (a) and (b) AFM topography image after tip-induced fracture over a nanohelix and the corresponding SEM image of the fractured part. Reproduced with permission from ref. 17. Copyright (2006) American Chemical Society.

The quasi-static uniaxial tensile testing of ZnO NWs was done using a designed MEMS test-bed.[19] The size effect on the fracture strains for ZnO NWs varied from 5–15% by tensile test, which was very high compared to bulk ZnO. It was proposed that the high fracture strain properties of ZnO NWs might make them a very viable and potential material for nanoscale sensors and actuators.

The mechanical properties of ZnO NWs with diameters from 85–542 nm were investigated by bending experiments in SEM.[20] It was found that the fracture strengths were close to the theoretical strength, large strains of up to 4–7% were obtained before the final elastic fracture and no detectable plastic deformation was observed. The large fracture strains and strengths may be attributed to a lack of defects and flaws.

The mechanical properties of ZnO NWs with diameters from 18–304 nm were measured by lateral force AFM and a brittle fracture phenomenon was found.[4] Figure 9.10 shows the AFM images and force–displacement (*F–d*) curve. The ultimate strength of ZnO NWs increased for small diameter wires. The researchers proposed that defect density would decrease with the decreasing of the diameter, which enhanced the strength of the 1D ZnO nanomaterials.

The effect of the surface morphology of the ZnO NWs on their mechanical properties was studied by AFM.[21] An average Young's modulus of 148 GPa was obtained and the size dependence was found to be unaffected by the detailed micro and macro surface morphology. The bending strain of 0.2–0.7% is one order of magnitude lower than that reported in the literature. It indicates that an irregular surface, for example cracks, flaws, a curved and neck-like surface and body defects dominates the fracture properties of ZnO NWs, rather than the elastic behaviour.

The fracture properties of [0001] ZnO NWs with diameters from 20–512 nm were investigated by *in situ* TEM uniaxial tensile experiments.[6,22] Figure 9.11 shows the TEM images and *F–d* curve. They found that the brittle fracture

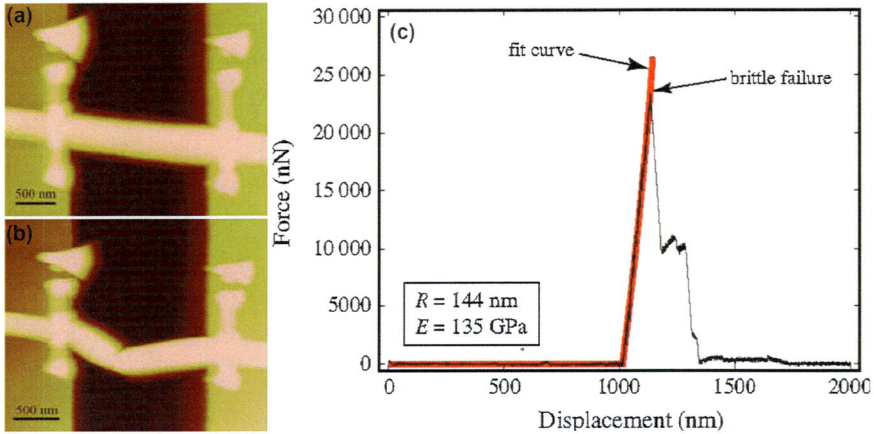

Figure 9.10 (a) and (b) Tapping-mode AFM images of a 144 nm radius ZnO NW before bending and after brittle failure; (c) *F–d* curve recorded during manipulation by AFM tip-induced lateral bending.
Reprinted (figure) with permission from B. Wen, J. E. Sader and J. J. Boland, Phys. Rev. Lett., 101, 175502, 2008. Copyright (2008) by the American Physical Society.[4]

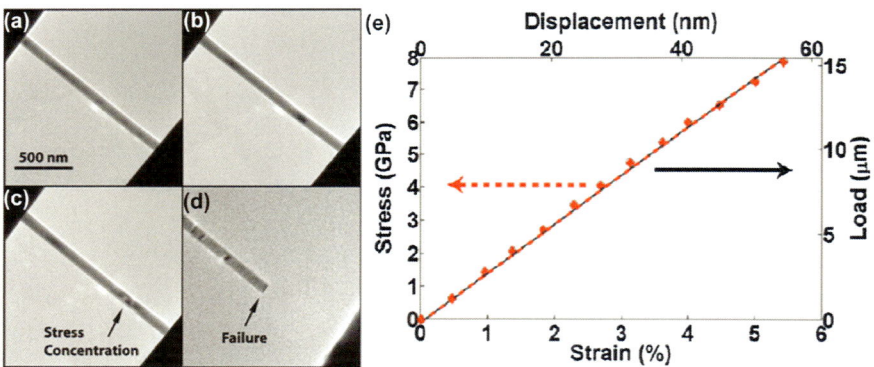

Figure 9.11 (a)–(d) A sequence of TEM images taken during the tensile testing of ZnO NW until fracture; (e) stress–strain (red dotted line) and *F–d* (black solid line) plot obtained experimentally for a 55 nm NW.
Reproduced with permission from ref. 6. Copyright (2009) American Chemical Society.

along the (0001) cleavage plane and the fracture stresses were as high as 9.53 GPa and failure strains were of about 6.2%.

The elastic and failure properties of ZnO NWs under different loading modes were studied with *in situ* SEM tension and buckling tests on single ZnO NWs along the [0001] polar direction.[16] Figure 9.12 shows the tensile SEM images and stress–strain curve. Both the tensile and the bending modulus was found to increase as the diameter decreased from 80 to 20 nm and the bending modulus increased more rapidly than the tensile modulus, which demonstrated that the elasticity size effects in ZnO NWs are

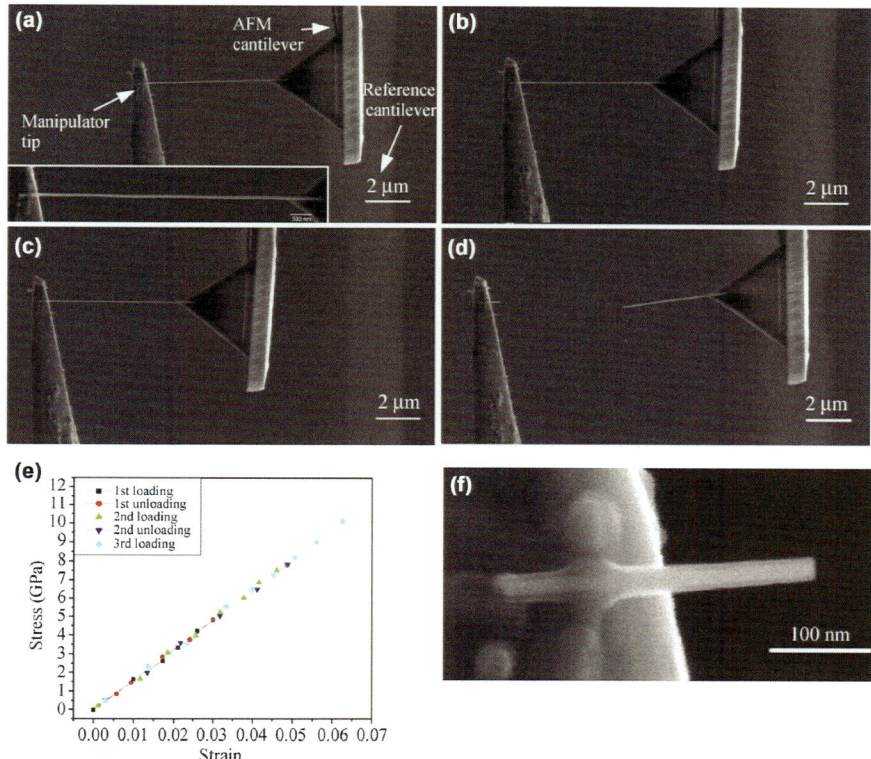

Figure 9.12 (a)–(c) A series of SEM images taken during the tensile test for an NW with a diameter of 20 nm. Inset of (a): high-resolution SEM image of the NW used for strain measurements; (d) SEM image showing that fracture occured on the NW when the load was applied to a certain value; (e) a typical stress–strain response of the specimens with a diameter of 20 nm under repeated loading and unloading; (f) enlarged SEM image of the broken end of the NW on the probe tip.

Reproduced from Nano Res., Mechanical properties of ZnO nanowires under different loading modes, 3(4), 2010, 271–280, F. Xu, Q. Q. Qin, A. Mishra, Y. Gu and Y. Zhu, (© Tsinghua University Press and Springer-Verlag Berlin Heidelberg 2010) With permission of Springer.[16]

mainly due to surface stiffening. The tension experiments also showed that fracture strains and strength of ZnO NWs increased as the diameter decreased.

We also have done lots of research on the mechanical nanodamage and nanofailure of 1D ZnO nanomaterials using AFM and *in situ* TEM electromechanical resonance in recent years. The mechanical behaviour of ZnO NBs with different cross-sections were investigated. Figure 9.13 shows the morphology changes of ZnO NBs with different cross-sections under external forces. It was found that the NBs with a triangular cross-section would produce mechanical nanodamage more easily under the same external force conditions than the NBs with a rectangular cross-section. The mechanical nanodamage mechanism may be due to the surface energy of the triangular NB being clearly lower than that of the rectangular NB.

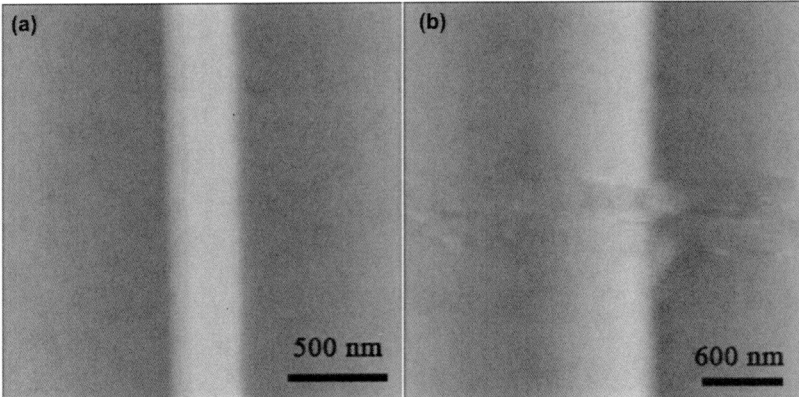

Figure 9.13 AFM images of ZnO NBs under external forces: the ZnO NB with (a) rectangular cross section and (b) triangular cross section.

We studied the mechanical service behaviour of ZnO NWs with diameters ranging from 67–201 nm and different scanning angles at a scanning rate of 14.8 $\mu m \, s^{-1}$ by AFM.[23] Figure 9.14 shows some testing results and relationship curves. The force calibration equation was established between the actual forces applied on the surface of ZnO NWs, applied forces and scanning angles. It was proved that the actual forces strengthened when scanning angles existed. The actual fracture threshold forces increase linearly with the increase of diameters of the ZnO NWs, but do not depend on the scanning angles. The investigation provides a method for studying mechanical service behaviour of 1D nanomaterial under a non-normal stress state and calibrating the actual force applied by AFM under different scanning angles.

During the investigation of mechanical properties of nanomaterials using AFM, a non-normal stress state with large scanning angles is a universal phenomenon. We studied the mechanical service behaviour of ZnO NWs with diameters ranging from 177–386 nm under a non-normal stress state using AFM at a constant scanning rate.[24] Figure 9.15 shows some AFM testing results and relationship curves. The application scope of the threshold force equation determining the threshold forces of ZnO NWs was expanded to a larger diameter of ~400 nm. The criterion for security service of the ZnO NWs was established by the combination of the threshold force equation and the force calibration equation. The criterion was used to predict the security service range of ZnO NWs successfully. The Young's modulus of the ZnO NWs obtained in the scanning process is far below the values measured under a static state in a previous report, but the fracture strength has good consistency. The success in predicting the security service range of ZnO NWs is very important and has meaningful consequences for practical applications of ZnO NWs under a non-normal stress state.

In addition, the investigation of fatigue behaviour of ZnO wires under high-cycle strain was systematically performed with an *in situ* TEM

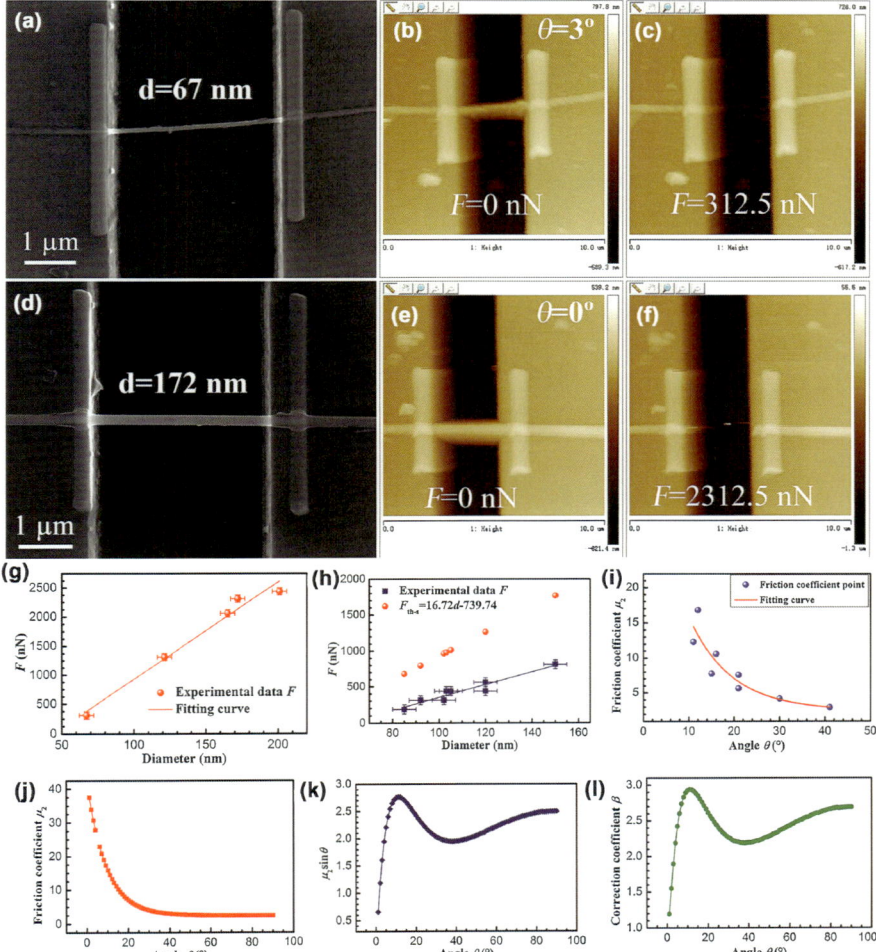

Figure 9.14 (a) and (d) SEM images of two ZnO NWs with diameters of about 67 and 103 nm. AFM images of the ZnO NWs with scanning angles of (b) and (c) 3° and (e) and (f) 15° before and after fracture; (g) relationship between the actual fracture threshold forces and diameters of ZnO NWs at small scanning angles; (h) relationship between the applied fractured forces F, actual fracture threshold forces F_{th} and diameters of ZnO NWs; (i) relationship between the lateral resistance friction coefficient μ_2 and the scanning angle θ; relationship between (j) lateral resistance friction coefficient μ_2, (k) $\mu_2\sin\theta$, (l) calibration coefficient β and the scanning angles.

Reproduced from ref. 23 with permission from The Royal Society of Chemistry.

electromechanical resonance method.[25] Figure 9.16 shows some TEM resonance testing results and relationship curves. The modulus of the ZnO NWs with diameters less than 100 nm approaches or exceeds the bulk modulus 140 GPa, while the modulus of the ZnO wires with diameters more

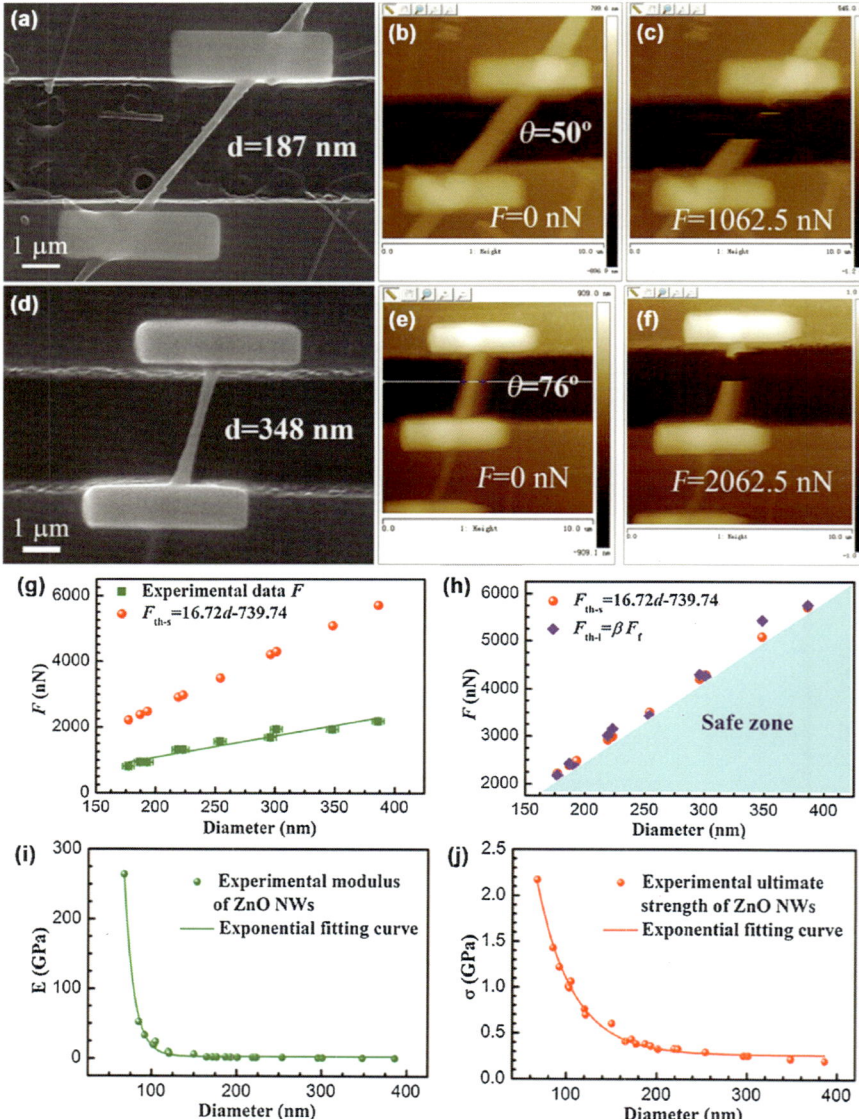

Figure 9.15 (a) and (d) SEM images of two ZnO NWs with diameters of about 187 and 348 nm. AFM images of the ZnO NWs with scanning angles of (b) and (c) 50° and (e) and (f) 76° before the test and after fracture under the applied forces; (g) relationship between the fracture force F, threshold forces F_{th} and diameters of ZnO NWs; (h) coincidence of the threshold forces of the ZnO NWs calculated by threshold force equation and force calibration equation, and the light blue triangle is the security zone; relationship between (i) Young's modulus, (j) fracture strength of ZnO NWs and diameters obtained in the experiments.

Reproduced from ref. 24 with permission from The Royal Society of Chemistry.

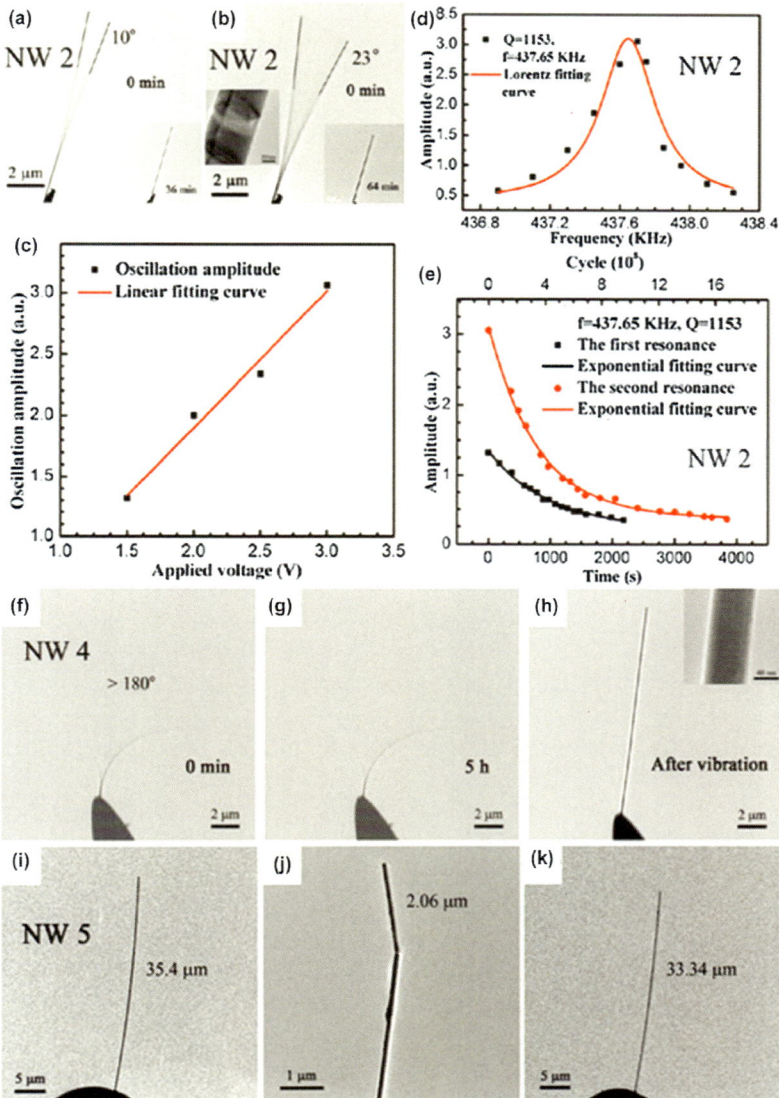

Figure 9.16 *In situ* TEM resonance images of the ZnO NW with $d_2 = 177$ nm: (a) $A_{21} = 1.319$ μm and (b) $A_{22} = 3.064$ μm; (c) oscillation amplitudes of the two ZnO NWs with approximate lengths (9735 and 9912 nm, respectively) as a function of the applied voltage of alternating current V_{ac} that shows a linear dependence; (d) the measured fwhm of the resonance peaks from the ZnO NW; (e) the amplitude changes of the ZnO NW with time and cycles increasing; (f)–(h) *in situ* TEM resonance images of the 70 nm ZnO NW with larger amplitude in various states; TEM images of a ZnO NW with $d_5 = 95$ nm: (i) original, (j) damaged after e-beam irradiation, (k) fractured after resonance for seconds.

than 100 nm is far below that of the bulk ZnO. No damage or failure was observed in the intact ZnO wires after resonance for about 10^8–10^9 cycles, while the damaged ZnO NW subjected to electron beam irradiation fractured after resonance of a few seconds. The research results will provide a useful guide for designing, fabricating, and optimising electromechanical nano-devices based on ZnO nanomaterials, as well as for future applications.

9.4 Electromechanical Nanodamage and Nanofailure

Some nanodevices based on ZnO nanomaterials were designed for certain functions, such as strain sensors and nanogenerators.[26,27] These nano-devices work under electrical and mechanical conditions at the same time. So, the electromechanical coupling effect should be considered in these electromechanical nanodevices. The investigation of electromechanical properties was important and necessary to provide us more theoretical support in practical applications.

Electromechanical nanodamage or nanofailure is the structural damage or functional failure of nanomaterials and nanodevices under the effect of coupling electricity and external forces/pressures. As typical semiconductor materials, ZnO nanomaterials and nanodevices cannot avoid the collective effect of electricity and external forces. So, it is necessary and important to study the service behaviour of ZnO nanomaterials and nanodevices under the electromechanical coupling conditions.

Studies of the electromechanical properties of 1D ZnO nanomaterials are few. We investigated the electromechanical nanodamage in ZnO NBs by using C-AFM.[7] Figure 9.17 show the measured schematic, testing results, and relationship curves. The measured damage threshold voltage was found to decrease from 12 to 6 V as the loading forces changed from 20 to 180 nN. The threshold of the voltage of nanodamage changes approximately linearly as the loading forces increase. The mechanism of the decrease in the damage threshold voltages is suggested to be attributed to the strain in-duced change in electric structures in ZnO.

We also found electromechanical nanodamage and nanofailure of ZnO nanodevices using AFM.[28] Figure 9.18 shows the measured schematic, testing *I–V* curves and energy band diagrams of the measured structure. A single ZnO NB with a thickness of about 500 nm was fixed on the 6 T film and the double diode was fabricated with the structure of PtIr tip/ZnO NB/6T film/ITO film, and the *I–V* characteristics were studied using a standard C-AFM with PtIr-coated tips. From the plots, the negative differential resistance (NDR) phenomenon was clearly observed in the *I–V* characteristics during the first measurement but disappeared during the second measurement. The NDR disappeared in the second measurement, which is also not consistent with the electron resonant tunnelling, but indicates that the 6 T film has broken down after the first measurement.

We studied the electromechanical coupling effect on service behaviour of ZnO NWs with different diameters by C-AFM. Figure 9.19 show some

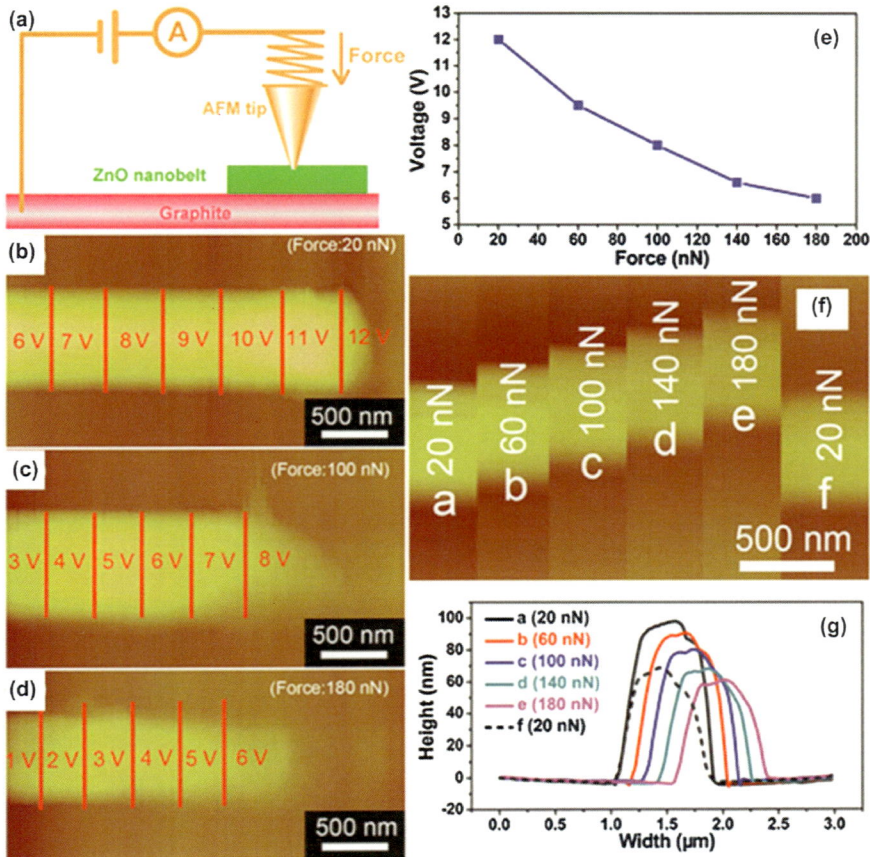

Figure 9.17 (a) A schematic diagram of nanodamage measurements. AFM images of single ZnO NBs under the different applied bias and the different loading forces of (b) 20 nN; (c) 100 nN; and (d) 180 nN; (e) relationship between the loading forces and nanodamage threshold voltages; (f) AFM images of single ZnO NBs under the different loading forces; (g) the corresponding line profiles of the ZnO NB.
Reproduced from ref. 7 with permission of AIP Publishing.

measured results, relationship V_{th}–F_{th} curves and mechanism schematics. The fracture threshold voltages of the ZnO NWs with equal diameters decreased linearly with the increase of external forces. Meanwhile the fracture threshold voltages increased linearly with the increase of diameters under the constant external force. Stress concentration effects strengthened the accumulation of electrons at the interface of the ZnO NW and AFM tip, which led to more Joule heat being generated and melted the ZnO NW at a lower threshold voltage. The investigation results provide the researchers and users with a helpful guide to designing, fabricating, and operating corresponding electromechanical and piezoelectric nanodevices based on ZnO NWs in practical applications.

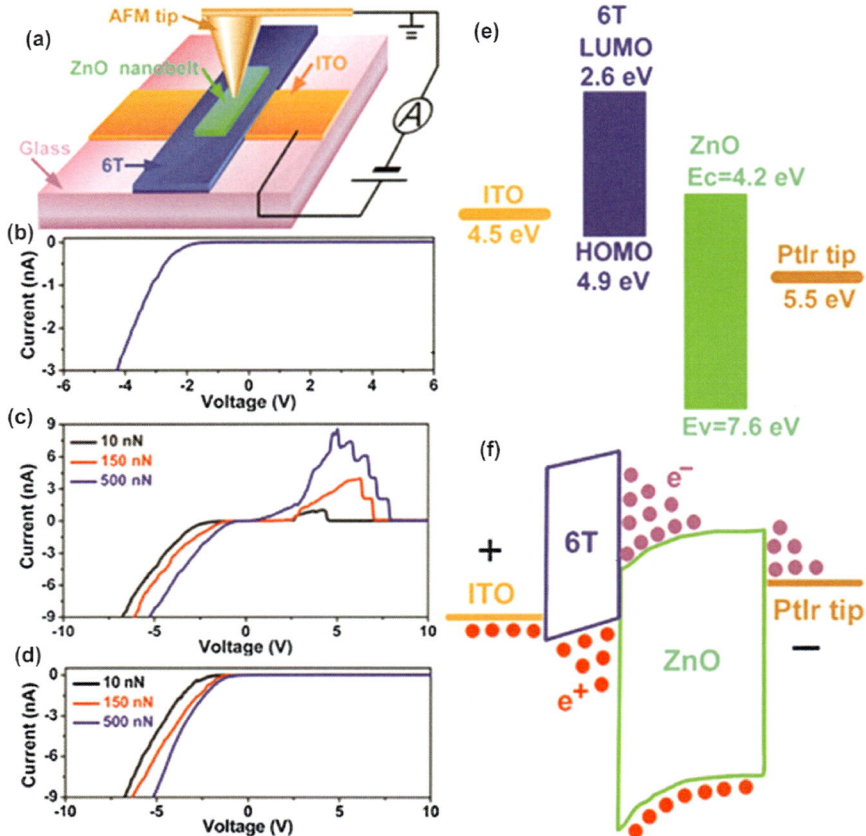

Figure 9.18 (a) A schematic diagram and (b) *I–V* characteristic of the structure of PtIr tip/ZnO NB/6T film/ITO double diodes; (c) *I–V* characteristics of the fabricated double diodes in the first measurement and (d) the second measurement under the different loading forces, which were measured on the different spots of the ribbon with the same height; (e) energy band diagrams of ZnO, 6T, PtIr, and ITO. E_c and E_v represent conduction band and valence band of ZnO, respectively; (f) energy band diagram of the fabricated double diodes under a positive bias. Reproduced from ref. 28 with permission of AIP Publishing.

9.5 Chemical–Mechanical Hybrid Nanodamage and Nanofailure

In view of practical applications, nanoscale devices, *i.e.* biosensors, must be exposed to different service conditions, such as an acidic or alkaline environment. Therefore, the chemical robustness of these low-dimensional working elements has drawn great attention because of their large aspect ratio.[2,6]

Chemical nanodamage or nanofailure is the structure damage or functional failure of nanomaterials and nanodevices under external chemical

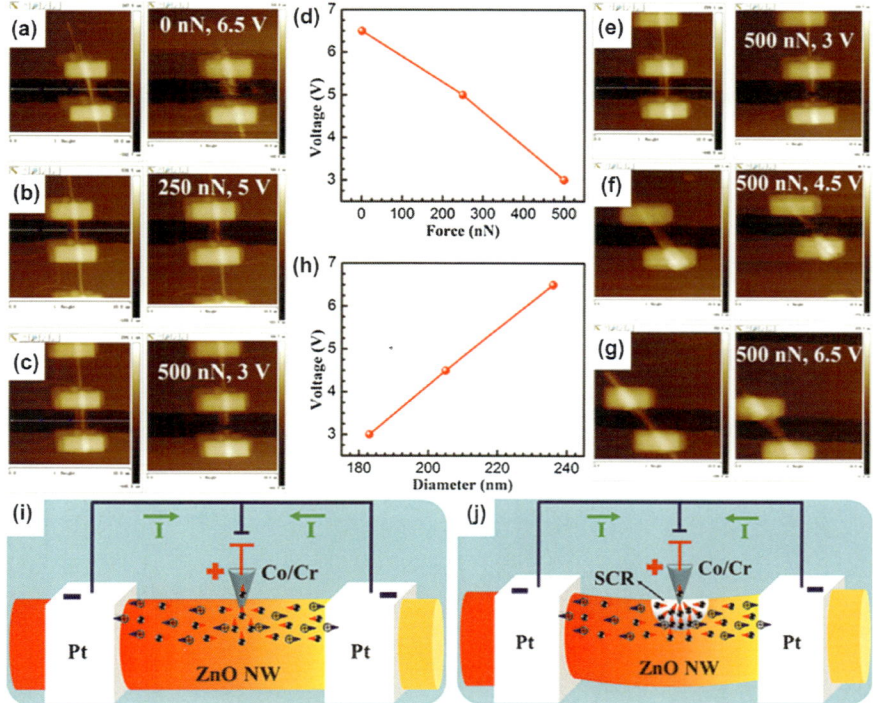

Figure 9.19 (a)–(c) AFM images revealing the electromechanical service behaviours of the three section ZnO NWs (left: initial; right: final); (d) relationship between fracture threshold voltages of the ZnO NWs and external forces; (e)–(g) AFM images revealing the electromechanical service behaviours of the three ZnO NWs (left: initial; right: final); (h) relationship between fracture threshold voltages of the ZnO NWs and diameters under constant external forces; (i) carriers transport in the ZnO NW under the applied external voltages; (j) carriers transport in the bending ZnO NW induced by external forces under the applied external voltages.

surroundings. As sensors or detectors, nanomaterials or nanodevices would be eroded by ambient surroundings like moisture and gases, especially chemical gases or solutions. In recent years, some chemical nanodamage or nanofailure phenomena of ZnO nanomaterials have been reported by some groups.

The interaction of hexagonal ZnO wires with different solutions, including deionised water, ammonia, NaOH solution and horse blood serum was systematically investigated.[2] Figure 9.20 shows SEM images of a ZnO wire that has interacted with pure horse blood serum for different lengths of time. The results showed that ZnO can be dissolved (the extreme form of nanodamage) by the above solutions. The etching process started from the edge of the ZnO wires. The interaction of ZnO wires with horse blood serum also showed that ZnO wires can survive in the fluid for a few hours, after

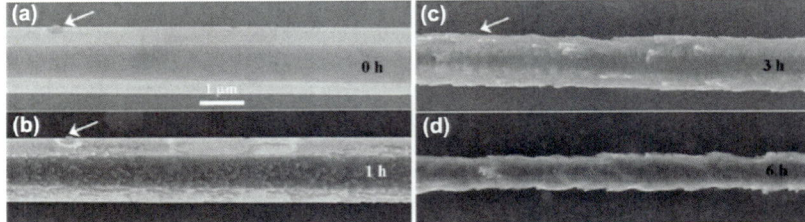

Figure 9.20   SEM images of a ZnO wire that has interacted with horse blood serum
(pH ≈ 8.5) for (a) 0; (b) 1; (c) 3; and (d) 6 h, respectively, with the
arrowhead indicating the same reference area.
Reproduced with permission from John Wiley and Sons Copyright ©
2006 WILEY-VCH Verlag GmbH & Co. KGaA, Weinheim.[2]

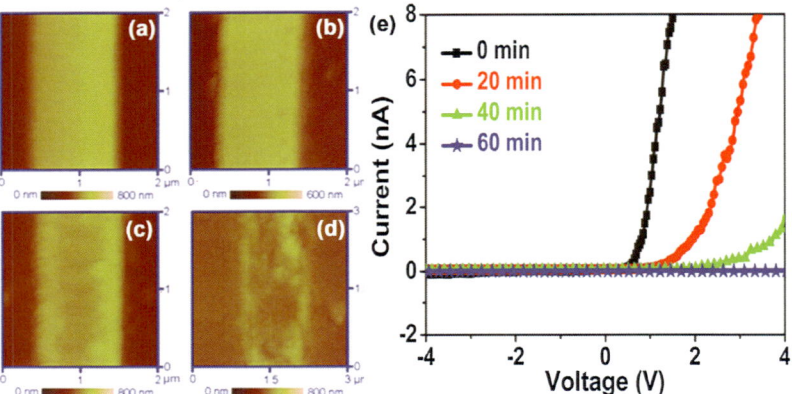

Figure 9.21   AFM images of ZnO NBs under the interaction with ammonia for
different times: (a) 0 min; (b) 20 min; (c) 40 min; (d) 60 min; (e) The
electrical transportation of the ZnO NB degrades as the increase of the
nanodamage.

which they degrade into mineral ions. The biodegradability and biocompatibility of ZnO wires could potentially allow their application in *in situ*
biosensing and biodetection. The study would set the foundation for ex-
panding the application of ZnO nanostructures in bioscience.

We also investigated the chemical nanodamage of a single ZnO NB in
weak alkaline conditions. Figure 9.21 shows the morphology and electrical
transportation changes under ammonia at different times. Etching pits appeared in the surface of the ZnO NB after interaction with ammonia for
20 min and they dissolved absolutely after 1 h. The electrical transportation
of the ZnO NB also degraded with the increase of nanodamage under
ammonia. This study would provide some meaningful guidance for the
applications of ZnO nanodevices in biological detection.

We also studied the influence of diameter on the dissolving behaviour
of ZnO wire in a HCl solution where pH = 6.[29] Figure 9.22 shows the
morphology, diameter and electrical property changes of the ZnO wire in the

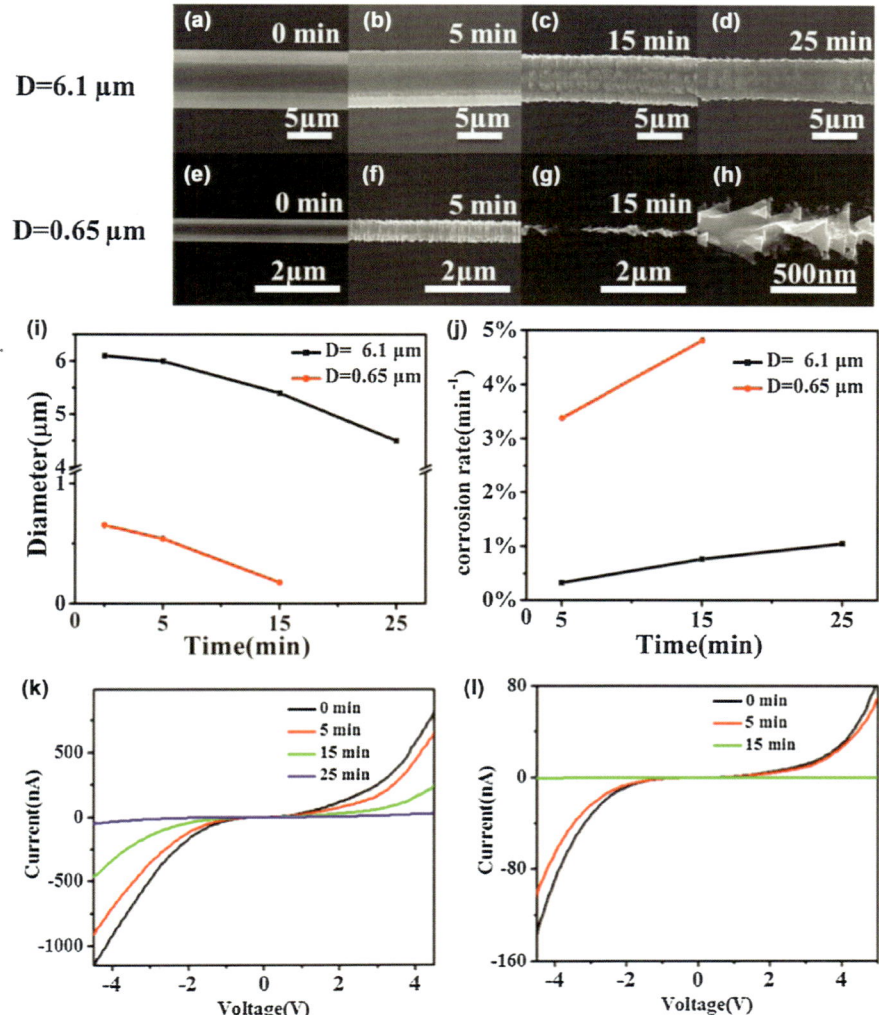

Figure 9.22 SEM images of the ZnO wires after interacting with HCl solution with pH = 6 for different time. (a) 0; (b) 5; (c) 15; (d) 25; (e) 0; (f) 5 and (g) 15 min; (h) the high-magnification SEM image of (g); (i) the diameters of the two ZnO wires after interacting with HCl solution of pH = 6 for different time; (j) the corrosion rates of the two wires at different time; (k) the *I–V* characteristics of the wires with a diameter of 6 μm after interacting with HCl for 0, 5, 10 and 25 min, respectively; (l) the *I–V* characteristics of the wires with a diameter of 1 μm after interacting with HCl for 0, 5 and 10 min, respectively.
Reproduced from ref. 29 with permission from The Royal Society of Chemistry.

HCl solution over time. The preferential etching plane and electrical properties upon treatment were also discussed in detail. The smaller diameter wire resulted in a more pronounced corrosion rate, which is most likely due

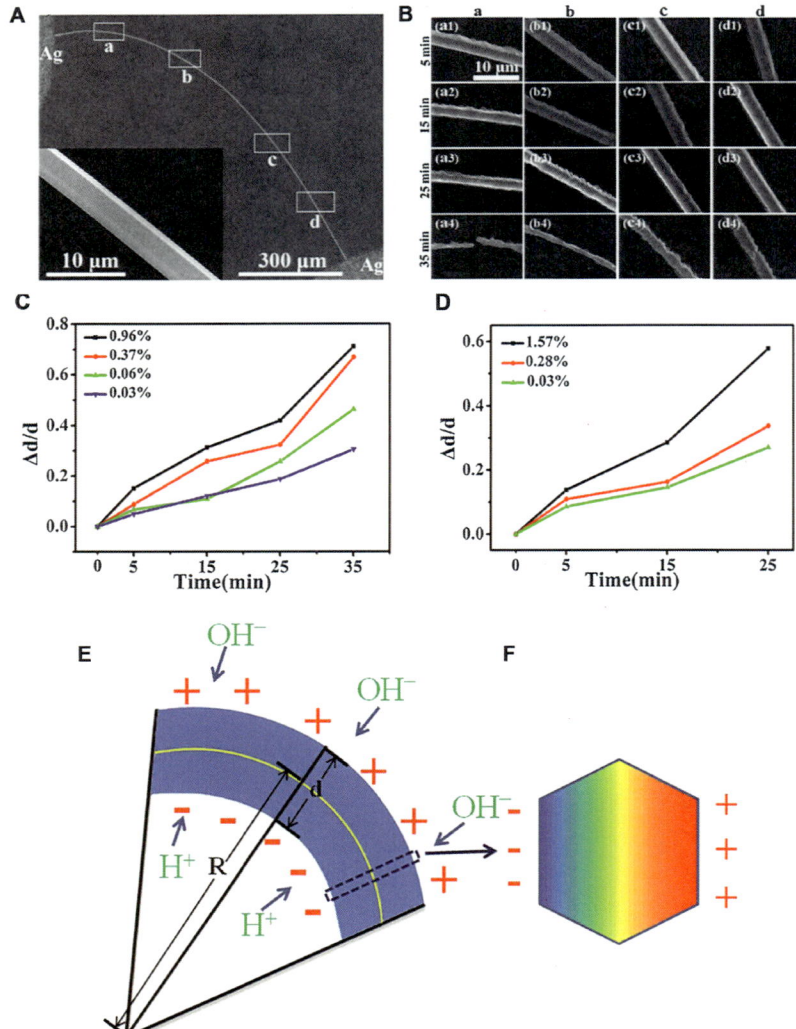

Figure 9.23 SEM images of a bent ZnO wire that interacted with 4 M KOH solution over time. (A) Low-magnification SEM image of the ZnO wire, the inset is the high-magnification SEM of the ZnO wire; (B) high-magnification SEM images of the different regions in (A) interacting with KOH solution for 5, 15, 25 and 35 min: (a1)–(a4) area a, (b1)–(b4) area b, (c1)–(c4) area c, and (d1)–(d4) area d. Diameter reduction rates of the two bent wires corresponding to the different regions in different solutions: (C) KOH; and (D) HCl; (E) piezoelectric effect of the inner and outer surfaces of the wire which attracts hydrogen and hydroxyl ions; (F) cross-sectional schematic diagram of bending-induced piezoelectric effect. The rainbow colours stand for the potential charge distribution in the wire induced by bending. Reproduced from ref. 30 with permission from The Royal Society of Chemistry.

to its higher specific surface area. The detailed morphological study demonstrated that ZnO wire had a preferential etching plane {10$\bar{1}$1} in the HCl solution. The etching also strongly reduced the electrical properties of the ZnO wires. These findings may provide valuable guidance for designing nanodevices based on ZnO wires.

In addition, we studied the piezoelectric effect on the corrosion behaviour of ZnO wires in acidic and alkaline environments for the first time.[30] Figure 9.23 shows the morphology, diameter changes and mechanism schematics of the ZnO wires under stress in the chemical solution for different times. KOH and HCl solutions were chosen as the simulated environments. A strain-free ZnO wire corroded almost symmetrically in solution, while the bent wire corroded quite differently and the failure phenomenon appeared faster under larger strain due to the higher piezoelectric potential. The corrosion behaviours of individually bent ZnO wires have been clearly observed under various strains estimated using the local curvature model. The corrosion phenomena of bent ZnO MWs in acidic and alkaline environments were different. The outer surface of the wire attracts free hydroxide ions and the inner one attracts hydrogen ions from the solution, which promotes the chemical reaction because of the piezoelectric potential which is generated by strain. The experimental results indicated that the corrosion rate is quite sensitive to strain, which provides a recommendation for the design and evaluation of nanodevices that serve in extreme environments.

Although various nanodamage and nanofailure forms were introduced by the categories discussed here, actual nanodamage and nanofailure may be a combination of several types due to the complexity of the actual working situation. So, we should distinguish the main factors in a specific situation that would be beneficial for investigating nanodamage and nanofailure in 1D ZnO nanomaterials and nanodevices. Through the above, and subsequent investigations on nanodamage and nanofailure of 1D ZnO nanomaterials and nanodevices, it would be useful to improve the properties of 1D ZnO nanomaterials and guarantee the stability, safety and long life of 1D ZnO nanodevices applied in daily life.

References

1. J. H. Zhan, Y. Bando, J. Q. Hu and D. Golberg, Nanofabrication on ZnO nanowires, *Appl. Phys. Lett.*, 2006, **89**(24), 243111.
2. J. Zhou, N. S. Xu and Z. L. Wang, Dissolving behavior and stability of ZnO wires in biofluids: a study on biodegradability and biocompatibility of ZnO nanostructures, *Adv. Mater.*, 2006, **18**(18), 2432–2435.
3. Q. Wan, J. Huang, A. X. Lu and T. H. Wang, Degenerate doping induced metallic behaviors in ZnO nanobelts, *Appl. Phys. Lett.*, 2008, **93**(10), 103109.
4. B. M. Wen, J. E. Sader and J. J. Boland, Mechanical properties of ZnO nanowires, *Phys. Rev. Lett.*, 2008, **101**(17), 175502.

5. Y. Yang, Y. Zhang, J. J. Qi, Q. L. Liao, L. D. Tang and Y. S. Wang, Electric-induced nanodamage in single ZnO nanowires, *J. Appl. Phys.*, 2009, **105**(8), 084319.

6. R. Agrawal, B. Peng and H. D. Espinosa, Experimental-computational investigation of ZnO nanowires strength and fracture, *Nano Lett.*, 2009, **9**(12), 4177–4183.

7. Y. Yang, J. J. Qi, Y. S. Gu, W. Guo and Y. Zhang, Electrical and mechanical coupling nanodamage in single ZnO nanobelts, *Appl. Phys. Lett.*, 2010, **96**(96), 123103.

8. Q. Zhang, J. J. Qi, Y. Yang, Y. H. Huang, X. Li and Y. Zhang, Electrical breakdown of ZnO nanowires in metal-semiconductor-metal structure, *Appl. Phys. Lett.*, 2010, **96**(25), 253112.

9. Y. Wei, P. Liu, K. L. Jiang, L. Liu and S. S. Fan, Breaking single-walled carbon nanotube bundles by Joule heating, *Appl. Phys. Lett.*, 2008, **93**(2), 023118.

10. Z. Xu, D. Golberg and Y. Bando, In situ TEM-STM recorded kinetics of boron nitride nanotube failure under current flow, *Nano Lett.*, 2009, **9**(6), 2251–2254.

11. A. M. Nie, J. B. Liu, C. Z. Dong and H. T. Wang, Electrical failure behaviors of semiconductor oxide nanowires, *Nanotechnology*, 2011, **22**(40), 405703.

12. P. F. Li, Q. L. Liao, Z. Zhang, Z. Z. Wang, P. Lin, X. H. Zhang, Z. Kang, Y. H. Huang, Y. S. Gu, X. Q. Yan and Y. Zhang, Investigation on the mechanism of nanodamage and nanofailure for single ZnO nanowires under an electric field, *ACS Appl. Mater. Interfaces*, 2014, **6**(4), 2344–2349.

13. Q. Zhang, J. J. Qi, J. Zhao, X. Li and Y. Zhang, Multi-zone light emission in a one-dimensional ZnO waveguide with hybrid structures, *Opt. Mater. Express*, 2011, **1**(2), 173–178.

14. X. W. Gu, M. Jafary-Zadeh, D. Z. Chen, Z. Wu, Y. W. Zhang, D. J. Srolovitz and J. R. Greer, Mechanisms of failure in nanoscale metallic glass, *Nano Lett.*, 2014, **14**(10), 5858–5864.

15. P. X. Gao, W. J. Mai and Z. L. Wang, Superelasticity and nanofracture mechanics of ZnO nanohelices, *Nano Lett.*, 2006, **6**(11), 2536–2543.

16. F. Xu, Q. Q. Qin, A. Mishra, Y. Gu and Y. Zhu, Mechanical properties of ZnO nanowires under different loading modes, *Nano Res.*, 2010, **3**(4), 271–280.

17. Z. L. Wang, Mechanical properties of nanowires and nanobelts, *Dekker Encycl. Nanosci. Nanotechnol.*, 2004, 1773–1786.

18. S. Hoffmann, F. Östlund, J. Michler, H. J. Fan, M. Zacharias, S. H. Christiansen and C. Ballif, Fracture strength and Young's modulus of ZnO nanowires, *Nanotechnology*, 2007, **18**(20), 205503.

19. A. V. Desai and M. A. Haque, Mechanical properties of ZnO nanowires, *Sens. Actuators A*, 2007, **134**(1), 169–176.

20. C. Q. Chen and J. Zhu, Bending strength and flexibility of ZnO nanowires, *Appl. Phys. Lett.*, 2007, **90**(90), 043105.

21. G. Y. Jing, X. Z. Zhang and D. P. Yu, Effect of surface morphology on the mechanical properties of ZnOnanowires, *Appl. Phys. A*, 2010, **100**(2), 473–478.

22. R. Agrawal, B. Peng, E. E. Gdoutos and H. D. Espinosa, Elasticity size effects in ZnO nanowires-a combined experimental-computational approach, *Nano Lett.*, 2008, **8**(11), 3668–3674.

23. P. F. Li, Q. L. Liao, Z. Z. Wang, P. Lin, Z. Zhang, X. Q. Yan and Y. Zhang, Calibration on force upon the surface of single ZnOnanowire applied by AFM tip with differentscanning angles, *RSC Adv.*, 2015, **5**(59), 47309–47313.

24. P. F. Li, Q. L. Liao, Z. Z. Wang, P. Lin, Z. Zhang, X. Q. Yan and Y. Zhang, AFM investigation of nanomechanical properties of ZnO nanowires, *RSC Adv.*, 2015, **5**(42), 33445–33449.

25. P. F. Li, Q. L. Liao, S. Z. Yang, X. D. Bai, Y. H. Huang, X. Q. Yan, Z. Zhang, S. Liu, P. Lin, Z. Kang and Y. Zhang, In situ transmission electron microscopy investigation on fatiguebehavior of single ZnO wires under high-cycle strain, *Nano Lett.*, 2014, **14**(2), 480–485.

26. J. Zhou, P. Fei, Y. F. Gao, Y. D. Gu, J. Liu, G. Bao and Z. L. Wang, Mechanical-electrical triggers and sensors using piezoelectric micowires/nanowires, *Nano Lett.*, 2008, **8**(9), 2725–2730.

27. Z. L. Wang and J. H. Song, Piezoelectric nanogenerators based on zinc oxide nanowire arrays, *Science*, 2006, **312**(5771), 242–246.

28. Y. Yang, J. J. Qi, Q. L. Liao, W. Guo, Y. S. Wang and Y. Zhang, Negative differential resistance in PtIr/ZnO ribbon/sexithiophen hybrid double diodes, *Appl. Phys. Lett.*, 2009, **95**(12), 123112.

29. J. J. Qi, K. Zhang, Z. X. Ji, M. X. Xu, Z. Z. Wang and Y. Zhang, Dissolving behavior and electrical properties of ZnO wire in HCl solution, *RSC Adv.*, 2015, **5**(55), 44563–44566.

30. K. Zhang, J. J. Qi, Y. Tian, S. N. Lu, Q. J. Liang and Y. Zhang, Influence of piezoelectric effect on dissolving behavior and stability of ZnO micro/nanowires in solution, *RSC Adv.*, 2015, **5**, 3365–3369.

CHAPTER 10

Field Emission and Electromagnetic Wave Absorption

QINGLIANG LIAO AND YUE ZHANG*

University of Science and Technology Beijing, Beijing, China
*Email: yuezhang@ustb.edu.cn

10.1 Field Emission Properties and Applications

There are many electrons in solids and the electrons cannot escape solids under normal conditions unless they are excited. The most common excitation methods are under a particular temperature, electric field, light radiation or high energy electrons. If the electrons get enough additional energy to overcome the barrier, they will escape from the solid. The electron emission mechanism can be divided into four modes: (1) thermionic emission, (2) field emission, (3) photoelectron emission and (4) secondary electron emission. Thermionic emission is where electrons obtain high energy and escape across the surface barrier by raising the temperature of the materials. Field emission is a totally different way and is provided by a strong external electric field which suppresses the surface of the object barrier instead of giving additional energy to the electrons. When the applied electric field is zero, electrons cannot escape because the energy is lower than the barrier. With an increase of the applied electric field, the height (and the width) of the barrier decrease. In this way, many electrons in emitters can escape by a tunnelling effect or even across the top of the barrier at absolute zero, forming field electron emission.

Nanoscience & Nanotechnology Series No. 43
ZnO Nanostructures: Fabrication and Applications
By Yue Zhang
© Yue Zhang 2017
Published by the Royal Society of Chemistry, www.rsc.org

In 1928, Fowler and Nordheim proposed the theory of emitting electrons into a vacuum from a metal surface at absolute zero by using quantum theory. An electric field is added to the metal and vacuum interface to bend the band structure of the metal and lead the electrons across the metal barrier.[1] Then the formula of the field emission current density is conducted, which is called the *F–N* formula. The formula can be simplified as follows when the electric field value is not too large:

$$J = A \frac{F^2}{\phi} \exp\left(-\frac{B\phi^{3/2}}{E}\right) \tag{10.1}$$

where J is the current density (A m^{-2}), F is electric field of emission region (V m^{-1}), ϕ is work function, E is the actual electric field strength (V m^{-1}), A and B are constants with the value of $A = 1.56 \times 10^{-10}$ (AV^{-2} eV), $B = 6.83 \times 10^9$ (V eV$^{-3/2}$ m^{-1}), respectively.

Because the voltage (V) and the distance (*d*) between the cathode and anode can be measured in a practical study, the relationship between the macroscopic electric field E and the actual surface electric field F of the emitters can be explained as follows:

$$F = \beta \frac{V}{d} \tag{10.2}$$

where β is the field enhancement factor.

It can be seen from the *F–N* formula that for the same material, when the electric field strength is greater, the emission current will be greater. Under the same electric field, the tip surface has greater field strength and thus selecting the tip as the emitter on the field emission display has obvious advantages. Nanostructures with an ultra-fine tip, up to tens of nanometres or even a few nanometres, can greatly reduce the field emission threshold voltage and turn-on electric field. Moreover, high-density nanostructures could be used as electron emitters for high intensity electron emission. In other words, the electron emission cathode consists of large numbers of nanostructure emitters for electron emission. During the work process, even a small number of emitters are damaged, but most of the other emitters also work well. The lifetime of the field emission devices will be improved greatly.

ZnO nanostructures have a negative electron affinity, high mechanical strength, high thermal stability and chemical stability and high specific surface area, so they are very suitable for field emission devices.

10.1.1 Field Emission Properties of Large Area Nanowires

The field emission of a large area nanowire is different from that of a single nanowire and it means that all the nanowires (NWs) on the substrate are used for the electron emitters. Under the investigation of field emission properties, the turn-on field is defined as the electric field value when the current density is 0.1 μA cm^{-2} and the threshold field is defined as the electric field value when the current density is *ca.* 1 mA cm^{-2}.

There are many reports on the field emission properties of large area ZnO NWs.[2–10] Well-aligned single crystal ZnO NWs were grown on silicon substrates by metal vapour deposition at low temperatures (550 °C) and their field emission properties were investigated first.[2] A ZnO nanowire array cathode with a 50 nm diameter has the following properties: the turn-on field is about 6 V μm^{-1}, the threshold field is 11 V μm^{-1}, and the field enhancement factor is about 847. It is proposed that the field enhancement factor increases due to the improvement of the crystallinity of NWs, which results from increasing the preparation temperature. Although the field enhancement factor of ZnO NWs is much lower than that of carbon nanotubes, the brightness for flat panel display use is sufficient. In fact, the ZnO nanowire array cathode can produce greater than 1000 cd cm^{-2} brightness at 0.1 mA cm^{-2} current density. In 2005, a uniform, large-scale and bilayered ZnO nanorod array on silicon substrate was synthesised using a catalyst and template-free chemical reaction in a dilute solution, which is shown in Figure 10.1.[3] Moreover, the growth mechanism of the bilayered ZnO nanorod

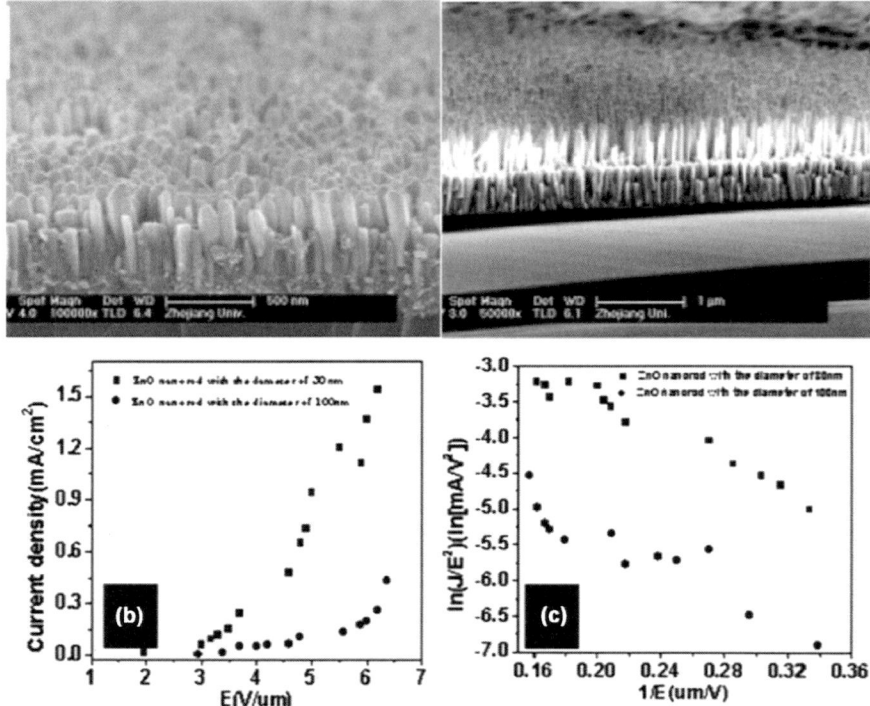

Figure 10.1 Bilayered ZnO nanorod arrays with different morphologies and corresponding field emission properties.
Reproduced with permission from ref. 3. Copyright (2005) American Chemical Society.

array has been proposed. The field emission properties of the ZnO nanorod array were optimised by modulating the diameters of the ZnO nanorods.

The field emission properties of ZnO nanoneedle arrays prepared by vapour deposition were investigated.[4] The results show that the turn-on field is about 2.4 V μm^{-1} and the emission current intensity is about 2.4 mA cm^{-2} when the electric field is about 7 V μm^{-1}. The results show that the sharp needle tip geometry is the reason for good field emission performance. There are many studies on the field emission properties of ZnO nano-structures with different morphologies.[5–9] Comparative studies on the field emission properties of ZnO nanostructures with different tip shapes were carried out. Needle-like ZnO nanostructures have the best field emission properties. Excellent field emission properties were obtained for other shapes, such as pencil-like and injector-like ZnO nanostructures. ZnO nanotubes were prepared by hydrothermal synthesis and the field emission properties were also studied. The results show that they have excellent field emission stability.[8] In addition, the densities of ZnO nanorod arrays have been modulated by controlling the growth position. High density ZnO nanorod arrays have excellent field emission properties.[9]

The tetraleg-ZnO (T-ZnO) nanostructure is an excellent field emission material. T-ZnO nanostructures were used as surface-conduction emitters, and a surface-conduction field-emission cathode was fabricated.[10] The effects of distance and thickness between the adjacent emitter on the electron emission efficiency of the T-ZnO thin film were investigated. When the inter-electrode distance is 0.1 mm and the film thickness is 8 μm, the optimum electron emission efficiency is 60%. When the emission current density reaches 0.6 mA cm^{-2}, the turn-on electric field is 1 V μm^{-1}. This can meet the demands of field emission display devices. The cathode has good emission stability and uniformity. A T-ZnO nanostructure cathode was fabricated by a screen-printing method.[11] The field emission properties were significantly improved by the addition of a carbon nanotube buffer layer between the ZnO and the Ag film. This is due to the formation of good mechanical and electrical contact between the ZnO and the substrate through the carbon nanotubes.

The bicrystalline ZnO nanowire arrays were prepared by thermal oxidation of a zinc film and have high-performance field emission properties.[12] The fabricated patterned ZnO nanowire arrays and corresponding field emission properties are shown in Figure 10.2. Field emission measurement results show that the prepared ZnO NWs have excellent field emission properties. Uniform emission can be obtained and the turn-on field is 7.8 V μm^{-1}. The reports indicated that the method is advantageous for large-scale synthesis of ZnO NWs for field emission applications.

The patterned growth of ZnO nanostructures is an efficient method to improve the field emission properties. The patterned growth of ZnO NWs combines the direct patterning of ZnO nanoparticle seeds and subsequent low-temperature hydrothermal growth, as shown in Figure 10.3.[13] By using this control of pattern geometry and printing time, radially grown ZnO NW

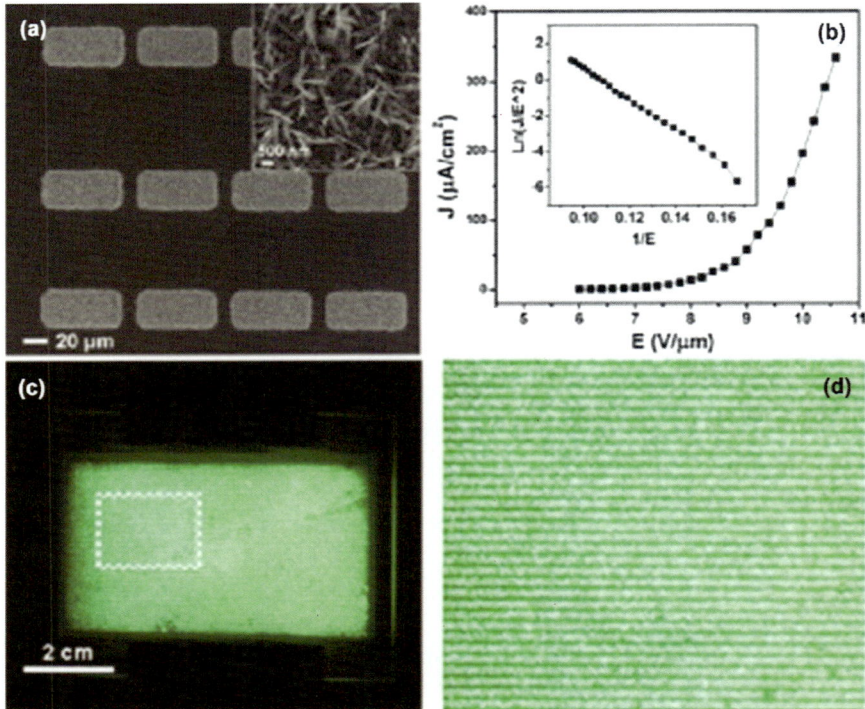

Figure 10.2 Typical SEM images of patterned ZnO nanowire arrays prepared on ITO glass and the corresponding field emission properties.
Reproduced with permission from ref. 12. Copyright (2013) American Chemical Society.

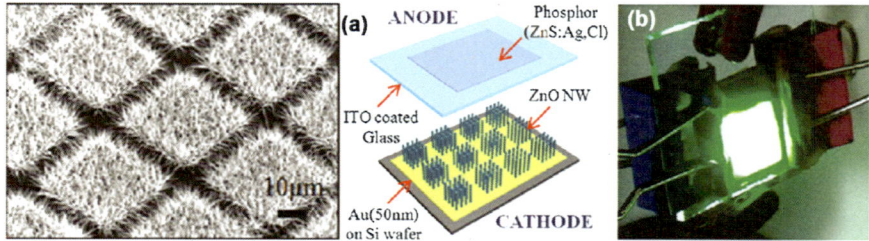

Figure 10.3 Patterned ZnO NW arrays and a luminous field emission device with a patterned ZnO NW array.
Reproduced with permission from ref. 12. Copyright (2011) American Chemical Society.

structures were created for use in the fabrication of an efficient field emission device. The optimum patterned ZnO NW field emission device fabricated exhibited a very low turn-on electric field, which is attributed to the decrease in the field emission screening effect that results from the radial structures of the patterned ZnO NW arrays.

10.1.2 Field Emission Properties of Single NWs

There are few reports on the field emission properties of single ZnO NWs.[14,15] The field emission properties of a single ZnO nanowire can accurately reveal the intrinsic emission properties of ZnO NWs and the factors that affect the emission properties of the NWs. In addition, the group also studied the single In-doped ZnO NWs field emission properties.

The field emission properties of single ZnO NWs could be investigated in a modified high vacuum transmission electron microscope (TEM). The vacuum degree is about 10^{-7} Pa, the temperature is room temperature, and the distance between the tip of the nanowire and the anode is adjustable. Figure 10.4 shows the *in situ* measurement system for the field emission properties of single ZnO nanowire.

The field emission current of single nanowires were measured using two NWs, and the inter-electrode distances can change from 1.5 µm to 200 µm. The field emission properties and the *F–N* curve are shown in Figure 10.5. The numbers labelled on the curve edges are the inter-electrode distance corresponded to each group of measurement.

The *V–I* curves of the nanowire field emission show that the field emission current increases rapidly with the increase of the voltage at different inter-electrode distances, and the turn-on voltage of the field emission increases with the increase of the inter-electrode distance. The field enhancement factors, calculated from the *F–N* curve, increased with the increasing of inter-electrode distance and exhibit a good linear relationship. The results are shown in Figure 10.6. In addition, the field emission of In-doped ZnO NWs

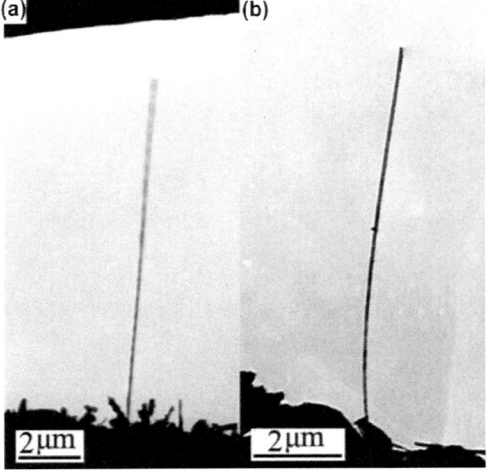

Figure 10.4   *In situ* field emission property measurement of a single nanowire.
Reproduced with permission from Y. Huang, Journal of Physics: Condensed Matter, 19, Field-emission properties of individual ZnO nanowires studied *in situ* by transmission electron microscopy, 176001, 2007, IOP Publishing, Ltd.[14]

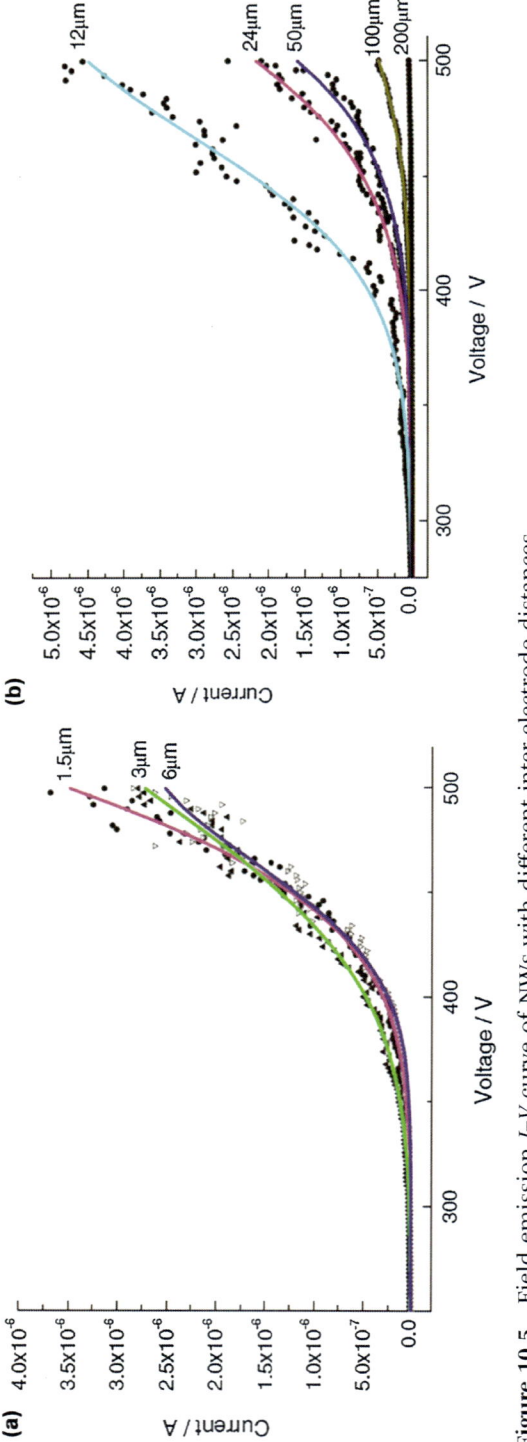

Figure 10.5   Field emission *I–V* curve of NWs with different inter-electrode distances. Reproduced with permission from Y. Huang, Journal of Physics: Condensed Matter, 19, Field-emission properties of individual ZnO nanowires studied *in situ* by transmission electron microscopy, 176001, 2007, IOP Publishing, Ltd.[14]

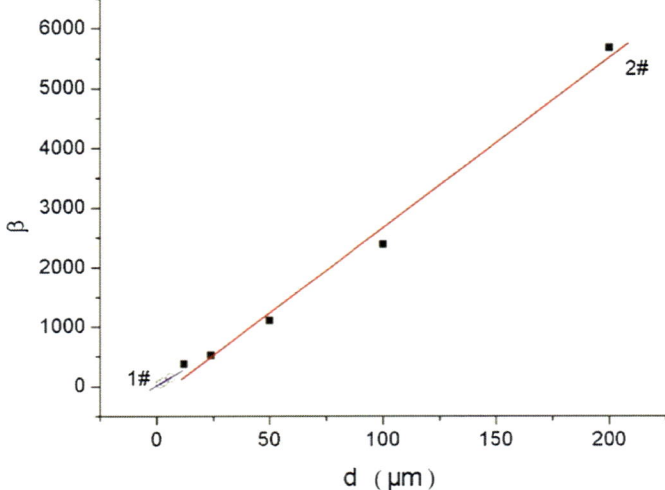

Figure 10.6   Relationship between the field enhancement factor β and the inter-electrode distance.
Reproduced with permission from Y. Huang, Journal of Physics: Condensed Matter, 19, Field-emission properties of individual ZnO nanowires studied *in situ* by transmission electron microscopy, 176001, 2007, IOP Publishing, Ltd.[14]

has two kinds of emission mechanism: conduction band emission and valence band emission.[15]

10.1.3   High Intensity Field Emission

The field emission performance described above is measured under conditions where the cathode is in the low-voltage DC electric field, in which the cathode emission current density is only up to several $A\,cm^{-2}$, which makes it difficult to meet the demands of the accelerator, travelling wave tubes and other high intensity electron beam device requirements. The so-called high density emission is where the cathode can achieve high density electron beam emission in the high-voltage pulsed electric field. High intensity electron beam emission is not the fundamental field emission but the plasma-induced field emission, which also known as explosive emission. The presence of the surface emitter in the cathode allowed the field strength of the emitter tip to increase several hundred times. At the same time, the emitter adsorbed gas and heated by electron beam. Finally, a plasma layer was formed on the emitter surface. In this way, the current density of electron beam in the electric field reached up to dozens of $A\,cm^{-2}$. The investigation of the emission properties of ZnO nanowire array under a high voltage pulse electric field were reported.[16,17]

The high intensity field emission performance of ZnO nanowire arrays was investigated by high-voltage pulsed electric field, which can be divided into single-pulse and double-pulse. In the double-pulse test, the pulse width of the pulsed power system was about 100 ns and the maximum voltage was 8×10^5 V, and the distance between anode and cathode was 98 mm. The cathode emission surface was a circle with a diameter of 50 mm, and the test environment was under a degree of vacuum of about 1×10^{-3} Pa. The voltage and current waveforms of ZnO nanowire arrays and a CCD photograph of the electron emission from the cathode are shown in Figure 10.7.

It is found that the emission current density of the ZnO nanowire arrays on the silicon substrate is 91.16 A cm^{-2} under a pulsed electric field of 8.16 V μm^{-1} and the value reaches 132.42 A cm^{-2} on stainless steel. During the process of the high current electron beam in ZnO nanowire arrays, the emitter first emits electrons under a high-voltage field, which is called field emission. Then, the emitter gas adopted by the emitter is ionised by the electron beam to form a plasma. Finally, the plasma emits a high current electron beam in the electric field.

The high intensity pulse field emission mechanism of nanostructures was investigated.[18] The emission of the nanostructure cold cathode under a high-voltage pulsed electric field is not purely field emission and there are two electron emission stages: field emission and plasma-induced emission. There was a transition from field emission to plasma-induced emission during cathode emission under single-pulse. In the case of multi-pulse, the first pulse emission process was the same as that of a single-pulse emission. The subsequent pulse field emission was entirely a plasma-induced emission, and the emission current was enhanced by the expansion of the plasma. Whether single or double pulse, the emission processes have a transition from field emission to plasma-induced emission. The high intensity electron beam emission results from the high voltage electric field and is enhanced by surface plasma.

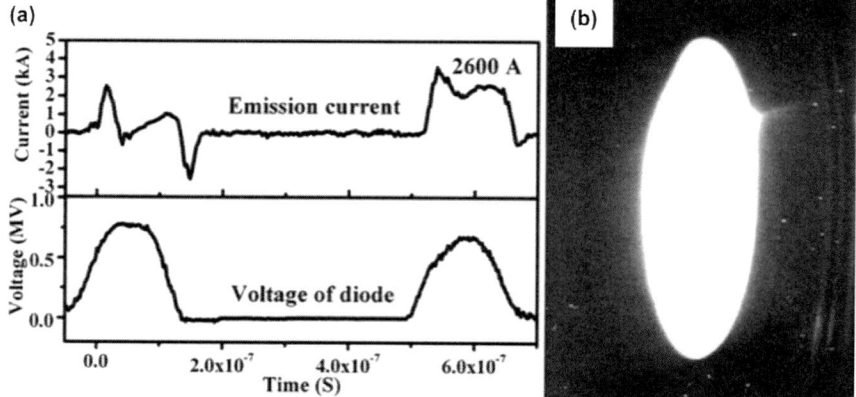

Figure 10.7 (a) The voltage and current waveforms of ZnO nanowire arrays cathode; (b) the CCD photograph of electron emission from cathode. Reproduced from ref. 16 with the permission of AIP Publishing.

10.1.4 Influencing Factors of Field Emission Properties

The field emission properties of ZnO nanostructures were affected not only by their own characteristics (work function), but also by some others such as geometric factors (nanowire size, tip morphology, density distribution, *etc.*), doping and so on. Additionally, the growth substrates of NWs and subsequent processing of NWs (such as annealing) have a great effect on the field emission properties.

10.1.4.1 *Morphology*

The influence of morphology on field emission properties by comparing needle-like ZnO nanostructures and hexagonal ZnO nano-arrays was investigated.[19] The field emission properties were investigated at room temperature with a vacuum of 3.5×10^{-7} Pa. The inter-electrode distance was 300 μm and the voltage could be modulated in the range of 0–1100 V. The morphologies of ZnO nanostructure for field emission property measurements are shown in Figure 10.8. The threshold fields of these two ZnO nanostructures were 2.3–2.4 V μm^{-1}. When the electric field reached 3.7 V μm^{-1}, the field emission density reached 4.31×10^{-5} A cm^{-2} and 1.94×10^{-5} A cm^{-2}, respectively.

The tip morphology of needle-like ZnO nanowire arrays is shown in Figure 10.9. This kind of nanostructure has a rough surface and the tip has an angle of 30°. The tip diameter is less than 20 nm, and the transition angle is about 117°, which is the intersection angle of the $(0\bar{1}13)$ and $(01\bar{1}3)$ surfaces. These two ZnO nanostructures have a similar morphology and density, but they have different tip geometries, which results in different field emission properties.

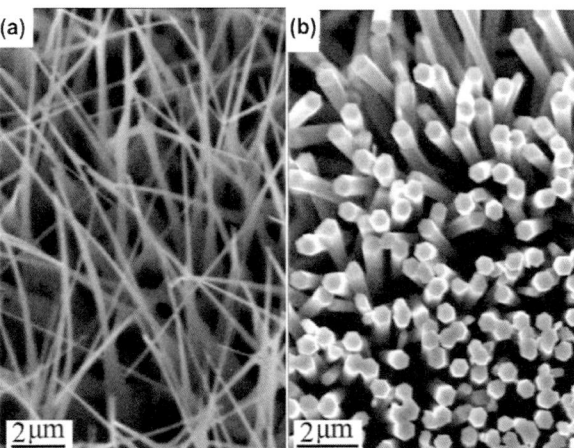

Figure 10.8 ZnO nanostructures with different morphologies for the investigation of field emission properties.
Reproduced with permission from ref. 19. Copyright (2007) American Chemical Society.

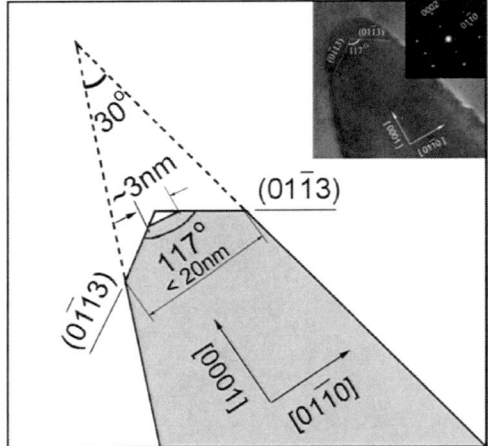

Figure 10.9 The morphology of the tip of a needle-like ZnO nanostructure and its schematic diagram.
Reproduced with permission from ref. 19. Copyright (2007) American Chemical Society.

Similarly, other reports also show that the tip geometric shape of the nanoneedles was responsible for good field emission properties and the field enhancement factor β is linearly related to the reciprocals of the emitter radius.[3,4] Other reports also have similar research results.[20]

10.1.4.2   Doping

Doping is an effective method for improving the field emission properties of ZnO nanostructures. The field emission intensity of ZnO NWs could be improved by In doping.[15] In-doped ZnO NWs have an In content of 17.2 at.% and the distance between anode and the nanowire emitter is 1.5 µm, and the maximum applied voltage is 420 V. The voltage-field emission current curves and F–N curves of In-doped ZnO NWs are shown in Figure 10.10.

The results show that the field emission properties of a single In-doped ZnO nanowire are better than those of pure ZnO nanowires and the values of $1/V \sim \ln(I/V^2)$ of pure ZnO NWs can be fitted to a perfect straight line while those of In-doped ZnO NWs can only be segmented into two straight lines with different slopes. As per the calculation, the corresponding work functions of these two lines are 4.89 eV and 1.70 eV, respectively and the energy difference is 3.19 eV, which is very close to the bandgap of ZnO (3.37 eV). The calculated results show that the In-doped ZnO has a lower work function than that of pure ZnO and thus has better field emission properties. The electron structure and density of states (DOS) of pure ZnO and In-doped ZnO could be calculated by using first principles based on DFT. The difference in the field emission characteristics between pure ZnO

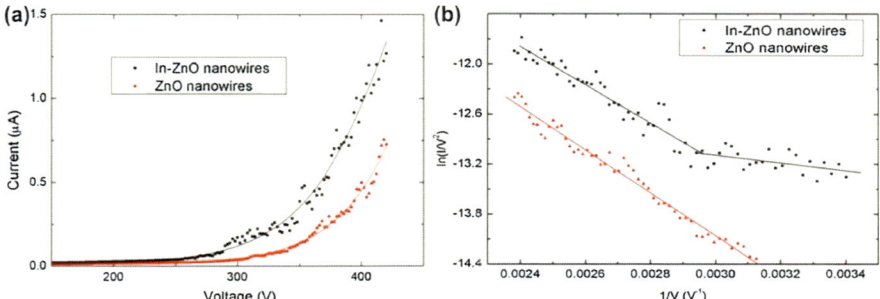

Figure 10.10 (a) Dependence of emission current on voltage for a single nanowire; (b) *F–N* curve of a single nanowire.
Reproduced with permission from ref. 15. Copyright (2007) American Chemical Society.

and In-doped ZnO was explained. The calculations showed that In-doped ZnO has a Fermi level in the conduction band and electrons for field emission came from two parts: electrons under the deep valence band and conduction band under the Fermi level. The electrons derived from the conduction band in the field emission current have a small work function, which resulted in a low threshold voltage. However, the emission current is very low due to the low-density electrons in the conduction band. When the applied voltage increases to a value, the electron emission from valence band becomes the dominant emission. Then, the emission current would be increased significantly. Therefore, the *F–N* curves of In-doped ZnO NWs are fitted with two straight lines with different slopes.

The field emission properties of ZnO nanofibre arrays were also improved by Ga-doping.[21] When the Ga dopant concentration is 0.73%, the work function is reduced from 5.3 eV for pure ZnO to 4.47 eV, which effectively reduces the emission threshold field of nanofibres and improves the field emission properties of nanofibres. The threshold field of T-ZnO was reduced effectively by Mg doping, and the field emission current density was also improved at the same time.[22]

10.1.4.3 Other Influencing Factors

The reductive atmosphere treatment can improve the field emission properties of ZnO nanostructures, because H_2 atmosphere treatment increases the oxygen vacancy and free electron concentration of ZnO nanoparticles, thus increasing the intensity of the field emission current. The work function of ZnO nanowires was reduced from 5.3 eV to 5.08 eV, and the turn-on field decreased correspondingly from 3.9 Vμm^{-1} to 3.6 Vμm^{-1} through annealing in the H_2 atmosphere.[23] The turn-on electric field of the ZnO nanotetrapods was decreased and the field emission current was increased through the H_2 atmosphere treatment at 500 °C for 1 h.[24]

It has been reported that annealing in an O_2 atmosphere also can improve the field emission properties of ZnO nanorod arrays.[25] The ZnO nanorod arrays prepared by chemical vapour deposition were annealed in oxygen, air and NH_3 atmosphere before testing. The results for the descending order of the field emission properties are as follows: oxygen atmosphere annealed nanorods, directly grown nanorods, air atmosphere annealed and NH_3 atmosphere annealed nanorods. It was concluded that annealing in an oxygen atmosphere may lead to a decline in the work function of ZnO nanorods.

Many studies show that the growth substrates of ZnO nanomaterials have a great impact on the field emission properties. A well-aligned ZnO nanorod array on an Au/Ti/n-Si substrate with a field emission turn-on voltage of only 0.85 V μm^{-1} and the emission current density of 1 mA cm^{-2} was achieved at the applied electric field of 5.0 V μm^{-1}.[26] ZnO NWs grown on carbon cloth have a very low field emission threshold electric field of ~0.7 V μm^{-1} and great field enhancement factor of ~41 100, which means when the electric field intensity is 0.7 V μm^{-1}, the field emission current density is as high as 1 mA cm^{-2}.[27] Under these conditions, the emission is not only related to the ZnO NWs, but also to the participation of carbon cloth.

Changing the growth substrate of ZnO nanostructures or treatment of substrate also can greatly improve the high-density field emission properties of ZnO nanostructures. The high density current emission performance of ZnO nanorod array cathode can be improved by using stainless steel as the cathode substrate. The emission current intensity of the ZnO nanorod array cathode can be increased from 1.47 kA (current density 74.87 A cm^{-2}) of Si substrate to 2.60 kA (current density 132.42 A cm^{-2}) of stainless steel substrate with similar density and morphology grown on a 50-mm diameter silicon substrate and a stainless-steel substrate.[16]

10.2   Electromagnetic Wave Absorption Properties and Applications

With the rapid development of modern radar and microwave electronic technologies, absorbing materials are widely used in microwave anechoic chambers, for electromagnetic shielding, to reduce the reflection of optical devices, to avoid interference of communication equipment, and in stealth weapons and many other fields. The so-called absorbing material is a kind of material that can absorb and attenuate the incident electromagnetic waves and convert electromagnetic energy into heat dissipation or make it disappear due to electromagnetic interference. Nanomaterials exhibit excellent absorbing properties due to their unique properties such as small size effect, surface effect, quantum size effect and macroscopic quantum tunnelling effect. They also have the characteristics of light weight, good compatibility and wide frequency band of absorption, which gives them the best development potential to be absorbing materials.

During the investigation of the microwave absorption properties of a nanostructure, the reflection loss (RL) and screen transmittance of the nanostructure are important parameters. The RL of the absorbing materials was investigated by vector network analysers at the 2–18 GHz band. Both the real and imaginary parts of the complex permittivity and permeability of samples were measured by a vector network analyser system in the frequency range of 2–18 GHz. In addition, the electromagnetic shielding properties were investigated by an arch reflecting method in the range of 2–18 GHz.

T-ZnO is one of the most unique ZnO nanostructures; it has a unique three-dimensional structure and excellent semiconducting and piezoelectric properties, and the combination of these features give it good properties for electromagnetic wave absorption. The electromagnetic wave absorption properties of a T-ZnO absorber were studied.[28] When the mass percentage of ZnO in the coating is 20%, the maximum absorption value of the sample reaches 15.02 dB at 17.9 GHz and the band width reaches 13.6 GHz at 4.4–18 GHz. The absorption of electromagnetic waves of T-ZnO is mainly due to the dielectric loss, and the forming of a conductive network.

The wave-absorbing properties of a cage-like coating that is composed of a ZnO nanotetrapod and granular SiO_2 were investigated.[29] When compared with the pure SiO_2 and ZnO–Si composite particles, the coating exhibits a strong electromagnetic attenuation. The maximum absorption of the coating reaches 10.68 dB at 12.79 GHz and the reflectance is less than −6 dB in the frequency range of 8–18 GHz.

ZnO NWs exhibit a certain electromagnetic wave absorption performance due to their high surface-to-volume ratio and a morphology that is like antenna. The ZnO nanowire/polyester absorber coating has strong electromagnetic attenuation characteristics in the X-band (8–12 GHz).[30] When the mass fraction of ZnO is 7%, the maximum absorption value reaches 12.28 dB. ZnO NWs and dendritic ZnO were fabricated, and their reflectivity was studied by measuring the electromagnetic parameters, in order to study the absorption properties of these two types of ZnO nanostructures, as shown in Figure 10.11.[31] The results show that the absorption of dendritic ZnO is better than that of ZnO NWs. When the thickness is 5 mm, the minimum reflectance of dendritic ZnO reaches −42 dB at 3.6 GHz.

Usually, non-magnetic wide-bandgap metal oxide semiconductors lack these μeV electronic transitions and applications. Good microwave absorbers were obtained by fabrication of metal oxides using 2D electron gas plasma resonance generated by a hydrogenation process.[32] High absorption with reflection loss values as large as −49.0 dB (99.99999%) is obtained in the microwave region by using ZnO nanoparticles. Calculated electric field distributions at the resonances of the void-like modes 2DEG and RL of pristine and hydrogenated ZnO are shown in Figure 10.12. The frequency of absorption can be tuned by the particle size and hydrogenation conditions. These results may pave the way for new applications for wide bandgap semiconductors, especially in the μeV regime.

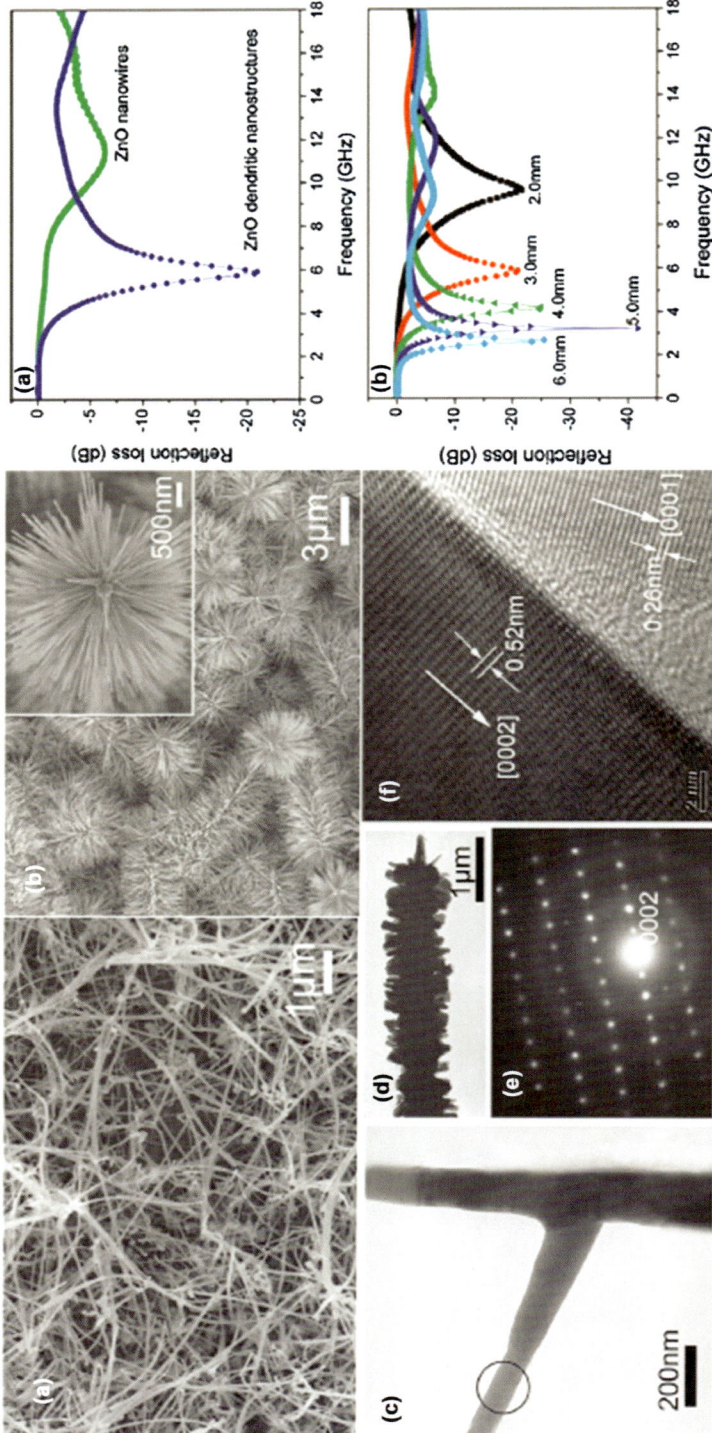

Figure 10.11 The morphologies of ZnO nanostructure and corresponding electromagnetic wave absorbing properties. Reproduced from ref. 31 with the permission of AIP Publishing.

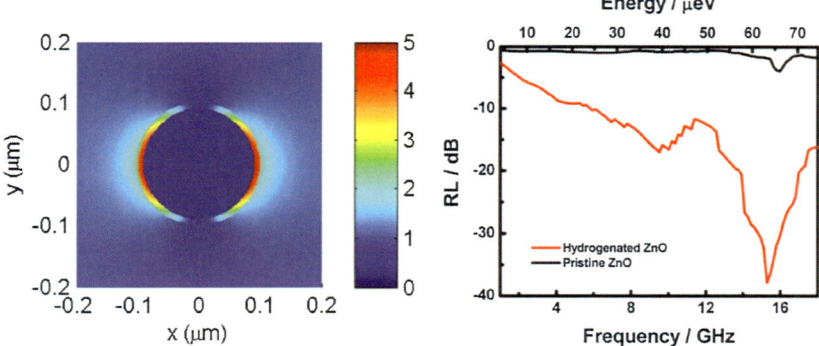

Figure 10.12 Calculated electric field distributions at the resonances of void-like modes 2DEG constrained in spherical shell and reflection loss (RL) in the microwave/μeV region of pristine and hydrogenated ZnO.
Reproduced with permission from ref. 32. Copyright (2015) American Chemical Society.

10.2.1 Absorption Properties of T-ZnO/Epoxy Resin Coatings

T-ZnO/epoxy resin (EP) wave-absorbing coatings were fabricated with T-ZnO as the absorbent and epoxy resin as the binder as follows: the T-ZnO was added into the EP resin, which was diluted by absolute ethyl alcohol, vibrated by ultrasonic wave for about 1 h and then the curing agent was put into the composite, and stirred gently. The mixture was sprayed layer-by-layer onto an aluminium plate with a square of 180 mm×180 mm and cured at 25–30 °C for at least 2 h.[28]

10.2.1.1 *Effects of the T-ZnO Concentration on Microwave Absorption Properties*

The microwave absorption properties of the T-ZnO/EP resin coatings with different ZnO concentrations and thicknesses are shown in Figure 10.13. The microwave absorption properties of the T-ZnO/EP resin coatings with different ZnO concentrations and a thickness of 1.5 mm are summarised in Table 10.1.

This shows that the absorption properties were improved as the concentration of T-ZnO increases. The minimum reflection loss was −1.74 dB when the concentration of T-ZnO is 11% and reduced to −3.23 dB under a concentration of 16%. When the concentration of T-ZnO increased to 20%, the minimum reflection loss decreased to −3.89 dB at 17.4 GHz. The difference in minimum reflection loss of the coatings results from the concentration of T-ZnO in the coating, which attenuates the electromagnetic wave energy mainly by forming conductive networks.

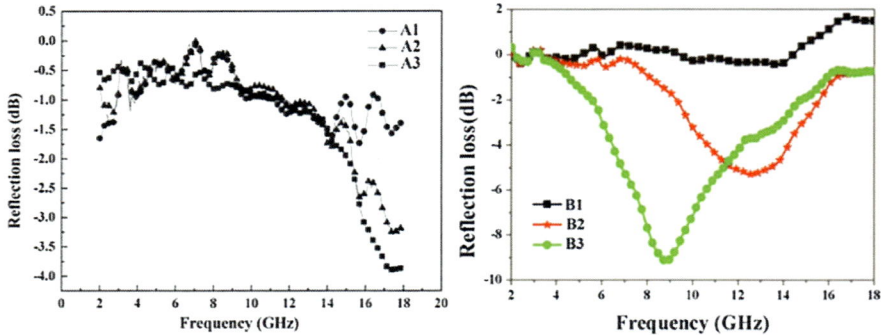

Figure 10.13 Absorption characteristics of ZnO/EP resin coatings under different ZnO concentrations and different thicknesses.

Table 10.1 Absorption properties of ZnO/EP resin coatings with different ZnO concentrations.[a]

Sample number	T-ZnO concentration (wt%)	Thickness (mm)	Minimum reflection loss (dB)	Corresponding frequency (GHz)
1	11	1.5	−1.74	15.7
2	16	1.5	−3.23	18.0
3	20	1.5	−3.89	17.4

[a]Reprinted with permission from *Acta Phys. Sin.*[28]

10.2.1.2 Effects of the Coating Thickness on Microwave Absorption Properties

The absorption properties of the ZnO/EP resin coatings with different thickness were investigated and the results are shown in Figure 10.14. When the coating thickness increases, the absorption properties were improved. The minimum RL was −0.38 dB at a thickness of 1.5 mm and the RL reduced to −5.30 dB at a thickness of 2.5 mm. When the thickness increased to 3.5 mm, the minimum RL reached −9.11 dB. However, it can also be seen that the maximum absorbing peak shifts toward a lower frequency as the thickness increased. When the coating thickness enlarged from 1.5 to 2.5 and 3.5 mm, the peak frequency is 14.0, 12.9 and 8.8 GHz, respectively.

Microwave absorption may result from dielectric loss or magnetic loss. These are characterised with the complex relative permittivity ε_r ($\varepsilon_r = \varepsilon' - j\varepsilon''$, where ε' is the real part, ε'' the imaginary part) and the complex relative permeability μ_r ($\mu_r = \mu' - j\mu''$, where μ' is the real part, μ'' the imaginary part).[33] To investigate the absorbing mechanism of the coating, the complex permittivity ε and permeability μ of the pure T-ZnO were measured. Figure 10.14 shows the frequency dependence of the permittivity and permeability of T-ZnO, respectively. It was found that

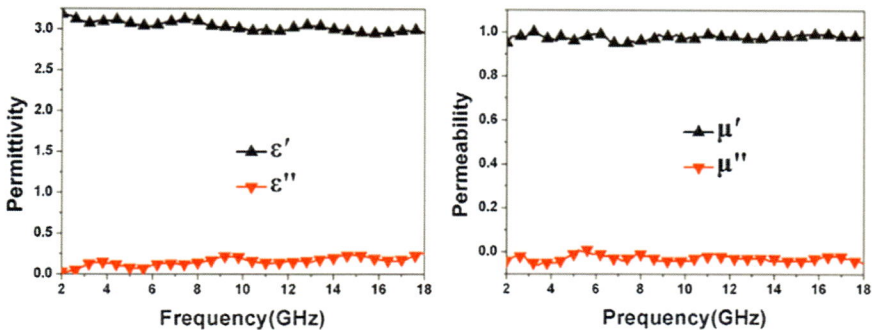

Figure 10.14 Frequency dependence of the permittivity (a) and permeability (b) of T-ZnO.
Reproduced with permission from *Acta Phys. Sin.*[28]

the values of the imaginary part of permittivity are larger than that of permeability. The value of the imaginary part of permittivity and permeability are close to 3.0 and 1.0, respectively. The results revealed that the value of dielectric loss $\tan \delta_E$ ($\varepsilon''/\varepsilon'$) was larger than that of magnetic loss $\tan \delta_M$ (μ''/μ'). Therefore, the electromagnetic wave absorption mechanisms of T-ZnO are mainly through dielectric loss rather than magnetic loss.

A wave absorption mechanism for T-ZnO could be proposed. First, the T-ZnO needle has a nanoscale diameter, and the quantum confine effect will improve the wave-absorbing properties of T-ZnO greatly. As per the Kubo theory, the energy levels of T-ZnO are not continuous but split because of the quantum confinement effect. When an energy level is in the range of microwave energy, the electron will absorb a photon in order to hop from a low energy level to a higher one. Meanwhile, the defects and suspending band can cause multiple scattering and interface polarisation, which results in the electromagnetic wave absorption. Second, the wavelength of the 2–18 GHz electromagnetic wave is larger than the size of T-ZnO, which can reduce the electromagnetic wave reflection. It can easily lead to Rayleigh scattering when the incident electromagnetic wave reacts with the T-ZnO, which results in electromagnetic wave absorption in all directions. Furthermore, the coating consists of networks resulting from the tetraleg-shaped structure of T-ZnO, and T-ZnO has good conductive properties. It is possible for the electromagnetic wave to penetrate the cellular material formed by the numerous conductive networks of T-ZnO, and the energy will be attenuated. The charge concentration at the needles' tip of the T-ZnO is distinct when the material is under an electric field. The tips of T-ZnO will act as multipoles that will be tuned with the incident electromagnetic waves and contribute to strong absorption.[34] Besides, the piezoelectric character of T-ZnO is also a factor in damaging the entered energy of microwave and reducing the reflectivity.

10.2.2 Absorption Properties of T-ZnO/Carbon Nanostructure Coatings

The fabrication process of T-ZnO/carbon nanostructure composites is as follows: the calculated amount of mixed carbon black particles or carbon nanotubes and tetrapod-like ZnO whiskers are prepared. Then the mixture is added into a two-part epoxy resin comprising a main agent and a curing agent. The mixture is diluted by absolute ethanol and stirred gently for 30 min. Thereafter, the resulted epoxy resin composition is poured into a mould and cured at room temperature for at least 24 h.

10.2.2.1 Absorption Properties of T-ZnO/Carbon Black Coatings

Carbon black (CB)/T-ZnO composites were prepared using CB and T-ZnO as absorbents and EP as the binder.[35] The electromagnetic parameters and microwave absorption properties of the composites were measured in the frequency range of 2–18 GHz. The influences of absorbents concentration and composites thickness on microwave absorption properties were studied. The results show that the minimum reflection loss for CB/T-ZnO/EP composites is −19.31 dB at 10.4 GHz and the bandwidth corresponding to reflection loss blow −5 dB is 9.68 GHz, when the content of CB and T-ZnO whiskers are 7 wt% and 13 wt%, respectively, and the composites thickness is 3 mm.

10.2.2.2 Absorption Properties of T-ZnO/CNTs Coatings

The typical fabrication process of T-ZnO/CNT composites is as follows: the calculated amount of mixed raw CNTs and T-ZnO nanostructures are sufficiently dispersed by ultrasonication for about 30 min. Then the mixture is added into EP, which is diluted by absolute ethyl alcohol, and dispersed by ultrasonication for about 30 min again, and then the curing agent is put into the composites and stirred gently. The mixture is sprayed layer-by-layer onto an aluminium plate and cured at room temperature for at least 24 h.[36]

The morphologies of absorbents and T-ZnO/CNTs /EP composites are shown in Figure 10.15. It can be found that CNTs are dispersed uniformly and form a network structure. T-ZnO nanostructures are arranged together with CNTs and EP. The reflectivity, complex permittivity and permeability of the composites were measured. The measured sample was made by mixing T-ZnO/CNTs with molten paraffin into a ring for electromagnetic parameter measurement.

A list of the microwave absorption properties of all manufactured samples is presented in Table 10.2.

The absorption properties of T-ZnO/CNTs/EP composites with a thickness of 1.2 mm were measured as shown in Figure 10.16.[36] It can be seen that CNT and T-ZnO concentration has an obvious effect on the microwave absorption properties. T-ZnO/EP and CNTs/EP composites have weak absorption performance. The minimum RL for sample **1** is −3.48 dB at 10.24 GHz. The

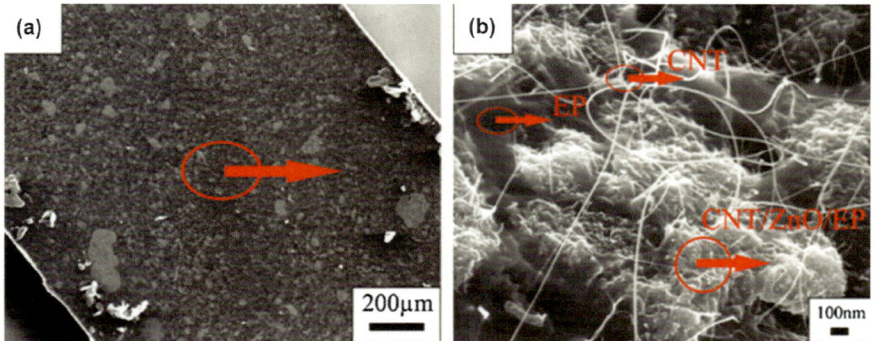

Figure 10.15   Typical SEM images of T-ZnO/CNTs/EP composites.
Reprinted from Mater. Sci. Eng. B, 175(1), H. Li, J. Wang, Y. Huang, X. Yan, J. Qi, J. Liu and Y. Zhang, Microwave absorption properties of carbon nanotubes and tetrapod-shaped ZnO nanostructures composites, 81–85. Copyright (2010), with permission from Elsevier.[36]

minimum RLs for samples **2**, **3** and **4** are −7.83 dB, −9.35 dB and −8.48 dB, respectively. The best RL of −11.21 dB at 16.16 GHz is obtained from T-ZnO/CNTs/EP composites, and reflection loss is over 10 dB between 15.52 GHz and 17.04 GHz. The minimum RL for sample **6** reaches 13.36 dB at 14.24 GHz, and the RL is over 10 dB between 13.28 GHz and 16 GHz. The results indicate that the CNTs with an appropriate concentration of T-ZnO nanostructures can improve the impedance matching and attenuation characteristics.[37] The microwave absorption properties of T-ZnO/CNTs/EP composites are improved significantly with the content of CNTs and T-ZnO nanostructures being 12 wt% and 8 wt%, respectively. The positions of the microwave absorption peaks move to the lower frequencies due to the increase of T-ZnO nanostructures. This result is like the previous reports on CNTs/Ag-NWs coatings, which show that the positions of absorption peaks move to the lower frequencies by filling the Ag NWs into CNTs.[38] This indicates that the absorption peak frequency of the T-ZnO/CNTs/EP composites can be modulated easily by changing the amount of CNTs and T-ZnO nanostructures.[39]

The mechanism of the T-ZnO/CNTs/EP is that the T-ZnO possesses isotropic crystal symmetry and plays an important role in the process of microwave absorption. It can form isotropic quasi-antennas and some discontinuous networks in the composites. Then, it is possible for the electromagnetic wave to penetrate the nanocomposites formed by the numerous antenna-like semiconductive T-ZnO nanostructures and the energy will be attenuated. However, there are many interfaces between the epoxy matrix and CNT outer surfaces. When compared with the CNTs, there are more interfaces between the T-ZnO and CNT inner surfaces in the composites. Therefore, interfacial multipoles contribute to the absorption of the T-ZnO/CNT/EP composites.[40] These results were confirmed by the theoretical calculation.[41] Furthermore, the size, defects, and impurities also have effects on the microwave absorption property of the T-ZnO.

Table 10.2 Microwave absorption properties of prepared T-ZnO/CNTs/EP composites.[a]

Sample Number	CNT concentration (wt%)	T-ZnO concentration (wt%)	Thickness (mm)	Minimum reflection loss (dB)	Corresponding frequency (GHz)	Absorption bandwidth (<10 GHz) (GHz)
1	0	20	1.2	-3.48	10.24	0
2	8	0	1.2	-7.83	18.00	0
3	12	0	1.2	-9.35	18.00	0
4	20	0	1.2	-8.48	17.78	0
5	8	12	1.2	-11.21	16.16	1.5
6	12	8	1.2	-13.36	14.24	2.8
7	12	8	1.5	-23.07	12.16	5
8	12	8	2.2	-23.23	12.8	4.4
9	12	8	2.7	-19.95	8.16	2.56

[a]Reprinted from Mater. Sci. Eng: B, 175(1), H. Li, J. Wang, Y. Huang, X. Yan, J. Qi, J. Liu and Y. Zhang, Microwave absorption properties of carbon nanotubes and tetrapod-shaped ZnO nanostructures composites, 81–85. Copyright (2010), with permission from Elsevier.[36]

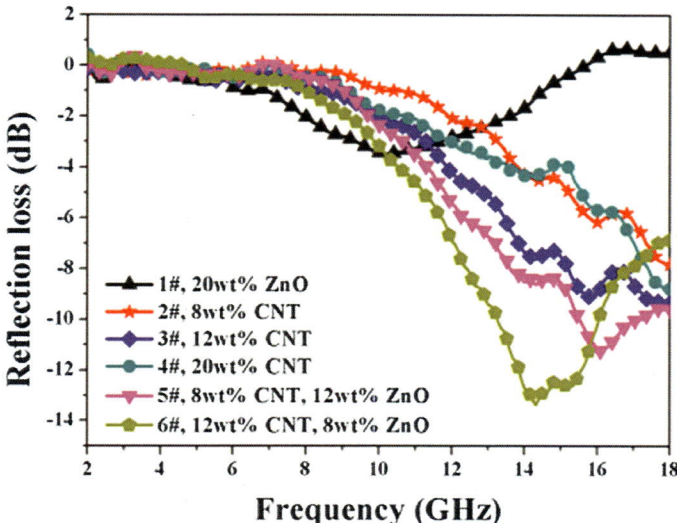

Figure 10.16 Absorption properties of CNTs/T-ZnO/EP composites with different CNT and T-ZnO nanostructure content.
Reprinted from Mater. Sci. Eng. B, 175(1), H. Li, J. Wang, Y. Huang, X. Yan, J. Qi, J. Liu and Y. Zhang, Microwave absorption properties of carbon nanotubes and tetrapod-shaped ZnO nanostructures composites, 81–85. Copyright (2010), with permission from Elsevier.[36]

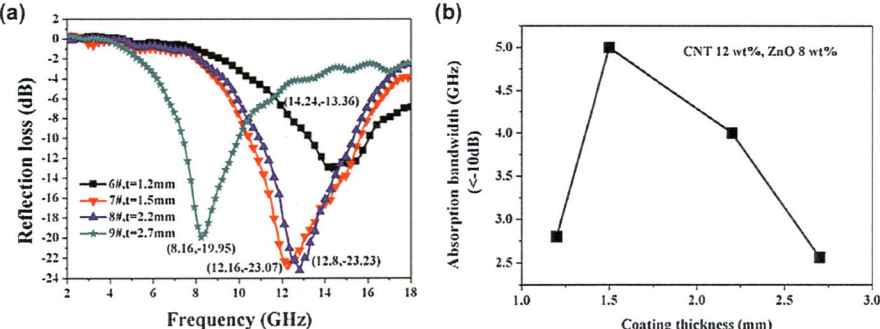

Figure 10.17 (a) Absorption properties of T-ZnO/CNTs/EP composites with different thicknesses; (b) frequency dependence of absorption bandwidth and coating thickness of T-ZnO/CNTs/EP composites.
Reprinted from Mater. Sci. Eng. B, 175(1), H. Li, J. Wang, Y. Huang, X. Yan, J. Qi, J. Liu and Y. Zhang, Microwave absorption properties of carbon nanotubes and tetrapod-shaped ZnO nanostructures composites, 81–85. Copyright (2010), with permission from Elsevier.[36]

To investigate the influence of the composite thickness on microwave absorption properties, T-ZnO/CNTs/EP composites with different thicknesses were prepared. Figure 10.17(a) shows that the minimum RL for sample **6** is −13.36 dB at 14.24 GHz with a thickness of 1.2 mm and the bandwidth below −10 dB is 2.8 GHz. When the thickness of sample 7 is 1.5 mm, the absorption properties have been improved obviously. The minimum RL for sample 7 is −23.00 dB at 12.16 GHz and the bandwidth of RL below −10 dB is 5 GHz. By increasing the thickness to 2.2 mm, the minimum RL for sample 8 is −23.24 dB at 12.71 GHz, and the bandwidth of RL below −10 dB is 4.4 GHz. When the composite thickness increases to 2.7 mm, the minimum RL for sample 9 is −19.95 dB at 8.16 GHz, and the bandwidth of RL below −10 dB is 2.56 GHz. It can also be seen that the maximum absorbing peaks shift toward a lower frequency with the increase of thickness, which is due to the interference resonance vibration.[33]

To study the effect of composite thickness on microwave absorption properties, a curve between the frequency dependence of absorption bandwidth and thickness of composites is presented in Figure 10.17(b). The frequency bandwidth of composites initially increases, and then decreases with the increase of the composite thickness. When the thickness is 1.5 mm, the frequency bandwidth reaches the maximum value of 5 GHz. This indicates that the CNTs mixed with T-ZnO nanostructures have potential application as broad frequency absorbing materials.

The RL of 20% CNTs and 12% CNTs mixed with 8% T-ZnO nanostructures at different thicknesses were calculated, as shown in Figure 10.18. 12% CNTs

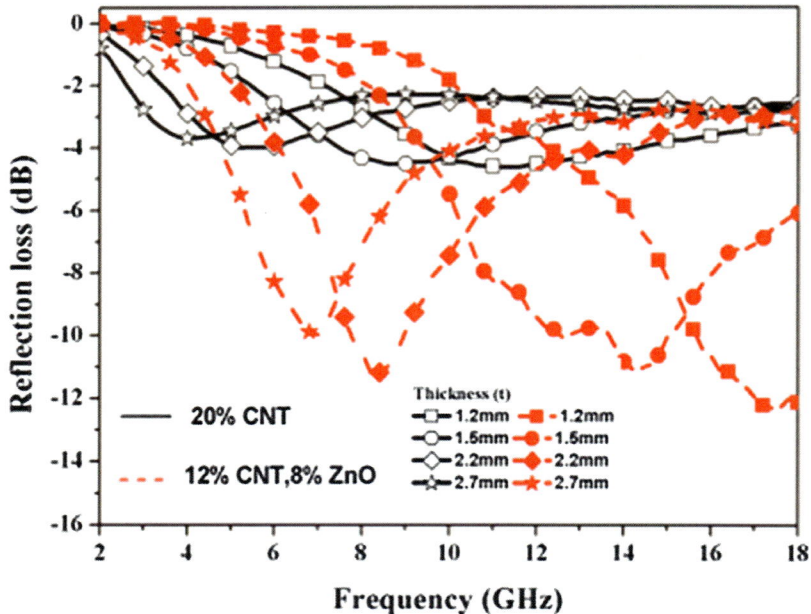

Figure 10.18   RL of 20% and 12% CNTs mixed with 8% T-ZnO at different thicknesses.
Reprinted from Mater. Sci. Eng. B, 175(1), H. Li, J. Wang, Y. Huang, X. Yan, J. Qi, J. Liu and Y. Zhang, Microwave absorption properties of carbon nanotubes and tetrapod-shaped ZnO nanostructures composites, 81–85. Copyright (2010), with permission from Elsevier.[36]

mixed with 8% T-ZnO nanostructures gives the optimum microwave absorption. The minimum reflection loss is -12.20 dB at 17.2 GHz with 1.2 mm thickness. The maximum bandwidth under -5 dB (68% absorption) is 6.7 GHz. with 1.5 mm thickness, which is consistent with the experimental results. The experimental and calculation results indicate that CNTs mixed T-ZnO nanostructures have excellent absorption properties.

As per the results mentioned above, T-ZnO/CNTs/EP composites exhibit excellent microwave absorption properties compared with those of CNTs/EP and T-ZnO/EP composites. A network formed by a small number of CNTs and some discontinuous networks formed by T-ZnO nanostructures coexist together with CNTs and EP in the composites. This special morphology makes it possible for the electromagnetic wave to penetrate the composites and the energy will be induced into a dissipative current and then the current will be consumed in the discontinuous networks, which leads to energy attenuation.[42] More importantly, the microwave absorption of T-ZnO/CNTs/EP composites is attributed to interfacial electric polarisation. The multi-interfaces between CNTs, T-ZnO nanostructures, EP and lots of agglomerates can be beneficial for the microwave absorption because of the interactions of electromagnetic radiation with charge multipoles at the interface.[41]

10.2.3 Absorption Properties of 3D ZnO Network Structures

Both 3D net-like ZnO and T-ZnO composites for microwave absorption were prepared by mixing the ZnO net-like nanostructures and the T-ZnO with paraffin wax with 50 vol.% of the ZnO, respectively. The composites were then pressed into a cylindrical-shaped compact for the measurement of the complex permittivity ε and permeability μ.

To investigate the microwave absorption properties of net-like ZnO micro-/nanostructures, the complex permittivity ε of the netlike micro-/nanostructures was measured.[43] For comparison, T-ZnO structures were correspondingly studied. Figure 10.19 shows the plots of the frequency *versus* the complex permittivity of the primary ZnO netlike nanostructure composites and T-ZnO netlike composites with 50 vol.% T-ZnO. The real permittivity of T-ZnO composite is about 3.1, and the imaginary permittivity is about 0.2 (corresponding green curve). However, for the ZnO netlike nanostructure composite, the ε' and ε'' values show a complex variation (corresponding red curve). The real part of relative permittivity (ε') declines from 15 to 6 in the frequency range of 2–18 GHz (Figure 10.19(a)). The imaginary part of permittivity (ε'') decreases from 4.4 to 3.6 and the curve exhibits two broad peaks in the 7–9 and 12–15 GHz ranges (Figure 10.19(b)). It is worth noting that the peaks of the ε'' curve appear at 8 and 14 GHz, suggesting resonance behaviour, which is expected when the composite is highly conductive and the skin effect becomes significant. The imaginary part ε'' of ZnO netlike nanostructures is relatively higher in contrast to that of T-ZnO composites, which implies the distinct dielectric loss properties arising from the morphological variation. It is reasonable that the dielectric loss is attributed to

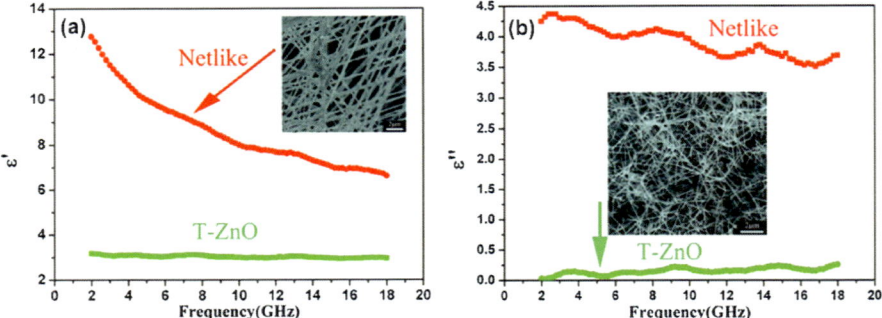

Figure 10.19 The frequency dependence of the real ε' and imaginary ε'' parts of relative complex permittivity for T-ZnO and net-like structure composites. Reproduced with permission from ref. 43. Copyright (2010) American Chemical Society.

the lags of polarisation between the 3D frame interfaces as the frequency is varied. ZnO net-like nanostructures possess more complicated interfaces than T-ZnO, resulting in better dielectric loss properties.

To explain the microwave absorption properties of the samples measured above, the RL of the netlike nanostructures and T-ZnO were calculated, respectively, using the relative complex permeability and permittivity at a given frequency and thickness layer as per the transmit-line theory, which follows the equations below:[44]

$$Z_{in} = Z_0(\mu_r/\varepsilon)^{1/2}\tanh[j(2\pi fd/c)(\mu_r\varepsilon)^{1/2}] \qquad (10.3)$$

$$RL(dB) = 20\log|(Z_{in} - Z_0)/(Z_{in} + Z_0)| \qquad (10.4)$$

where f is the microwave frequency, d is the thickness of the absorber, c is the velocity of light, Z_0 is the impedance of air and Z_{in} is the input impedance of the absorber. The relative complex permeability and permittivity were tested on a network analyser in the range 2–18 GHz. The simulations of the reflection loss of the two composites with a thickness of 2.0 mm are shown in Figure 10.20(a). The ZnO netlike micro-/nanostructure composite possesses a strong microwave absorption property, and the value of the minimum reflection loss for the composite is −30 dB at 14.4 GHz. However, the T-ZnO composite has almost no absorption. Figure 10.20(b) shows simulations of reflection loss of ZnO netlike micro-/nanostructure composites with different thicknesses. The value of the minimum reflection loss for the ZnO netlike micro-/nanostructure composite is −37 dB at 6.2 GHz with a thickness of 4.0 mm. When compared with the previous report,[10] in which the value of minimum reflection loss for the composite with 50 vol.% ZnO dendritic nanostructures is −25 dB at 4.2 GHz with a thickness of 4.0 mm, the ZnO net-like micro-/nanostructures have more excellent properties.

The absorption mechanism of ZnO nanotrees was explained by using an isotropic antenna mechanism.[40] The random distribution of the isotropic

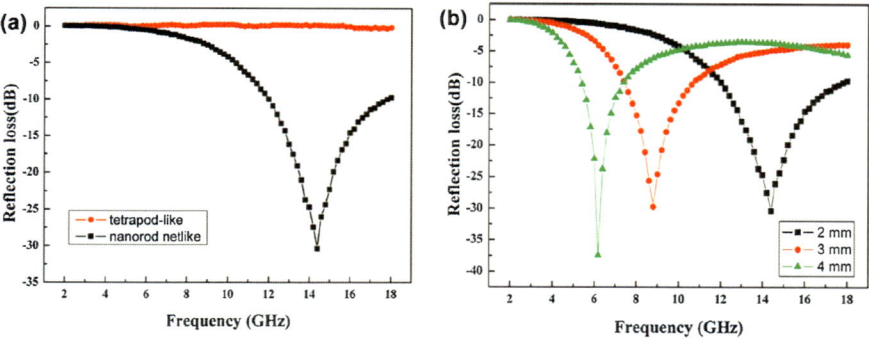

Figure 10.20   (a) Simulation of RL of 50 vol.% net-like and T-ZnO nanostructure composites; (b) simulation of RL of 50 vol.% ZnO net-like nanostructure composites with different thicknesses.
Reproduced with permission from ref. 43. Copyright (2010) American Chemical Society.

quasi-antenna ZnO semiconductive crystals not only leads to diffuse scattering of the incident microwaves, which results in the attenuation of electromagnetic energy, but also means that they act as receivers of microwaves, which can produce vibrating microcurrent in the local networks. Here, the ZnO netlike nanostructures have a special geometrical morphology. Such isotropic crystal symmetry can form continuous isotropic antennas networks in the composites. Moreover, it is possible for the electromagnetic wave to penetrate the composites formed by the numerous conductive ZnO networks. The energy will be induced into dissipative current, and then part of the current will generate electromagnetic radiation and the rest will be consumed in the discontinuous networks, which leads to the energy attenuation. Thus, the network frame of ZnO nanostructures in the composite will induce a certain amount of conductive loss. Compared with ZnO netlike nanostructures, no complex frame exists in T-ZnO. In short, ZnO net-like nanostructures, acting as receiving antenna, can receive electromagnetic energy and transform it into dissipative current. They also act as sending antenna transforming the vibrating current into electromagnetic radiation. Besides, the interfacial electric polarisation should also be considered. There are multi-interfaces between the isotropic antenna frame, and air bubbles can benefit the microwave absorption because of the interactions of electromagnetic radiation at the interfaces.[42]

ZnO nanostructure have been applied widely for functional nanodevices. The performance of various nanodevices is also improving. Novel and high performance devices based on ZnO nanostructures will be an important research field worldwide. With the improvement of fabrication technology and optimisation of performance, ZnO nanostructures will play an important role in the fields of new energy sources, environmental protection, biomedicine, information science and technology, security and defence.

References

1. R. H. Fowler and L. W. Nordheim, Electron emission in intense electric fields, *Proc. R. Soc. London, Ser. A*, 1928, **119**(781), 173–181.
2. C. J. Lee, T. J. Lee, S. C. Lyu, Y. Zhang, H. Ruh and H. J. Lee, Field emission from well-aligned Zinc Oxide NWs grown at low temperature, *Appl. Phys. Lett.*, 2002, **81**(19), 3648–3650.
3. H. Zhang, D. Yang, X. Ma and D. Que, Synthesis and field emission characteristics of bilayered ZnO nanorod array prepared by chemical reaction, *J. Phys. Chem. B*, 2005, **109**(36), 17055–17059.
4. Y. W. Zhu, H. Z. Zhang, X. C. Sun, S. Q. Feng, J. Xu, Q. Zhao, B. Xiang, R. M. Wang and D. P. Yu, Efficient field emission from ZnO nanoneedle arrays, *Appl. Phys. Lett.*, 2003, **83**, 144–146.
5. Q. Zhao, H. Z. Zhang, Y. W. Zhu, S. Q. Feng, X. C. Sun, J. Xu and D. P. Yu, Morphological effects on the field emission of ZnO nanorod arrays, *Appl. Phys. Lett.*, 2005, **86**(20), 203115.
6. R. C. Wang, C. P. Liu, J. L. Huang, S. J. Chen, Y. K. Tseng and S. C. Kung, ZnO nanopencils: Efficient field emitters, *Appl. Phys. Lett.*, 2005, **87**(1), 013110.
7. C. Li, Y. Di, W. Lei, Q. Yin, X. Zhang and Z. Zhao, Field emission from injector-like ZnO nanostructure and its simulation, *J. Phys. Chem. C*, 2008, **112**(35), 13447–13449.
8. A. Wei, X. W. Sun, C. X. Xu, Z. L. Dng, M. B. Yu and W. Huang, Stable field emission from hydrothermally grown ZnO nanotubes, *Appl. Phys. Lett.*, 2006, **88**(21), 213102.
9. Y. Zhang and C. Lee, Site-controlled growth and field emission properties of ZnO nanorod arrays, *J. Phys. Chem. C*, 2009, **113**(15), 5920–5923.
10. C. Li, K. Hou, W. Lei, X. Zhang, B. Wan and X. W. Su, Efficient surface-conducted field emission from ZnO nanotetrapods, *Appl. Phys. Lett.*, 2007, **91**(16), 163502.
11. C. Li, K. Hou, X. Yang, K. Qu, W. Lei, X. Zhang, B. Wang and X. W. Su, Enhanced field emission from ZnO nanotetrapods on a carbon nanofiber buffered Ag film by screen printing, *Appl. Phys. Lett.*, 2008, **93**(23), 233508.
12. C. X. Zhao, Y. F. Li, J. Zhou, L. Y. Li, S. Z. Deng, N. S. Xu and J. Chen, Large-scale synthesis of bicrystalline ZnO nanowire arrays by thermal oxidation of zinc film: growth mechanism and high-performance field emission, *Cryst. Growth Des.*, 2013, **13**(7), 2897–2905.
13. H. W. Kang, J. Yeo, J. O. Hwang, S. Hong, P. Lee, S. Y. Han, J. H. Lee, Y. S. Rho, S. O. Kim, S. H. Ko and H. J. Sung, Simple ZnO NWs patterned growth by microcontact printing for high performance field emission device, *J. Phys. Chem. C*, 2011, **115**(23), 11435–11441.
14. Y. Huang, X. Bai, Y. Zhang, J. Qi, Y. Gu and Q. Liao, Field-emission properties of individual ZnO NWs studied in situ by transmission electron microscopy, *J. Phys.: Condens. Matter.*, 2007, **19**(17), 176001.

15. Y. Huang, Y. Zhang, Y. Gu, X. Bai, J. Qi, Q. Liao and J. Liu, Field-emission of a single In-doped ZnO nanowire, *J. Phys. Chem. C*, 2007, **111**(26), 9039–9043.

16. Q. Liao, Y. Yang, L. Xia, J. Qi, Y. Zhang, Y. Huang and Z. Qin, High intensity, plasma-induced emission from large area ZnO nanorod array cathodes, *Phys. Plasmas*, 2008, **15**(11), 114505.

17. Q. Liao, J. Qi, Y. Yang, Y. Huang, Y. Zhang, Z. Zhang and L. Xia, Morphological effects on the plasma-induced emission properties of large area ZnO nanorod array cathodes, *J. Phys. D: Appl. Phys.*, 2009, **42**(21), 215203.

18. Q. Liao, Y. Zhang, L. Xia, X. Yan, J. Qi, Y. Huang and Z. Gao, Intense electron beam emission from carbon nanotubes and mechanism, *J. Phys. D: Appl. Phys.*, 2007, **40**(21), 6626–6630.

19. Y. H. Huang, Y. Zhang, L. Liu, S. S. Fan, Y. Wei and J. He, Controlled synthesis and field emission properties of ZnO nanostructures with deferent morphologies, *J. Nanosci. Nanotechnol.*, 2006, **6**(3), 787–790.

20. I. Yao, P. Lin and T. Y. Tseng, Nanotip fabrication of zinc oxide nanorods and their enhanced field emission properties, *Nanotechnology*, 2009, **20**(12), 183–187.

21. C. X. Xu, X. W. Sun and B. J. Chen, Field emission from gallium-doped Zinc Oxide nanofiber array, *Appl. Phys. Lett.*, 2004, **84**(9), 1540–1542.

22. H. Pan, Y. Zhu, H. Sun, Y. Feng, C. Sow and J. Lin, Electroluminescence and field emission of Mg-doped ZnO tetrapods, *Nanotechnology*, 2006, **17**(20), 5096–5100.

23. H. S. Jang, S. O. Kang, S. H. Nahm, D. H. Kim, H. R. Leeb and Y. I. Kim, Enhanced field emission from the ZnO NWs by hydrogen gas exposure, *Mater. Lett.*, 2006, **61**(8–9), 1679–1682.

24. H. Chen, J. Qi, Y. Zhang, Q. Liao, X. Zhang and Y. Huang, Field emission characteristics of ZnO nanotetrapods and the effect of thermal annealing in hydrogen, *Chin. Sci. Bull.*, 2007, **52**(9), 1287–1290.

25. Q. Zhao, X. Y. Xu, X. F. Song, D. P. Yu, C. P. Li and L. Guo, Enhanced Field emission from ZnO nanorods via thermal annealing in oxygen, *Appl. Phys. Lett.*, 2006, **88**(3), 033102.

26. C. J. Park, D. Choi, J. Yoo, G. Yi and C. J. Lee, Enhanced field emission properties from well-aligned Zinc Oxide nanoneedles grown on the Au/Ti/n-Si substrate, *Appl. Phys. Lett.*, 2007, **90**(8), 083107.

27. S. H. Jo, D. Banerjee and Z. F. Ren, Field emission of zinc oxide NWs grown on carbon cloth, *Appl. Phys. Lett.*, 2004, **85**(8), 1407–1409.

28. J. Cao, Y. Huang, Y. Zhang, Q. Liao and Z. Deng, Research on electromagnetic wave absorbing properties of nano tetraleg ZnO, *Acta Phys. Sin.*, 2008, **57**(6), 3641–3645.

29. M. Cao, X. Shi, X. Fang, H. Jin, Z. Hou and W. Zhou, Microwave absorption properties and mechanism of cagelike ZnO/SiO_2 nanocomposites, *Appl. Phys. Lett.*, 2007, **91**(20), 203110.

30. Y. Chen, M. Cao, T. Wang and Q. Wan, Microwave absorption properties of the ZnO nanowire-polyester composites, *Appl. Phys. Lett.*, 2004, **84**(17), 3367–3369.

31. R. Zhuo, H. Feng, J. Chen, D. Yan, J. Feng, H. Li, B. Geng, S. Cheng, X. Xu and P. Yan, Multistep systhesis, growth mechanism, optical, and microwave absorption properties of ZnO dendritic nanostructures, *J. Phys. Chem. C*, 2008, **112**(31), 11767–11775.

32. T. Xia, Y. Cao, N. A. Oyler, J. Murowchick, L. Liu and X. Chen, Strong microwave absorption of hydrogenated wide bandgap semiconductor nanoparticles, *ACS Appl. Mater. Interfaces*, 2015, **7**(19), 10407–10413.

33. L. Zhang, H. Zhu, Y. Song, Y. Zhang and Y. Huang, The electromagnetic characteristics and absorbing properties of multi-walled carbon nanotubes filled with Er_2O_3 nanoparticles as microwave absorbers, *Mater. Sci. Eng. B*, 2008, **153**(1–3), 78–82.

34. Z. Zhou, L. Chu, W. Tang and L. Gu, Studies on the antistatic mechanism of tetrapod-shaped zinc oxide whisker, *J. Electrostatics*, 2003, **57**(3), 347–354.

35. H. Qin, Q. Liao, G. Zhang, Y. Huang and Y. Zhang, Microwave absorption properties of carbon black and tetrapod-like ZnO whiskers composites, *Appl. Surf. Sci.*, 2013, **286**, 7–11.

36. H. Li, J. Wang, Y. Huang, X. Yan, J. Qi, J. Liu and Y. Zhang, Microwave absorption properties of carbon nanotubes and tetrapod-shaped ZnO nanostructures composites, *Mater. Sci. Eng. B*, 2010, **175**(1), 81–85.

37. A. Yusoff, M. Abdullah, S. Ahmad, S. Jusoh, A. Mansor and S. Hamid, Electromagnetic and absorption properties of some microwave absorbers, *J. Appl. Phys.*, 2002, **92**(2), 876–882.

38. D. Zhao and X. Li, Z. Shen.Electromagnetic and microwave absorbing properties of multi-walled carbon nanotubes filled with Ag NWs, *Mater. Sci. Eng. B*, 2008, **150**(2), 105–110.

39. Z. Fan, G. Luo, Z. Zhang, L. Zhou and F. Wei, Electromagnetic and microwave absorbing properties of multi-walled carbon nanotubes/polymer composites, *Mater. Sci. Eng. B*, 2006, **132**(1–2), 85–89.

40. D. Zhao and Z. Shen, Preparation and microwave absorption properties of carbon nanocoils, *Mater. Lett.*, 2008, **62**(21–22), 3704–3706.

41. X. Fang, M. Cao, X. Shi, Z. Hou, W. Song and J. Yuan, Microwave responses and general model of nanotetraneedle ZnO: Integration of interface scattering, microcurrent, dielectric relaxation, and microantenna, *J. Appl. Phys.*, 2010, **107**(5), 054304.

42. R. Zhuo, H. Feng, J. Chen, D. Yan, J. Feng, H. Li, B. Geng, S. Cheng, X. Xu and P. Yan, Growth mechanism, optical, and microwave absorption properties of ZnO dendritic Nanostructures, *J. Phys. Chem. C*, 2008, **112**(31), 11767–11775.

43. H. Li, Y. Huang, G. Sun, X. Yan, Y. Yang, J. Wang and Y. Zhang, Directed growth and microwave absorption property of crossed ZnO netlike micro-nanostructures, *J. Phys. Chem. C*, 2010, **114**(22), 10088–10091.

44. G. Sun, X. Zhang, M. Cao, B. Wei and C. Hu, Facile synthesis, characterization, and microwave absorbability of CoO nanobelts and submicrometer spheres, *J. Phys. Chem. C*, 2009, **113**(17), 6948–6954.

Subject Index